Contributions in Petroleum Geology & Engineering Volume 2

Applied Open-Hole Log Analysis

Volume 1: Geologic Evaluation of Naturally Fractured Reservoirs

Volume 2: Applied Open-Hole Log Analysis

Gulf Publishing Company
Book Division
Houston, London, Paris, Tokyo

Contributions in Petroleum Geology & Engineering Volume 2

Applied Open-Hole Log Analysis

James Brock

Contributions in Petroleum Geology & Engineering

Series Editor: George V. Chilingar, University of Southern California

Volume 2:

Applied Open-Hole Log Analysis

Acknowledgments

The author wishes to thank Schlumberger, Gearhart, Dresser Atlas, and Welex for the use of their charts and examples.

Dedication

I wish to dedicate this work to:
First and foremost, the person who made it possible, Jannye Brock.
The little person who made it almost impossible, Roxanne Brock.
The founders of this technology, Conrad and Marcel Schlumberger.
John and Dannye Bristow, and Louis and Carmen Scharfenstein.

Library of Congress Cataloging-in-Publication Data

Brock, James G., 1951–
Applied open-hole log analysis.

(Contributions in petroleum geology & engineering; 2)
Bibliography: p.
Includes index.
1. Oil well logging. I. Title. II. Series.
TN871.35.B764 1986 622′.18282 85-30604

ISBN 0-87201-638-2

Contents

Preface

The scope and purpose of this book is to instruct. When I first started to learn the art of log analysis, I found myself completely lost. The key to the learning process is the fact that this technology is not all science. If you wish to avoid frustration, then I suggest you take learning lightly at first. If you don't understand something very well, press on and come back to it later.

Log analysis is a "big picture" concept. Each log curve provides a small clue or picture element. As you read on, keep this in mind and try to build a mental picture of what is going on.

This book is organized into stages. First we build a common language. Second, we study the concept or foundation. Third, we study the individual curves on a log. Fourth, we analyze a complete log and build the "big picture." Fifth and last, we look at the supplementary curves. By this organization, the student has the opportunity to use and practice the fundamental skills of log analysis.

As you read this book, keep in mind that some techniques will work and some will not. This is entirely dependent upon a given area. The quality of the data is of great importance. When you find that one technique works well, use it, but don't assume that the same will hold true for other areas. For this reason, this work presents a wide scope of techniques.

James Brock, Ph.D.

1

The Petroleum Reservoir

A petroleum reservoir is a rock bed capable of containing gas, oil, or water. A commercially productive reservoir must have sufficient thickness and pore space to contain a large volume of hydrocarbons and must yield fluids at a satisfactory rate when the reservoir is penetrated. In this chapter we shall discuss the three primary parameters used to determine a formation's production. These are: porosity, permeability, and water saturation.

Before embarking on the subject of analysis, the reader should become acquainted with log scales. Figure 1-1 shows various combinations of scales, called tracks. The leftmost track is numbered 1. This scale, or track, is always linear. Immediately to its right is the depth track, which is not numbered. Tracks 2 and 3 may be linear, logarithmic, or split. When a log scale spans across tracks 2 and 3, it is said to be a "track 4" presentation. Both tracks combined as one will form the fourth track.

Porosity

Porosity is defined as the capacity of a formation to contain fluids and is denoted by the Greek symbol *phi* (ϕ). By definition, porosity is the void volume of the rock divided by the total rock volume. The porosity of a rock may be primary, such as the intergranular porosity of a sandstone, or secondary. Secondary porosity may be from dolomitization, solution channels, or fracturing.

Factors affecting porosity are sorting, packing, cementation, and angularity or roundness of the grains. In sandstones, porosity is controlled primarily by sorting. The term "poorly sorted" refers to a collection of grains of different sizes and shapes (Figure 1-2).

When sand grains are spherical and all one size, porosity is at its maximum, regardless of the size of the grains. Porosity then progressively decreases as the grains become more angular, because such grains pack together more closely. To better understand these factors, picture identical spheres stacked in a cubic structure, as shown in Figure 1-3. Calculations show that this represents a geometric maximum porosity of 47.6%. Porosity remains at 47.6% no matter what the grain's size is. The spheres may be basketball size or they may be the size of marbles.

If the same sand grains are now arranged in the closest possible manner, where the top sphere sits in the valley between the two lower spheres, (Figure 1-4), the porosity will be reduced by 25.9%. Here again, sphere size is unimportant as long as all of the spheres are the same. The term for this type of packing is *rhombic*.

Clean sands can be artificially mixed so that they are extremely well sorted. Such sands would exhibit about 43% porosity. Porosity then decreases to about 25% for poorly sorted medium-to-coarse grain sands.

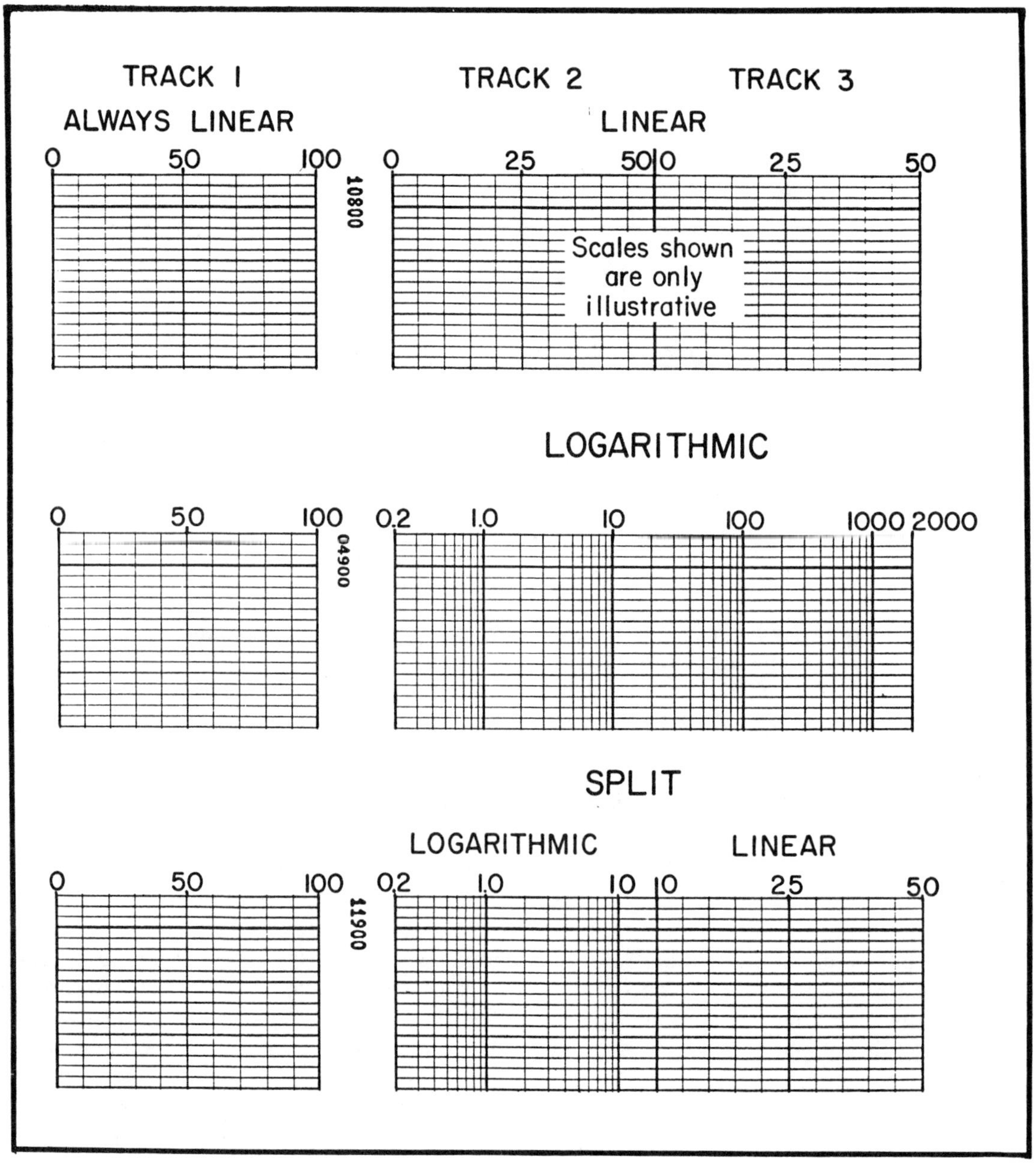

Figure 1-1. Example of log scales. (Courtesy of Schlumberger.)

Figure 1-2. Illustration of poorly sorted sand grains.

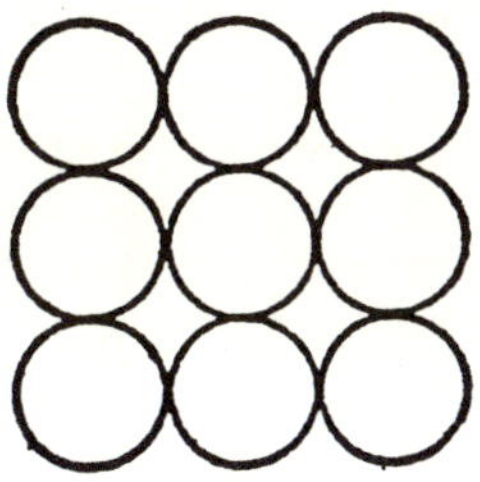

Figure 1-3. Illustration of cubic arrangement of sand grains (ϕ = 47.6%).

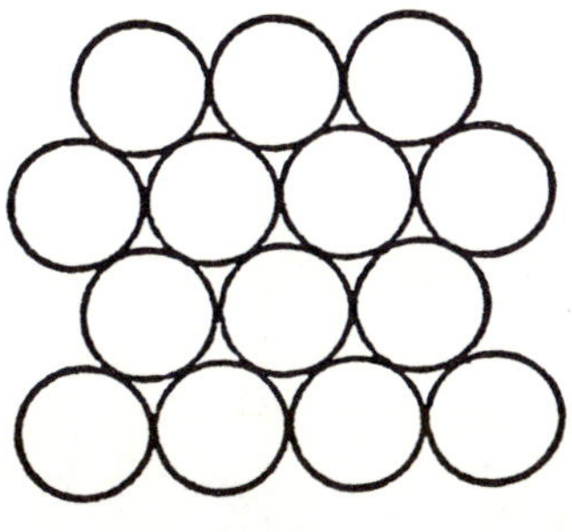

Figure 1-4. Illustration of rhombic arrangement of sand grains (ϕ = 25.9%).

Fine sand grains can exhibit a porosity of 30% or more, regardless of sorting. Packing refers to the geometrical configurations of the grains as discussed previously.

The angularity of the grain, or the lack of roundness, together with the sorting and packing, will affect porosity. This is due to the grains interlocking and thus reducing the available void space. Rock grain must be cemented in order to form a stone. Cementation adheres individual grains to each other. Very often this cementing agent will be quartz or calcite. Porosity will be influenced by the degree of cementation. A poorly cemented rock will have a higher porosity than a well-cemented one. The overlaying pressure frequently alters the size and shape of the reservoir rock. This compaction effect thus lowers the porosity of the reservoir. While this principle is true in most cases, it does have its exceptions. Generally, porosity will tend to be lower in deeper, older rocks.

Of additional importance is secondary porosity. Secondary porosity results from geologic agents such as leaching, fracturing, or fissuring, which occur after the rock-forming process is complete. For example, the dissolving of limestone or dolomite by ground waters is a leaching process which creates vugs or caverns as depicted in Figure 1-5.

Due to the brittle nature and chemical composition of carbonate formations, they become excellent exhibits of secondary porosity. In limestone reservoirs, the pore openings may be individually quite large. However, when averaged over a section, the porosity is generally lower than a sandstone. Dolomite reservoirs, because of the shrinking of solid volume as the material transforms from limestone to dolomite, frequently exhibit higher porosities. Both clastic and car-

Figure 1-5. Illustration of vuggy formations.

bonate reservoirs produce hydrocarbons from primary porosity. Carbonates producing from secondary porosity tend to have localized porosity values.

The last factor affecting porosity is connate water, or irreducible water saturation, S_{wirr}. Connate water adheres by surface tension to the sand grain and will not come loose, hence the term irreducible. Irreducible water has the effect of reducing the useful porosity of a bed.

Permeability

Permeability is defined as the ability of a reservoir to permit flow or passage of reservoir fluids. For a reservoir to be commercially productive, it is not sufficient to merely have oil or gas contained within the formation. It is the flow of these hydrocarbons from the reservoir to the surface that makes the accumulation valuable. Obviously fluid movement is possible only where pore spaces are interconnected. The unit of permeability is the **darcy,** named after the French engineer Henry D'Arcy, who in 1865 devised a method for quantifying permeability. This property is mathematically defined as follows: permeability is one darcy when one square centimeter of rock surface releases one cubic centimeter of fluid of unit viscosity in one second under a pressure differential of one atmosphere. The darcy is a large unit. Most reservoir rocks have permeabilities that are measured in millidarcies, or thousandths of a darcy. Although the concept of permeability appears quite simple, permeability determination is somewhat complex. Flow rate will increase as more pressure is exerted on the fluid or as fluid viscosity decreases. Gas, for instance, flows more easily than water. Water flows more easily than oil. Several factors must be known in order to determine the relative permeability of a rock formation to a given fluid.

Usually, the larger the pores become, the higher the permeability will be. In a well-sorted sandstone, pore and grain size generally increase together. In some cases, grain size increases while porosity stays the same. As porosity stays the same, permeability increases. This argument does not carry over to carbonates, as grain size is not really significant. What matters in carbonates is pore size. In many carbonate reservoirs, pore size or permeability is large and porosity is small.

Fractures can increase the permeability of a formation considerably. The permeability of fractures has been estimated to be a function of fracture width, as shown:

$$K = 50{,}000{,}000 \times (\text{width})^2$$

where K is permeability in darcies and width is in inches.

This approximation of fracture permeability indicates the reason why fractured rock fluid flow is controlled by fractures, if they are present. These fractures only contribute .5% to 1.5% porosity but completely control the flow. It can be seen that if a formation has one fracture .01 inches in width, over 90% of the fluid flowing to the well will be via the fracture.

Permeability with only one fluid in the pore is "absolute" permeability. "Effective" permeability is a permeability with more than one fluid present in the pores. Effective permeability is less than absolute permeability, as the fluid which wets the solid part of the rock reduces the hydraulic area available for the other fluid to flow through. "Relative" permeability is a ratio of the effective permeability of a specific fluid and the absolute permeability.

In Figure 1-6, we see the relative permeability of a water system with hydrocarbons present. At low water saturations, only oil will flow. As the water saturation increases, the relative permeability to oil de-

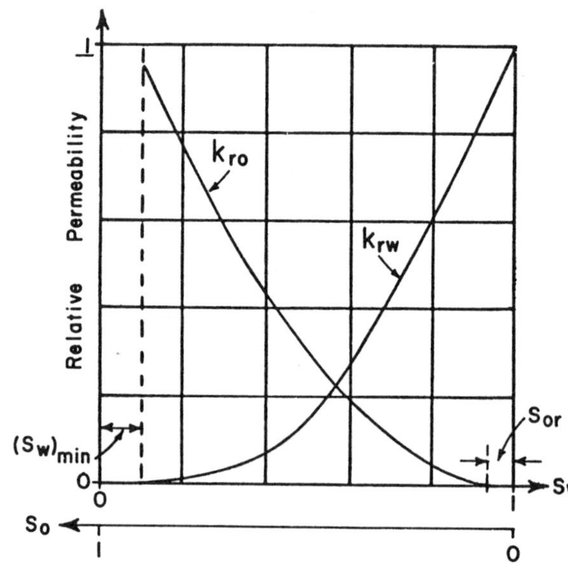

Figure 1-6. Plot of the relative permeability of oil to water.

creases until at some point water and oil will both flow. The water saturation at which the water starts to flow is often referred to as the "critical" water saturation.

From Figure 1-6, it can be seen that there is a range of water saturations at which only water will flow, even though the water saturation is less than 100%. Relative permeability curves are different for diverse rocks and fluids. So are the critical water saturations. The concept of relative permeability explains why a well which initially produces only oil will at some later point in its producing life begin to produce water with the oil. As the reservoir becomes depleted, its overall water saturation increases until, at some time, the critical water saturation is reached and only water will be produced.

Water Saturation

The fluid saturation of a rock is the ratio of the volume of the fluid within the pores of the rock to the total pore volume. Very simply, a water saturation (S_w) of 30% means that three-tenths of the pore space is filled with water. Water, oil, and gas may be found simultaneously in reservoirs. However, due to specific gravity, fluids tend to segregate within the reservoir.

When considering a possible productive interval, the fractional portion of the pore space which does not contain formation water is assumed to contain hydrocarbons. This can be expressed mathematically using the relationship:

$$S_h = 1 - S_w$$

where S_h = hydrocarbon saturation
S_w = the water saturation

Before concluding this chapter, let us examine the schematic representation of a well bore (Figure 1-7). R_{mc} is the mudcake resistivity, h_{mc} is the thickness of the mudcake, D_i is the diameter of invasion, h is the bed thickness, D_h is the hole diameter, R_{xo} is the resistivity of the flushed portion of the invaded zone, R_{mf}

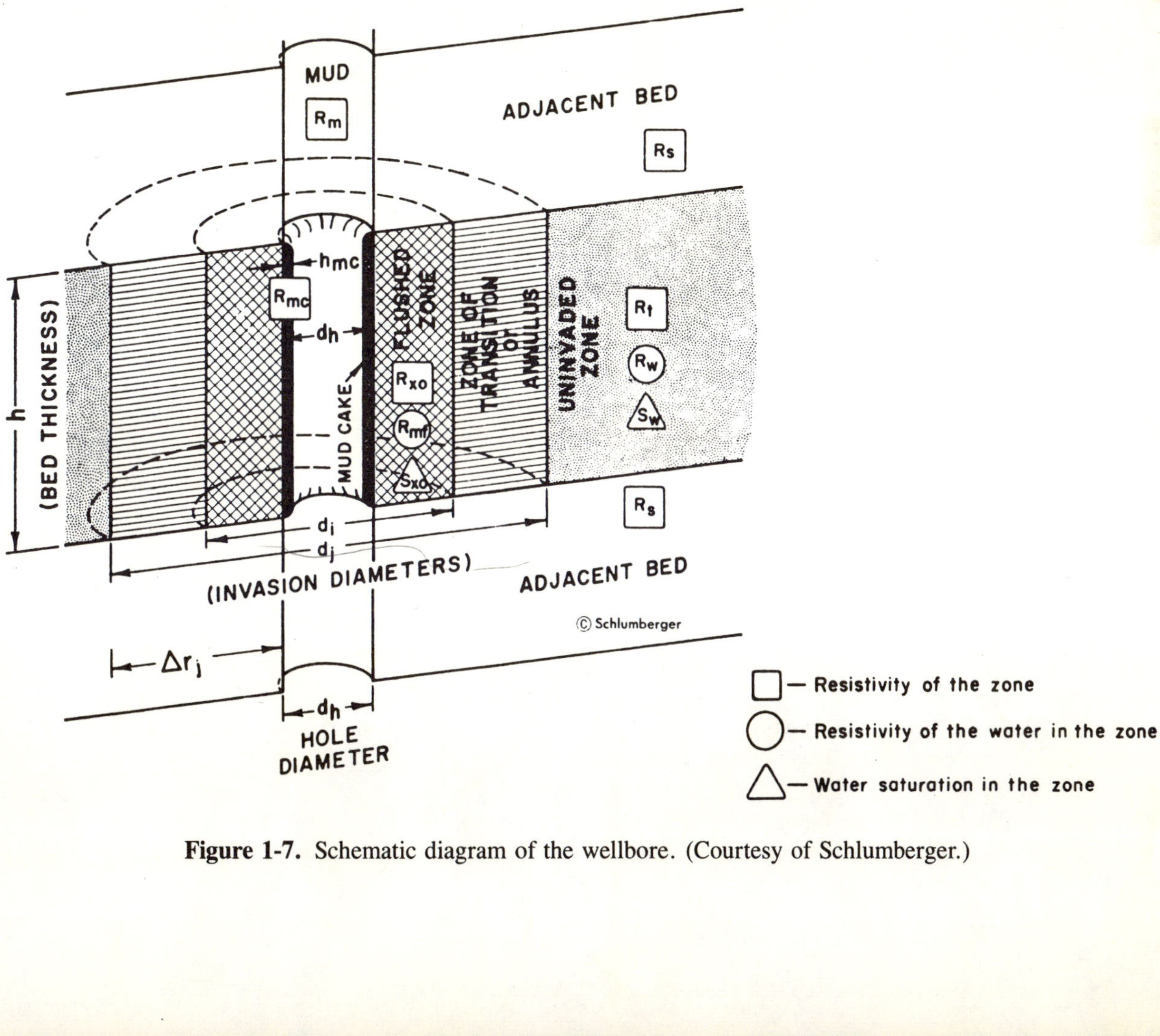

Figure 1-7. Schematic diagram of the wellbore. (Courtesy of Schlumberger.)

is the resistivity of the mud filtrate, and S_{xo} is the flushed zone water saturation.

In the uninvaded zone, R_t is the true virgin formation resistivity. R_w is the formation water resistivity, and S_w is the water saturation of the undisturbed formation. Water saturation in the undisturbed or uninvaded zone is the parameter we are most interested in. Its determination allows the log analyst to estimate the amount of hydrocarbons present in a given formation. This information, along with porosity and a permeability estimate, can be used to predict the well's ultimate production.

Summary

Porosities in sandstones generally will be less than 40%. When porosities are less than 7% in a gas-bearing sandstone or 8% in an oil bearing sandstone, the permeability is typically so low that nothing will be produced from the formation. 9% is often used as the lower limit of porosity for a good commercial producer. These porosities are assumed to be true porosities and devoid of measurement inaccuracies or improper calibrations. There are sands which produce below this limit, but these zones are usually fractured.

As a rule of thumb, 60% water saturation is used as the cut-off for production from a sandstone. Above this point, water will be produced in too great a quantity to be commercially economical. If the sandstone has a high shale content, corrections must be applied to porosity and water saturation calculations in order to achieve accurate values.

Porosities in carbonates are also usually less than 40%. When a porosity of 6% or greater is encountered in carbonate rocks, they can be quite productive. The relation between water saturation and porosity for productive carbonates is considerably more variable than in sandstones. A 50% water saturation is usually considered the cut-off point. However, some carbonates may produce hydrocarbons at 70% S_w, while others will produce water at only 30% S_w. Experience in a particular area is necessary to establish realistic cut-offs.

2

The Basics of Resistivity

The resistivity of a substance is its ability to impede the flow of electrical currents. The ability of a formation to conduct electricity is directly related to the amount of water in the formation. This is true since the formation grains (matrix) have negligible conductivity (high resistivity). By knowing bed resistivity, it is possible to determine the "water saturation" of a formation and hence its hydrocarbon saturation. Formation resistivity can also be related to the porosity of the formation.

Resistivity

Electrical resistivity of any material is related to the resistance by the following equation:

$$R = (r \times a)/L$$

where a = area in square meters of rock material exposed to the current flow
L = length of the material in meters
r = electrical resistance in *ohms*
R = electrical resistivity expressed in *ohm-meters*

Formation resistivities usually range from .2 to 1,000 ohm-meters. Resistivities higher than 1,000 ohm-meters are uncommon in permeable formations. Most formations are made up of rocks which when dry will not conduct electrical currents. Current flow in a formation is through interstitial water made conductive by salts in solution. These salts dissociate into positively charged sodium cations and negatively charged anions. Ions move under the influence of an electrical field, carrying an electrical current through the solution. Greater salt concentration lowers the formation water resistivity.

The concept of water saturation and its relation to porosity and resistivity is really quite simple if you form it with one element at a time. Imagine a cube one meter by one meter by one meter, as in Figure 2-1. Now let's fill the cube with salt water. Taking a test meter capable of measuring resistance, we measure the resistance of the water cube. Let us call this measurement R_w. Since we have a specific amount of water, we can now say that the resistivity of the cube of water is so many ohms per unit of volume. We have one cubic meter of volume, thus we can say that R_w is so many *ohms per meter/meter²*.

The resistivity of the water is:

$$\text{Resistivity} = \frac{R_w \times L}{\text{area}}$$

We measured R_w with a test meter. L is the length of the cube, which is one meter. The area is one meter by one meter. Solving the equation:

$$\text{Resistivity} = \frac{R_w \times 1}{1 \times 1} = \frac{R_w}{1} = R_w$$

In this case the resistivity of the water is the same as its measured resistance.

Now let us take one cubic meter of formation rock and insert it into the cube filled with water. What happens to some of the water? Yes, it spills over as the rock occupies the volume. Looking at Figure 2-2, we now have a cube with formation rock that is 100% saturated with water. The resistivity of this cube is called R_o.

When the cube had only water, it was simple to see that an ion could travel from one side of the cube to the other in a straight path. The length of this travel was one meter. Now that we have added the formation rock to the cube, the ion must find a meandering path around the rock grains to go from one side to the other. Naturally the length of this path will be greater than one meter.

The area is also altered by the rock. Instead of area occupied only by water, we have water occupying the pore spaces of the formation. Thus porosity becomes area.

$$\text{Resistivity } (R_o) = \frac{R_w \times L}{\Phi}$$

We can measure R_w and it is possible to measure porosity. This leaves L, or the length of the path of the ion to be determined.

The length of this ion cannot be measured. The best that can be done is simply to estimate it based on the texture of the rock, the cementation, and the amount of pore space available. We will consider this in more detail later in the text.

To complete our cubic reservoir, let's take some water out and replace it with oil. Because the oil is lighter than the water it will float on top. We still have the length, L; the porosity factor; and the resistivity of the water to consider. The only difference is that we no longer have 100% water. We took some out and replaced it with oil (Figure 2-3). If we took out 40% of the water, then we would have 60% left in the cube. This 60% is called the *water saturation,* or S_w. Now the total resistivity of the cube is:

$$\text{Resistivity (total)} = \frac{R_w \times L}{\Phi \times S_w}$$

In reality, the water saturation is what we want to know. If we can measure the total resistivity, the porosity, the water resistance, and estimate the length L, then we can solve the equation for S_w. It then follows that the amount of hydrocarbons is $1 - S_w$.

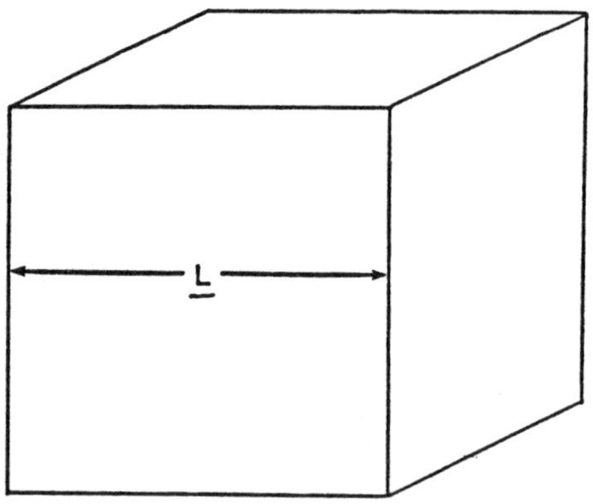

$$\text{Resistance of water} = \frac{R_w \times L}{\text{area}}$$

where L = 1 meter
R_w = Ohm-meter
area = 1 square meter
thus Resistance = resistivity

Figure 2-1. Model of water resistivity.

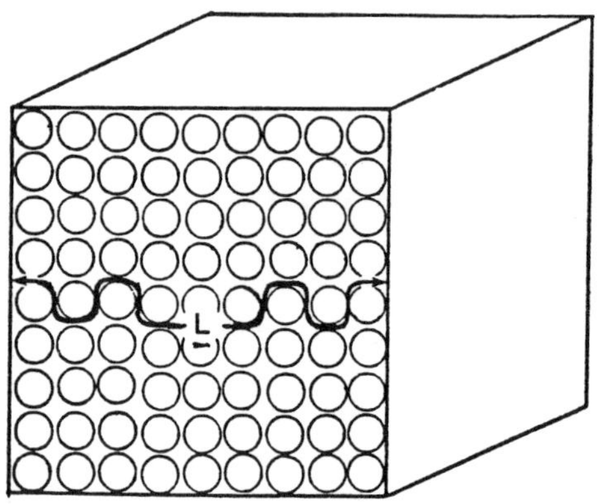

R_o = water saturated resistivity

$$R_o = \frac{R_w \times L}{\Phi}$$

Figure 2-2. Model of a 100% water-saturated formation.

It should be mentioned here that the resistivity of the water is not only a function of salinity but also of temperature. The higher the temperature is, the lower is the resistivity at a given salinity, so that water resistivity will vary according to temperature.

Equation 2-1 shows the general method for calculating formation temperature so that it can be applied towards adjusting the resistivity of the formation itself.

$$T_f = T_o + (G_g \times D_f)/100 \qquad (2\text{-}1)$$

or

$$T_f = (T_{td} - T_o)\, D_f D_{td} + T_o$$

where T_f = the formation temperature
T_o = the mean annual surface temperature
G_g = the geothermal gradient expressed in °F per 100 ft
D_f = the formation depth in feet
D_{td} = total depth
T_{td} = bottom hole temperature

Surface temperature is defined as the average annual temperature at some depth, typically from 70 to 100 feet below the surface of the earth, where seasonal temperature variations average out. This surface temperature may vary according to location. According to the United States Department of Agriculture, the following are average surface temperatures:

- Gulf of Mexico Coastal Regions—70°F
- Oklahoma—60°F
- Colorado—50°F
- Wyoming—40°F
- California—60°F

For the purpose of log analysis, however, surface temperatures 5 to 10 degrees higher than these produce more satisfactory results. Formation temperature may also be determined from the chart in Figure 2-4.

Because mud filtrate invades the formation, it is important for us to know what the mud filtrate resistivity is. The mud filtrate resistivity is measured at the surface with a cup-type device, and that resistivity is adjusted for temperature at the formation depth.

$$R_2 = R_1 \times \frac{T_1 + 6.77}{T_2 + 6.77} \qquad (2\text{-}2)$$

Equation 2-2 is the *ARPS formula* for determining resistivity at formation temperature. T_1 is the mean surface temperature expressed in °F. T_2 is the formation temperature in °F. R_1 is the resistivity at T_1, and R_2 is the resistivity at T_2.

As discussed earlier, the unknown factor is the length of the path of the ionic current. The length of this path is one meter for R_w and greater than one for R_o. This unknown quantity is called the *formation resistivity factor* or simply F. This factor is defined as:

$$F = R_o/R_w \qquad (2\text{-}3)$$

Since R_w is known, theoretically the ratio of R_o to R_w should give us their difference.

In a formation containing oil or gas, both of which are insulators, resistivity is a function of the formation factor F, the water resistivity R_w, and the water saturation S_w. S_w is a fraction of the pore volume occupied by formation water. We can then say that $1 - S_w$ is the fraction of the pore volume occupied by oil or gas. Archie determined experimentally that the water saturation of a clean formation can be expressed in terms of its true resistivity.

$$S_w^2 = FR_w/R_t \qquad (2\text{-}4)$$

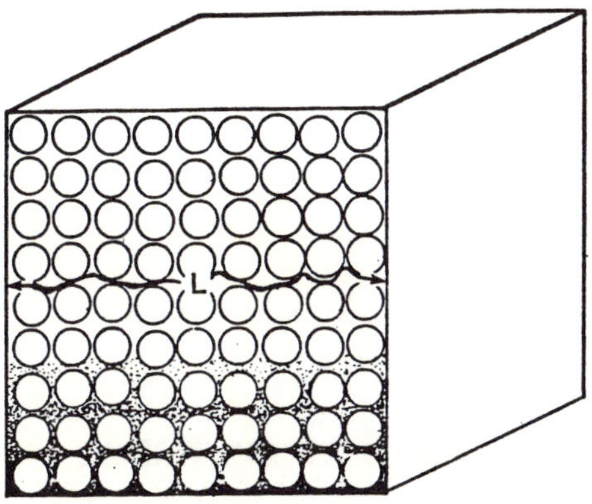

$$R_t = \frac{R_w \times L}{\Phi \times S_w}$$

Figure 2-3. Model of a formation containing oil and water.

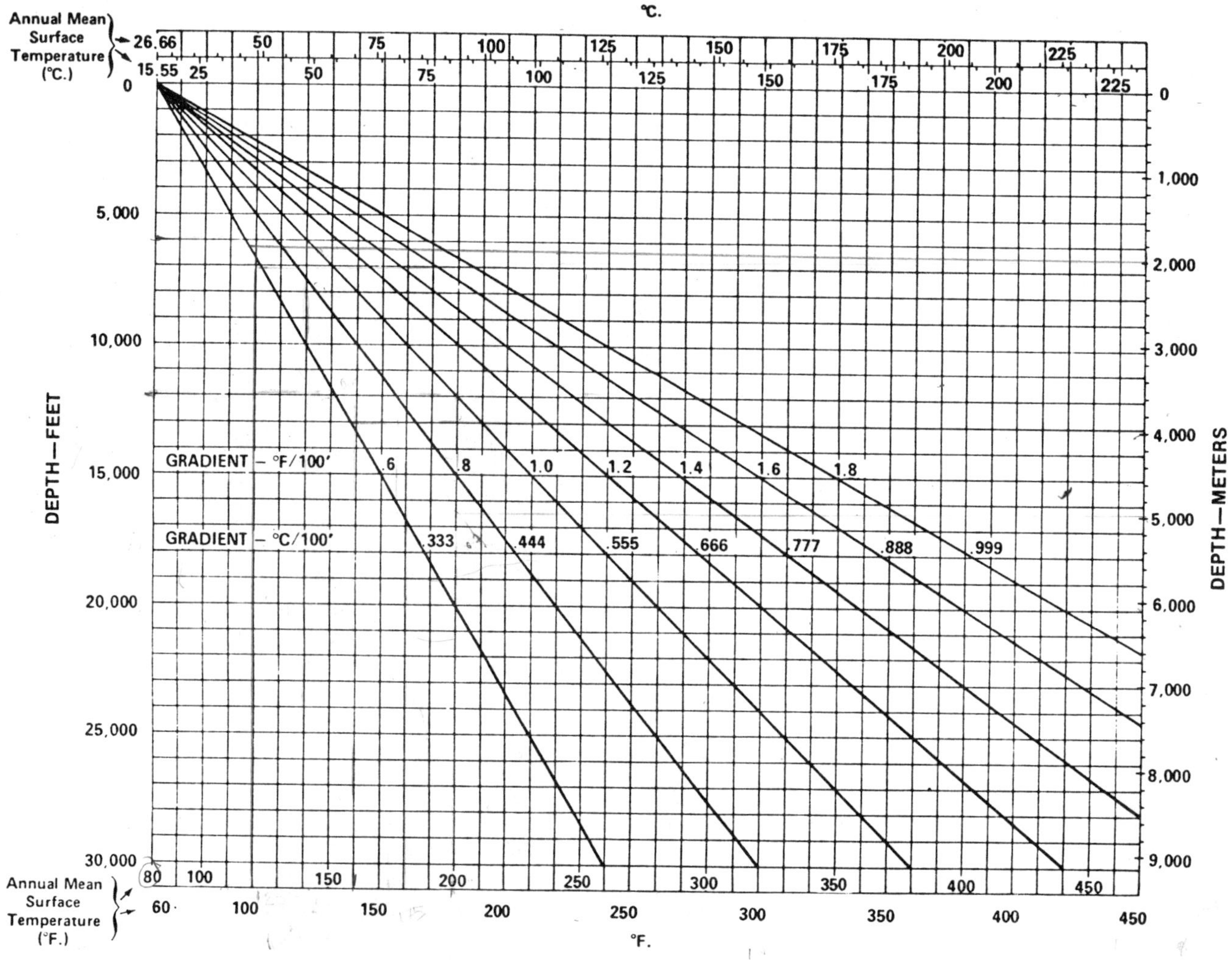

Example: Given: Temperature at 12,000′ from a maximum reading thermometer = 200°F.; average annual surface temperature = 80°F. Find the temperature at 6,000′.

Solution: Temperature gradient is 1.0°F. per 100′; temperature at 6,000′ is 140°F., assuming an annual mean surface temperature of 80°F.

Figure 2-4. Formation temperature-depth relationship. (Courtesy of Gearhart.)

since $F = R_o/R_w$ solving for R_o:

$$R_o = FR_w$$

thus water saturation can be expressed as:

$$S_w = \sqrt{R_o/R_t} \qquad (2\text{-}5)$$

and hydrocarbon saturation $= 1 - S_w$

Porosity Determination

For a constant porosity, the ratio of R_o to R_w remains nearly constant for all values of R_w below 1 ohm-meter. Experiments show that in more resistive waters, the value of F is reduced as R_w rises. This phenomenon is due to a greater influence of the surface conductance of the grains in fresh waters.

The value of R_o should be calculated, rather than picked from the log. Finding a clean zone that is

100% saturated with water is a difficult task. This places us in a difficult position because we must know the value of the formation factor.

Archie experimentally determined that the formation factor could be determined from three factors:

- Porosity Φ
- Cementation m
- Rock texture a

Thus:

$$F = \frac{a}{\Phi^m} \tag{2-6}$$

Through extensive use of this relationship, the following values have been used with great success:

$$F = .81/\Phi^2 \quad \text{in sands} \tag{2-7}$$

$$F = 1/\Phi^2 \quad \text{in carbonate rocks} \tag{2-8}$$

These two formulas are similar to the *Humble formula* developed from an intensive study of sandstone samples in the 1950s, i.e.:

$$F = .62/\Phi^{2.15} \tag{2-9}$$

Conclusion

It is not important to fully understand each concept presented in this chapter. The purpose is to acquaint you with the basic terms and relationships of analysis. Everything presented here will be covered in great detail in chapters to follow.

Proceed reading through Chapter 5, and review this chapter before reading Chapter 6.

3

Spontaneous Potential

The spontaneous potential (or SP) curve is a recording of naturally occurring earth potentials between a moveable electrode in the borehole and a fixed surface electrode. Such potentials are in no way induced by the passive measuring equipment. For this reason, SPs are also referred to as "self potentials." The SP is generally recorded in the left-hand track of the log. The SP is used to:

- Detect permeable beds.
- Locate their boundaries and permit correlation of such beds.
- Determine values of formation water resistivity (R_w).
- Give qualitative indications of bed shaliness.

The readings of the SP in shale are usually fairly constant and tend to follow a straight line on the log, called the *shale base line*. In permeable formations, the SP curves show excursions from the shale base line. In thick beds it tends to reach an essentially constant deflection defining a sand line. The SP may deflect either to the left (negative) or to the right (positive) of the shale base line, depending on the relative salinities of the formation water and of the mud filtrate. Figure 3-1 illustrates the typical SP response in shale, sand, and limestone.

The SP Log

In Figure 3-2, an electrode A, located at the end of an insulated cable C, is moved up or down the mud-filled borehole. The insulated conductor is connected to one terminal of a recording galvanometer; the other terminal of the galvanometer is connected to a potentiometer circuit P, and then to electrode B, usually placed in the mud pit or in a mud-filled hole dug for this purpose. Electrode B is called the SP ground.

The movement of film in the recorder is synchronized with the movement of the electrode along the borehole. The recorder registers the SP proportional to the depth of electrode A.

The boreholes in which the SP is recorded are usually filled with mud having a water base. The mud density is such that at each depth the hydrostatic pressure in the hole is greater than in the formations; as a result, the fluid contained in the permeable beds cannot contaminate the mud. Also, the mud is in constant circulation during the drilling operation, prior to logging, and therefore is homogeneous.

The recording galvanometer measures all the differences of potential appearing between electrodes A and B. The deflections on the SP log correspond to phenomena occurring at the contacts between the mud and the different beds, and also at the contacts between the beds themselves. These phenomena produce an electric current, which uses the mud as its return path. In doing so, it creates in the mud, by ohmic effect, potential differences which can be measured.

The position of the shale base line on the log recording has no useful meaning for interpretation purposes. SP scale sensitivity is chosen and the shale base line is positioned by the logging engineer so that the SP curve deflections remain in the SP track.

The SP cannot be recorded in holes filled with nonconductive muds because such muds do not provide

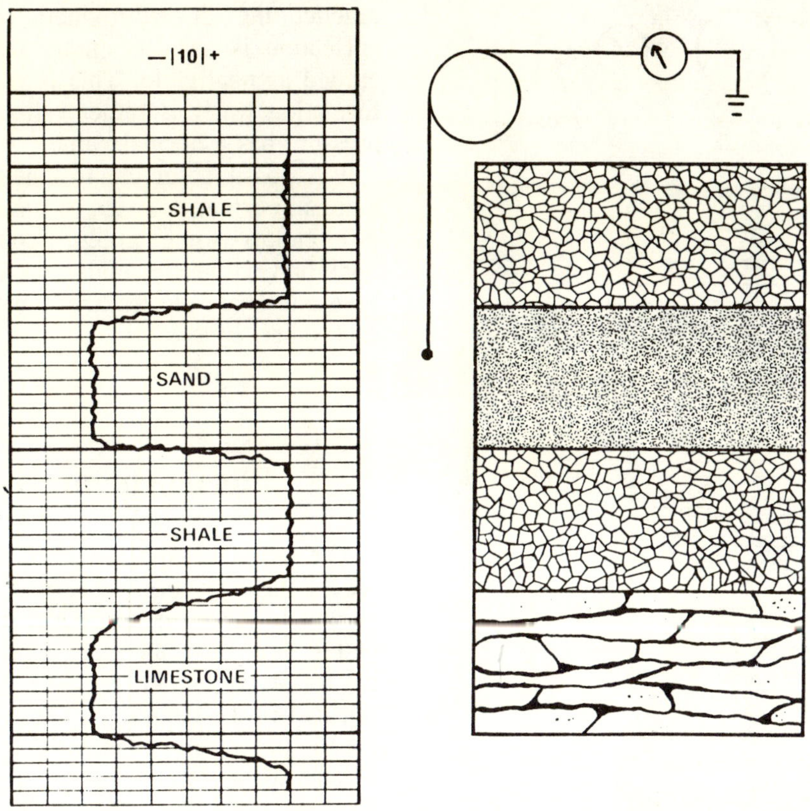

Figure 3-1. Spontaneous potential. (Courtesy of Gearhart.)

electrical continuity between SP electrodes and formation.

Normal SP deflections occur only in porous, permeable beds. The SP log, although indicating permeability, cannot be used to quantify either permeability or porosity.

SP deflections are due to earth currents which develop an electromotive or electrical force in the mud column. These SP currents are generated by two sources called electrochemical and electrokinetic.

Greater significance is often attached to each variation in the SP curve, so the presence of spurious potentials, or potentials induced by outside sources, on the log is undesirable. If present and bothersome, proper steps by the logging engineer should be taken to remove them. Some spurious potentials have no effect on the magnitude of SP deflections. The electrodes are chosen to be stable insofar as their contact potential with the mud is concerned. In practice, lead electrodes are used. A constant difference of potential may normally appear between the surface and downhole electrodes in the absence of any SP current. This difference of potential is not recorded on the SP log; it is counterbalanced by means of the surface potentiometer circuit used to set the shale base line. Accordingly, the potential of the downhole electrode is measured on the SP log with reference to an arbitrary constant. However, variations in potential (that is, the deflections on the SP log) do not depend on the arbitrary constant.

Electrokinetic Potential

An electrokinetic potential, E_k (also known as streaming or electrofiltration potential), is produced by the flow of an electrolyte through a nonmetallic, porous medium. Its magnitude is determined by several factors, among which are the differential pressure producing the flow and the resistivity of the electrode in the porous medium. Physical movement of electrolitic ions represents a moving charge and constitutes an electric current, much like the more familiar move-

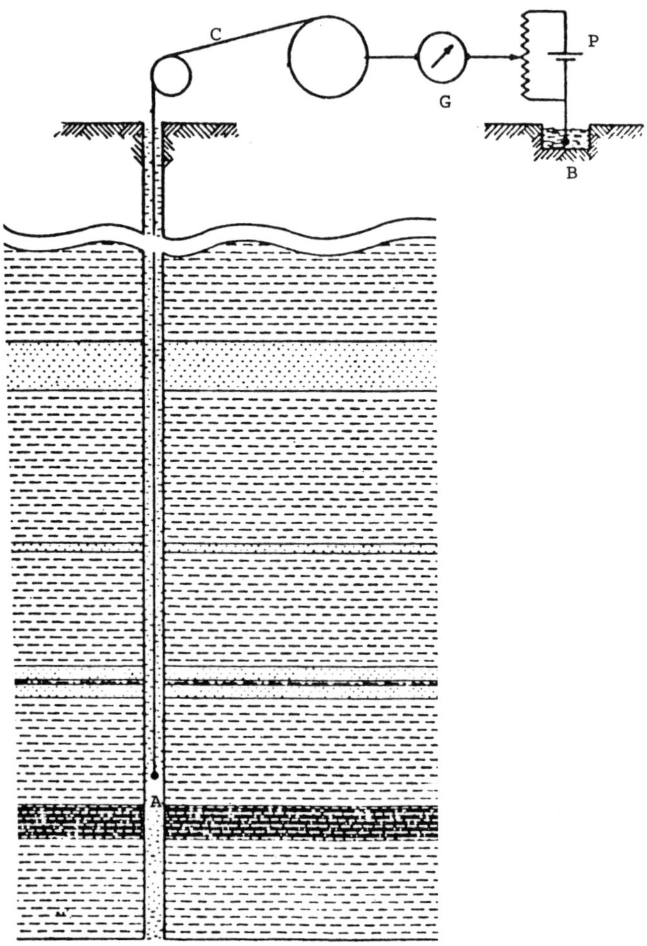

Figure 3-2. Schematic of the SP tool.

ment of electrons in a metallic conductor. Such an electrokinetic force is produced by the movement of filtrate through the mudcake on a permeable formation. Usually little or no electrokinetic force is generated across the permeable formation itself. This is because practically all of the differential pressure between borehole and formation is expended across the much less permeable mudcake. The remaining differential pressure across the formation is normally not great enough to produce an appreciable force.

An opposing electrokinetic force (E_k) is produced across the shale which often has sufficient horizontal permeability to permit a tiny amount of filtration flow from the mud. Horizontal permeability of the mudcake on the sand is essentially the same as that of the shale which has little or no mudcake. Thus, the net electrokinetic contribution to the SP deflection measured from the shale base line is the difference between the sand, mudcake, and shale potentials. In practice, the net electrokinetic contribution to the SP deflection is usually small, and it is normally regarded as negligible. This is particularly true if the formation water is rather saline and the differential pressure has a normal value.

It is possible for electrokinetic effects to become more important in cases of unusually large pressure differentials (for example, depleted-pressure formations or very heavy muds). In these cases, the mudcake and shale electrokinetic effects may not cancel each other.

Electrochemical Potential

The second component of the SP is mostly responsible for the deflections observed on the log. This component is the electrochemical portion of the SP. To better understand the electrochemical force, consider the following model. In 1943, Mounce and Rust duplicated in the laboratory the essential ingredients of the SP. In their model (Figure 3-3) a circular dish was filled with shale and waters of two different salinity concentrations, each separated with permeable membranes to prevent diffusion. A current was found to flow in the electrochemical cell thus created. Voltage generated was found to increase as the salinity contrast between the two solutions increased. When both

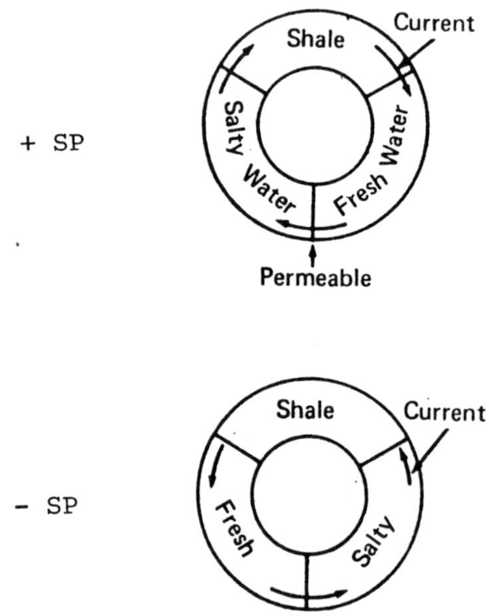

Figure 3-3. SP model after Mounce and Rust.

solutions were identical, no current flowed. The direction of current movement is shown by the arrows of Figure 3-3. Current flows from the fresh to the saline water and then to the shale. If the two waters are reversed, current flows in the opposite direction.

Consider a permeable formation, with thick, shale beds above and below; assume too, that the two electrolytes present, mud filtrate and interstitial water, contain only sodium chloride ions. Due to their layered clay structure and the electrical charges normally present on the layers, shales behave like ion-selective membranes, passing only the sodium cations but blocking the chlorine anions. When a shale separates sodium chlorine solutions of different salinities, the sodium cations (positive charges) move through the shale from the more-concentrated to the less-concentrated solution. This movement of charged ions is an electrical current, and the osmotic force causing them to move constitutes a potential across the shale (see Figure 3-4).

The curved arrow shows the direction of current flow in the well-bore, corresponding to passage of sodium cations through the adjacent shale from the salty formation water in the bed to the less-saline mud and back into the formation. Current flow from the sand to the shale is responsible for that component of the electrochemical SP known as membrane potential.

The other component of the electrochemical SP is produced at the edge of the invaded zone, where the mud filtrate and formation water are in direct contact. Sodium cations and chlorine anions can transfer from one solution to the other with ease. Since chlorine ions are more mobile than sodium ions, the net result is a flow of negative charges (chlorine anions) from the more-concentrated solution to the less-concentrated solution. This is equivalent to a conventional current flow in the opposite direction, indicated by the straight arrow A in Figure 3-5. The current flowing across the junction between the solutions of different salinity is produced by the liquid-junction potential. Liquid junction potential is only about one-fifth of the membrane potential.

R_w from the SP

If the permeable formation is not shaly, the total electrochemical force corresponding to the liquid junction and membrane potential, is equal to:

$$SP = K_{log} \frac{a_w}{a_{mf}} \tag{3-1}$$

Where a_w and a_{mf} are the chemical activities of the two solutions (formation water and mud filtrate) at formation temperature. K is a coefficient proportional

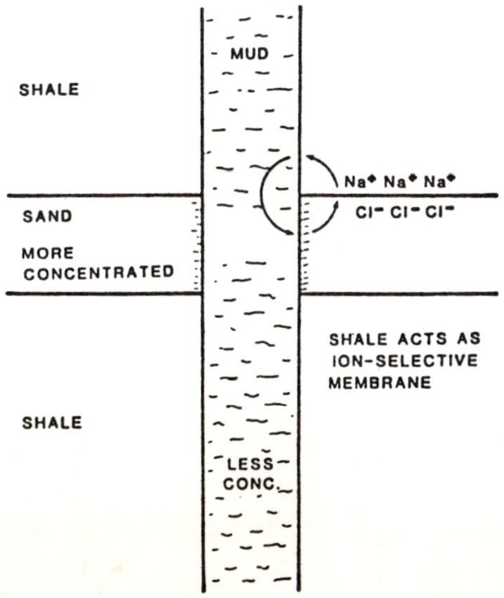

Figure 3-4. Illustration of membrane potential.

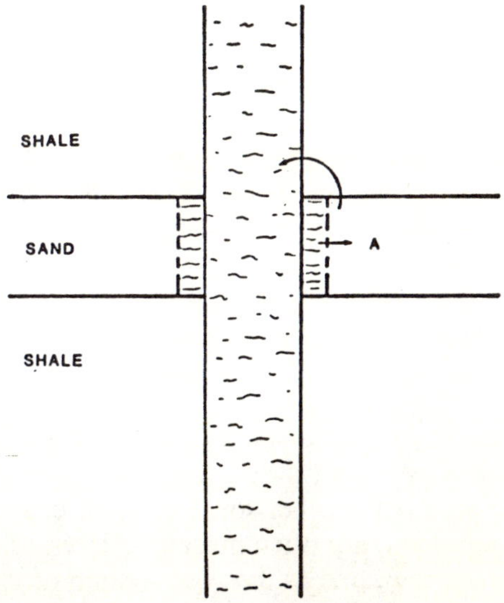

Figure 3-5. Illustration of liquid junction potential.

to the absolute temperature. Sodium chloride, formation water, and mud filtrate are equal to 71 at 77°F.

Since chemical activities are of little value to the log analyst, Equation 3-1 can be expressed in terms of water resistivities. Resistivity is also dependent on temperature.

Equation 3-2 solves for SP, given the resistivity of the formation water, the mud filtrate, and the formation temperature.

$$SP = -(60 + 0.133T) \log_{10} \frac{R_{mf}}{R_w} \quad (3\text{-}2)$$

where $-(60 + 0.133T)$ = the constant $-K$
T = the formation temperature in °F
R_{mf} = the resistivity of the mud filtrate
R_w = the water resistivity

Equation 3-2 is the most popular form of the SP relationship. Note that an empirical relation has been substituted for the constant K, which enables its value to be calculated. Thus, if the total SP deflection can be found from the log and K can be calculated and R_{mf} can be measured from a flowline sample, the desired quantity R_w can be found.

Several qualifications regarding the use of Equation 3-2 are in order. R_{mf} and the temperature at which it was found must be determined accurately. Such is not commonly the case in the field. Mud tests are typically performed by crew members while the engineer busies himself with other tasks. Crew members are not usually adequately trained and are unaware of the factors causing errors in their measurements. Time spent by the engineer to ensure good values is well spent. The R_{mf} measurement is the single most common source of difficulty in proper R_w determination.

Secondly, the expression $-K(60 + .133T)$ is, as stated, empirical and rests on no thermodynamic foundation; that is, K is not systematically temperature dependent. The higher the values of R_{mf} and R_w are, the more valid it becomes to use a temperature-dependent K. The value T corresponds to formation temperature in °F at the point where the SP deflection is read from the log.

Finally, it should be noted that actual filtrates and formation waters are rarely composed of pure sodium chloride solutions. Other ions, such as calcium and magnesium ions, are often present. Activities of such ions are not the same as those of sodium or chlorine. To account for their influence, an expression relating complex solution resistivity to an equivalent sodium chloride resistivity is used in a modification of Equation 3-2.

$$SP = -(60 + .133T) \log \frac{(R_{mf})_e}{(R_w)_e} \quad (3\text{-}3)$$

where $(R_{mf})_e$ and $(R_w)_e$ are equivalent resistivities. Equivalent resistivities are defined as proportional to the reciprocal of activities: $(R_w)_e = C/a_w$. The proportionality constant C is chosen so that $(R_w)_e = R_w$ for pure NaCl solutions with resistivities larger than about .3 ohm-meter. Where fresh water contains other ions, $(R_w)_e$ is different than R_w, corresponding to the resistivity of a pure NaCl solution having the same activity as the water in question.

In practice, $(R_{mf})_e$ is often assumed to equal R_{mf}; that is, the drilling mud system is assumed to contain only sodium chloride salts. This is not true when lignosulfonate muds or muds containing potassium, calcium, or magnesium are employed.

In such cases, R_{mf} must be converted to R_{mfe}. Either Equation 3-6 or 3-7 can be used, or a chart such as Schlumberger's SP-2 can be used to accomplish this task (Figure 3-6).

R_{we}, to be of value, must then be related to true water resistivity. The chart in Figure 3-7 is used for this purpose.

Where equations are preferred over charts, $(R_w)_e$ may be solved for directly:

$$\frac{R_{mfe}}{R_{we}} = 10^{-SP/K} \quad (3\text{-}4)$$

where $K = (60 + .133T)$
T = formation temperature in °F

The term $-SP/K$ is a mathematical term. If a negative SP is read from the log, then the term changes to positive.

$$-(-SP)K = +SP/K$$

or

$$10^{+SP/K}$$

Solving for R_{we} we transpose Equation 3-4 to:

$$R_{we} = \frac{R_{mfe}}{10^{-SP/K}} \quad (3\text{-}5)$$

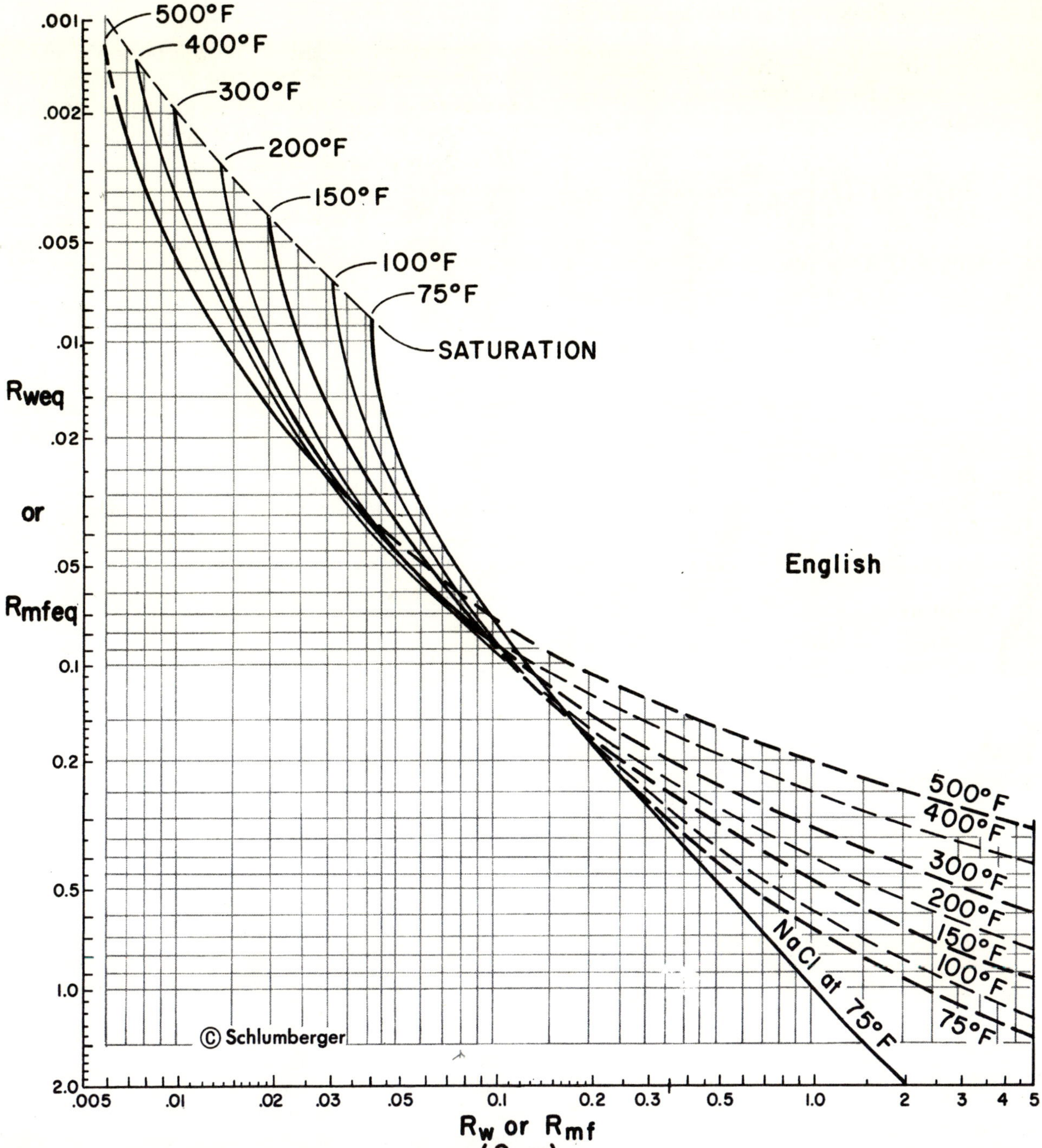

Figure 3-6. R_w versus R_{weq} and formation temperature. (Courtesy of Schlumberger.)

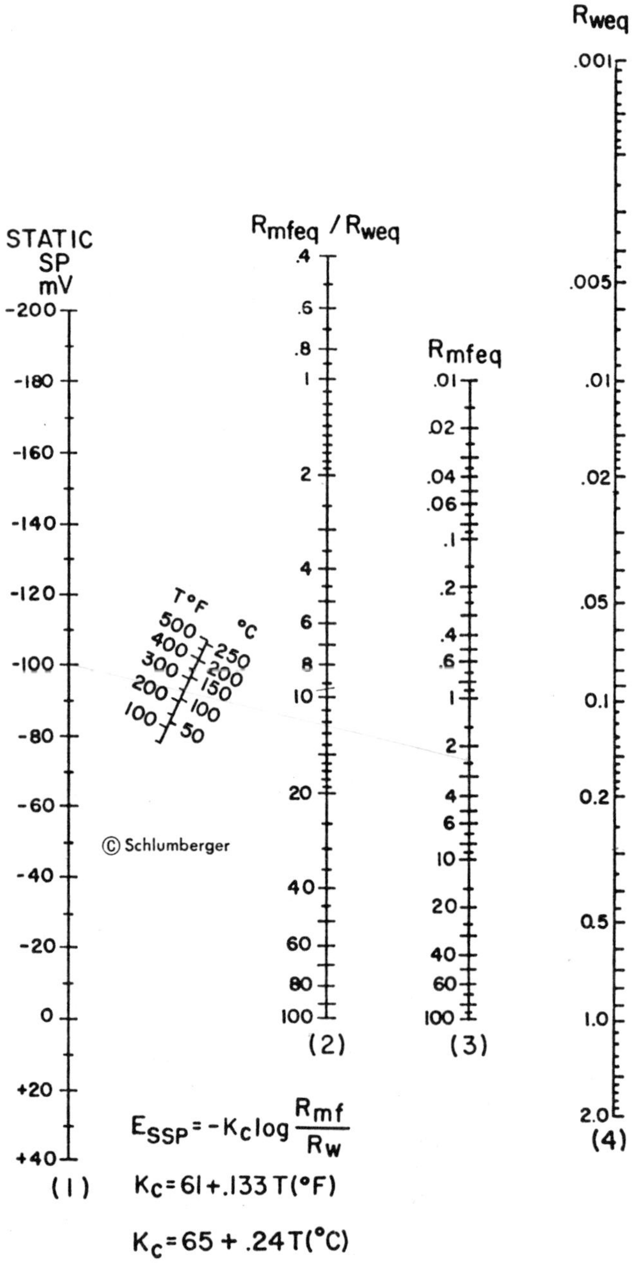

Figure 3-7. R_{weq} determination from the SP. (Courtesy of Schlumberger.)

In order to solve Equation 3-5 R_{mf} is first converted to R_{mfe} at 75°F. Either Equation 3-6 or 3-7 is used, depending on the following test:

If R_{mf} @ 75°F is more than .1, use:

$$R_{mfe} = 0.85\ R_{mf} \quad (3\text{-}6)$$

or if R_{mf} @ 75°F is less than .1, use:

$$R_{mfe} = \frac{146\ R_{mf} - 5}{337\ R_{mf} + 77} \quad (3\text{-}7)$$

The numerical constants of Equations 3-6 and 3-7 have been chosen for 75°F. The decision to use either one assumes R_{mf} to be at this temperature, therefore care must be taken to ensure all parameters are at 75°F. Now calculate K, using:

$$K = (60 + .133T)$$

where T = formation temperature

Read the SP from the log in a clean water-bearing interval and solve Equation 3-5.

Once R_{we} has been found, it is converted to R_w. If R_{we} @ 75°F is greater than .12 ohm-meter, use:

$$R_w = -0.58 + 10^{(0.69\ R_{we} - 0.24)} \quad (3\text{-}8)$$

If R_{we} @ 75°F is less than .12 ohm-meter, use:

$$R_w = \frac{77\ R_{we} + 5}{146 - 377\ R_{we}} \quad (3\text{-}9)$$

The Sp value used in the preceding equations or charts may require corrections. If the permeable bed is thin or contains some shale or dispersed clay, for example, SP deflections may be reduced. We shall now explore the influence of some of these factors as we discuss electromotive forces (emf) and the static SP.

The Static SP (SSP)

As stated before, the emfs generating the SP current, which affects the SP log, arise from two types of phenomena. The first one is of electrokinetic nature, producing an emf of filtration at the drilling mud/permeable bed contact. Since the hydrostatic pressure of the mud in the borehole is greater than the pressure in the permeable formation, some mud fluid filters slowly through the mudcake into the permeable bed. This causes an emf to appear primarily where the pressure difference is maximum, that is, across the mudcake. The emf depends on the nature of the filtrate and the filter and on the pressure difference. As a result, for a given formation, the emf will be uniform all along the drilling mud and the permeable bed contact. If the difference in pressure is about the same for various permeable formations traversed by the borehole, the emf of filtration will also be the same.

The second phenomenon is an electrochemical one. It occurs at the contact of media of different natures, and creates an emf at each such contact. For instance, there are three contacts or boundaries shown in Figure 3-8 at A, B, and C, as follows:

A = boundary between mud and salt-water sand
B = boundary between salt-water and clay
C = boundary between clay and mud

An emf of electrochemical origin exists at each of the boundaries, A, B, and C. As each medium is fairly homogeneous, each emf is uniform along the corresponding boundary.

At the boundary A, an electrokinetic phenomenon as well as an electrochemical one is present; thus, there is an emf corresponding to the algebraic sum of two forces of different origins.

In order to get a better understanding of the effect of the emf, it is convenient to consider first an idealized case where the SP currents are prevented from flowing. In this case (Figure 3-9) a borehole section traverses two identical, thick beds of clay separated by a rather thin salt-water sand. Although this would not easily be feasible in practice, it may be conceived that two insulating plugs are placed in the hole to interrupt the electrical continuity of the mud column at the two boundaries between sand and clay.

The presence of the plug does not affect the emf; at the boundaries A, B, and C there are emfs designated respectively by a, b, and c. Since the plugs prevent any current flow, the potential within each single medium, enclosed by boundaries or plugs, is then constant. However, the potential varies from medium to medium, the difference of potential between two adjacent media being equal to the emf existing at their common boundary.

The algebraic addition of the emfs A, B, and C, and the potential V_o determines their potential from one medium to the next. The values of the potentials are as follows:

- Potential in the upper section of the mud is V_o
- Potential in the upper shale is $(-V_o) + (-C)$
- Potential in the salt water sand is $(-V_o) + (-b) + (-c)$
- Potential at interface A between the plugs is $-(V_o) + (-a) + (-b) + (-c)$
- Potential in the lower shale is $-(V_o) + (-c)$
- Potential in the lower section of the mud is $-V_o$

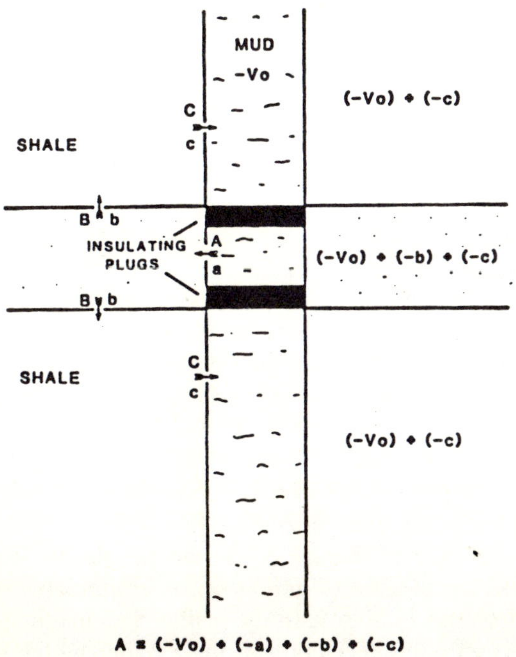

Figure 3-8. Illustration of static SP charges.

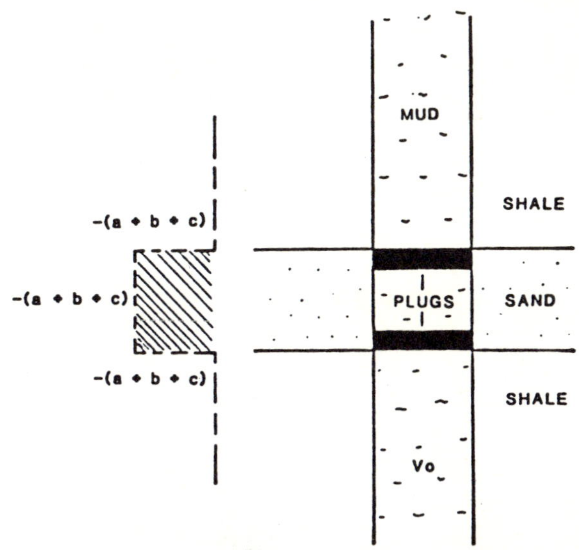

Figure 3-9. Illustration of static SP curve.

From the point of view of the SP log, it is, of course, essentially the potentials in the different sections of the mud which are of interest to us. In that respect, it is very important to remark that the potential of the mud section in front of the lower clay is the same as in the upper section. It is also fundamental to observe that the potential of the mud in front of the sand differs from the potential $-V_0$ in front of the clays by the quantity (a + b + c), which is the sum of the emfs of the three boundaries a, b, and c.

At this point, it is pertinent to remark that the boundary emfs show their effect through a combination of three boundaries having a common junction. This is quite fundamental in the analysis of SP logs. The combination of the three emfs at the boundaries, along any closed path traversing all three media, establishes what is conveniently called a three-link chain emf. The left hand (dotted curve) part of Figure 3-10 represents the SP of the ideal case.

On this diagram the potential in the mud versus depth is plotted. As it corresponds to the case where no current is flowing, or in other words, to a static case, it will be designated hereafter as the *static SP diagram.* Although a purely theoretical concept, the static SP is of great interest. It represents, in a convenient manner, the values of the emfs which produce the SP currents, and which therefore determine the SP log.

Once the insulating plugs have been removed to reestablish the continuity of the mud column, there is no longer a static equilibrium, but rather a dynamic state. The SP current can flow in the borehole through the mud, as well as in the formations. The three emfs a, b, and c add their effects to generate the SP current which follows the paths on Figure 3-10 by the solid lines. Each line corresponds to a line of flow, the current circulating in the direction of the arrows.

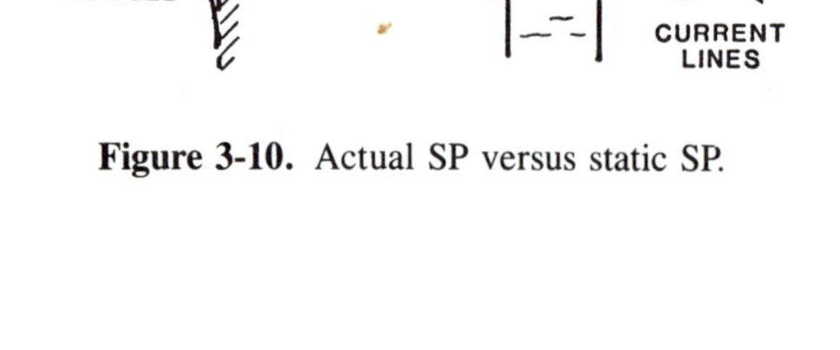

Figure 3-10. Actual SP versus static SP.

Each current line must necessarily cross the three boundaries, A, B, and C. Furthermore, that part of the current generated by each of the three emfs, a, b, and c, follows the same path; the current lines are independent of the repartition of the emfs between the three boundaries. In other words, the intensity of the current circulating in the mud of the borehole depends only upon the algebraic sum of all the partial emfs in the circuit, and does not depend upon the allocation of these partial emfs to each boundary, provided that each emf is uniform everywhere on its corresponding boundary.

Along its path, the SP current has to force its way through a series of resistances, both in the ground and in the mud. In so doing it produces potential differences according to Ohm's law. Along a given line of flow, the potential falls down continuously in the direction of the current, as indicated by the arrows, but at each boundary where an emf occurs, the potential is raised by the amount of the value of the emf. Along a closed line of current flow, the total drop of potential is equal to the sum of the emfs encountered.

Also, the intensity of the current being constant along its path, the potential drop varies according to the resistivity of the section through which it flows. This means that the total potential drop (which is equal to the sum of the emfs) is divided between the different formations and the mud in proportion to the resistivities encountered by the current in each medium. Accordingly, the potential drop in the borehole is only part of the emf, unless the electrical resistance offered by the mud is very large compared to the one in the formation.

The SP log records the potential drop occurring in the mud. It follows that the amplitude of the peak of the SP log approaches the amplitude of the static SP. When the resistance of the mud to the flow of the SP current is not large compared to the resistance in the formation, then the SP log will show a peak of lesser amplitude than the static SP.

It may also be seen that the current circulates in the mud, not only opposite the permeable formation, but also part way beyond the boundaries of the formation. As a result, though the static SP diagram indicates a sharp break corresponding to the boundaries of the permeable bed, the SP log exhibits a more progressive change in potential, extending along the borehole beyond the boundaries of that bed.

In the case of Figure 3-10, the permeable bed is thin, so the resistivity in that bed is appreciable compared to the total resistivity in the SP current path. This is why the deflection of the SP log, which measures the potential drop in the mud, is only a fraction of the total emfs. To make that point clearer, the SP log is presented as a solid line together with the static SP diagram (dashed line) whose deflection characterizes the total emfs involved.

As shown, the deflection of the SP log is not only smaller than the one of the static SP, but it is also much more progressive. It is interesting to note that the slope of the SP log measures the potential drop per unit length in the hole, which is proportional to the intensity of the SP current in the mud at the corresponding level. Starting from the top part of the log and going down, the slope increases progressively because the current in the hole increases, until the level X (contacted clay-sand) is reached. At that level, the intensity of the current in the hole is maximum, and this corresponds on the SP log to a maximum slope, or in other words, to an inflection point. Below that level, the current progressively decreases until it becomes nil in the middle of the sand; this corresponds to the point of maximum deflection. Farther down, the current flows in the opposite direction so the slope of the SP log is reversed. The slope increases gradually, until a new maximum is reached at the level Y (lower sand-clay contact), which corresponds to another inflection point on the SP; and still farther down, the slope progressively decreases again because the current itself decreases.

Keep in mind that the circular paths at sand-shale junctions are current flows only. The SP is developed across a resistance which is in the mud. Since the mud contains salt, it can be considered an electrolyte. Electrolytes are conductors of electricity, but, as all conductors have a certain amount of resistivity, so does the mud.

To illustrate the effect, consider the formation as an electrical circuit (Figure 3-11). As the SP electrode moves into the hole, it comes in contact with the first junction emf and deflects to its maximum negative level. Keep in mind that what is being measured is the potential drop or voltage across R_m, the mud resistivity. This drop is measured differentially across the sand and is thus responsible for the shape of the SP curve.

The Shape and Amplitude of the SP Curve

As the electrode moves closest to the maximum emf, you will note an inflection point on the curve. An inflection point is that part of a curve where the slope changes (Figure 3-12). The inflection points of the SP log are important for their interpretation.

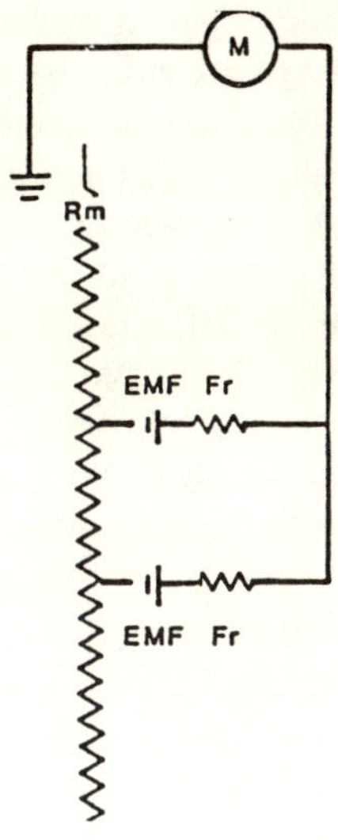

Figure 3-11. Electrical schematic of the SP.

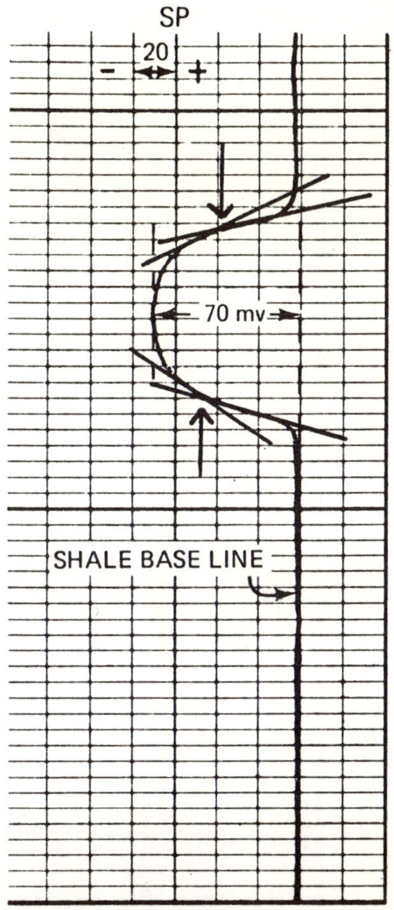

Figure 3-12. Illustration of bed boundaries. Two lines, or "ski slopes," are drawn. Where the lines cross is the point of inflection. This corresponds to the bed boundary. (Courtesy of Dresser Atlas.)

There has sometimes been a tendency to place the boundary between a permeable and an impervious bed at the point which corresponds to a half deflection on the SP log. This can be substantially wrong in some cases. The contact level should be taken as corresponding to the inflection point on the SP log.

When the SP curve deflects towards the left, it is considered negative. This is the case when formation water is more saline than the mud. If the reverse is true (Figure 3-13), that is, if the mud contains higher concentration of sodium chloride than the formation, then the SP deflection will be to the right, or positive. Although it may not have been clear earlier, it was stated that sodium chloride cations move through a shale or clay from the more-concentrated solution to the less-concentrated solution. In the case of a positive SP, the mud concentration is higher, and the sodium chloride cations flow from the mud through the shale and into the less-concentrated water sand. This constitutes a current flow opposite in direction from our previous examples.

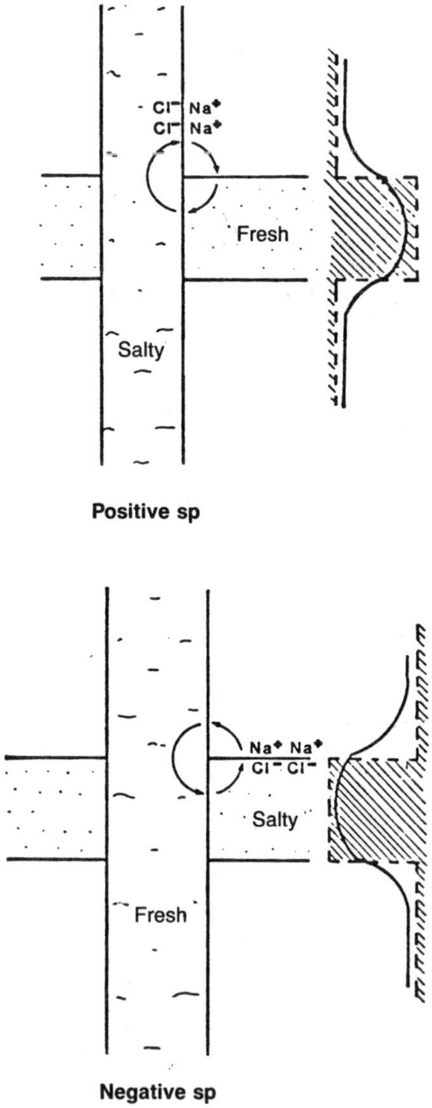

Figure 3-13. Illustration of positive and negative SP.

The shape and amplitude of the peak on the log opposite a given bed may be influenced by the following factors:

1. The total emf involved.
2. The thickness of the bed.
3. The resistivity of the bed, of the surrounding formations, and of the mud.
4. The diameter of the borehole.
5. The depth of penetration of the mud filtrate in the permeable beds.

Where the permeable beds contain some impervious and conductive material, such as shale, the SP log may also be affected by the presence of that material.

The SP log would be influenced additionally by the lack of homogeneity in the mud. A change in salinity of the mud at a certain level would result in a base line shift at that level. However, it has been found through experience that such changes in salinity are seldom encountered.

In this discussion, the expression "impervious bed" or "impervious formation" is always used to designate formations like, for example, shale or clay, which are at the same time impervious and porous. As they contain water, they are also conductive. Compact or hard formations which contain extremely little water, and have no permeability, are designated by the expression "hard formation." Finally, formations such as sands, porous limestones, and the like, which are permeable even though the permeability might be very low, are designated by the expression "permeable bed" or "permeable formation."

Thick Beds

All other factors remaining the same, a change of total emf affects the amplitude but does not otherwise modify the shape of the SP log. In practice, the emfs involved may vary from one hole to another. Either the salinity of the mud or of the formations may change or differential pressure between the mud and the formations may be different. In a given hole, however, there is a definite tendency for the total emfs to be the same for all beds of the same type. This is clearly shown by logs of thick, conductive, impervious and permeable beds. In front of the former, the SP log gives a good, straight base line. Peaks corresponding to the latter have generally the same value, which implies that the emfs are the same for all of them. True enough, there may be a number of thin, permeable beds which give peaks of smaller amplitude. Conversely, there might be, within thick, permeable formations, thin, impervious beds for which the SP log does not come back to the base line. This effect of the thin beds can be explained by other causes, as will be covered next. There is, therefore, no reason to assume that they produce smaller emfs. On the contrary, the fact that there is a good base line shows that all conductive impervious beds are of much the same nature. The fact that all large peaks are of the same amplitude shows that salinity and differential pressure are the same for all thick permeable beds. In all probability, these conditions should be the same for the thinner beds in between. It is quite logical to look for other factors, in particular, bed thickness, to explain the smaller deflections observed.

Permeable beds of different porosity, or with different dimension of grains, give the same emf if other factors are unchanged. The emfs are also independent of the permeability value, even down to fractions of one millidarcy. Electrochemical emfs depend only on the salinity of the permeable beds, on the nature of the impervious formation with which they are in contact, and on the nature of the mud. Electrofiltration emfs depend only on the differential pressure and on the nature of the mud, which itself determines the mudcake where that emf is generated.

It has been observed that salinity and differential pressure are not always constant for all permeable beds, especially at widely different depths or in very different formations. Fresh water sands or very salty sands will show respectively abnormally low or large amplitude peaks. The polarity of the peak even reverses, if the water in the sand is less salty than the mud. Depleted sands, where the pressure is very low, give peaks of large amplitude, especially with muds of very low salinity which favor the electrofiltration emfs. Generally these can be detected without too much difficulty in the thick permeable beds because the amplitude of the peaks on the SP log for the thick beds is equal to the total emf. In the thin beds, for which the peak amplitude on the SP depends on the thickness and on the resistivity, it is practically impossible to separate the effect of the different factors on the SP log, and therefore, to evaluate the emfs.

Thin Beds

For the purpose of explaining the influence of bed thickness on the SP, we will assume that the resistivity of all beds is the same and equal to the resistivity of the mud. This is not as theoretical as it may seem, since it occurs rather frequently in some areas.

Figure 3-14 shows a succession of saltwater sands separated by thick layers of shale. The SP log is somewhat rounded at each boundary, but its peaks are practically equal to those of the static SP diagram whenever the thickness of the sand is more than twice the hole diameter. For smaller thickness, the maximum value given by the peaks is definitely reduced. The SP log no longer reaches the static SP diagram. For sands having a thickness smaller than one-half of the diameter of the hole, the amplitude of the peak is approximately proportional to the thickness. With the assumption that all resistivities are equal, the proportionality factor is such that a thin bed whose thickness is X percent of the borehole diameter shows a deflection on the SP log which is about X% of the maximum possible.

Figure 3-15 represents an inverse case of the previous example. Shale beds of varying thicknesses are separated by thin saltwater sands. As can be seen, the log is simply a symmetrical image of the one previously shown. For thin shales, the deflection is approximately proportional to the thickness.

Figure 3-16 is a composite log, illustrating the SP log where bed thicknesses vary. The bottom portion is of particular interest, as it shows a succession of thin beds of sand and shale, hereafter referred to as a *sandwich*. Comparing the SP log of the sandwich with that of homogeneous sand A, it can be seen that the sandwich is characterized by:

1. A smaller average deflection.
2. Ripples about the average deflection.

When all resistivities are the same, as in this example, the average deflection represents a percentage of the total emf approximately equal to the percentage of sand in the sandwich. The ripple is a function of the thickness of the individual beds and decreases very quickly when individual thicknesses fall below one half the diameter of the hole. The discussion of thin interbedded layers in connection with the SP log is of particular importance when analyzing shaly sands.

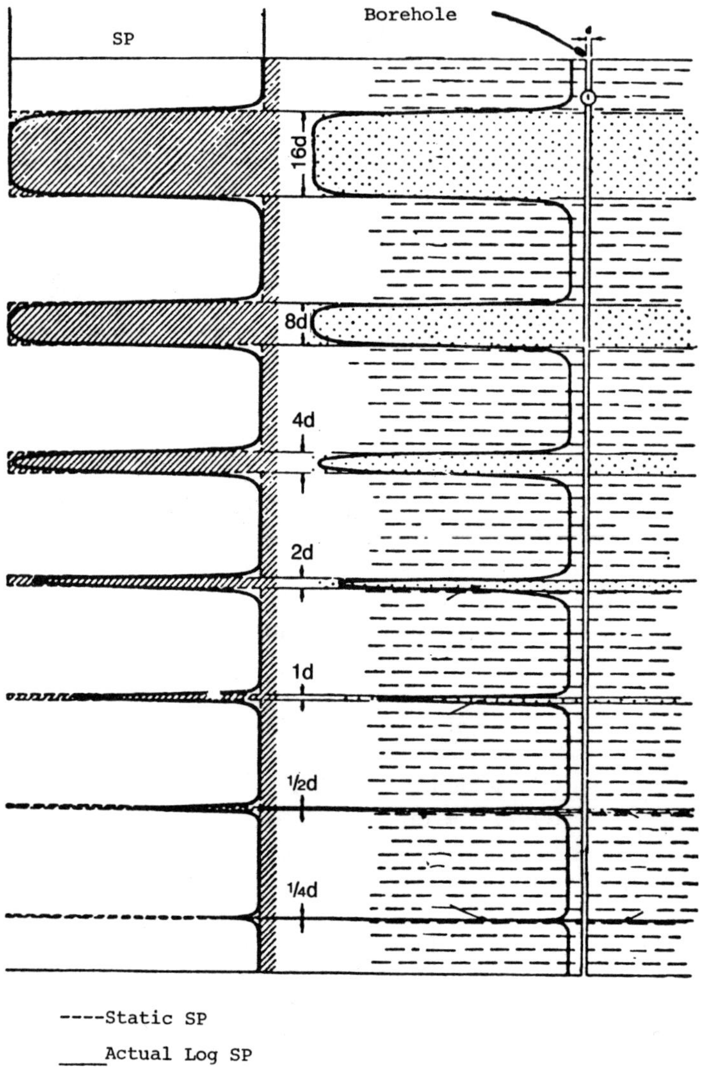

Figure 3-14. Illustration of the effect of bed thickness on the SP curve.

Resistivity Contrasts

In the previous examples, we assumed that formation and mud resistivities were equal. Now we are going to study the influence of resistivity of the mud and formations.

The effect of the resistivities of the media on the SP log is better considered as a function, not of their absolute values, but rather of the ratio between the resistivity of the formations R_t and the resistivity of the mud R_m. The value R_t/R_m is called the *resistivity ratio*. Figure 3-17 illustrates the influence of the resistivity ratio on the SP log.

In order to permit an accurate computation in each of the previous examples, the resistivity was chosen uniform for all beds, but now it is assumed greater than the mud resistivity. The SP log is similar in character to that where the formation resistivity is the same as the mud resistivity, except that:

1. The SP log is more rounded at the boundaries.
2. The peaks in corresponding thin beds are reduced in amplitude.

These two effects are more pronounced as the resistivity ratio becomes greater. This is illustrated by Figure 3-17, which corresponds to the same formation ar-

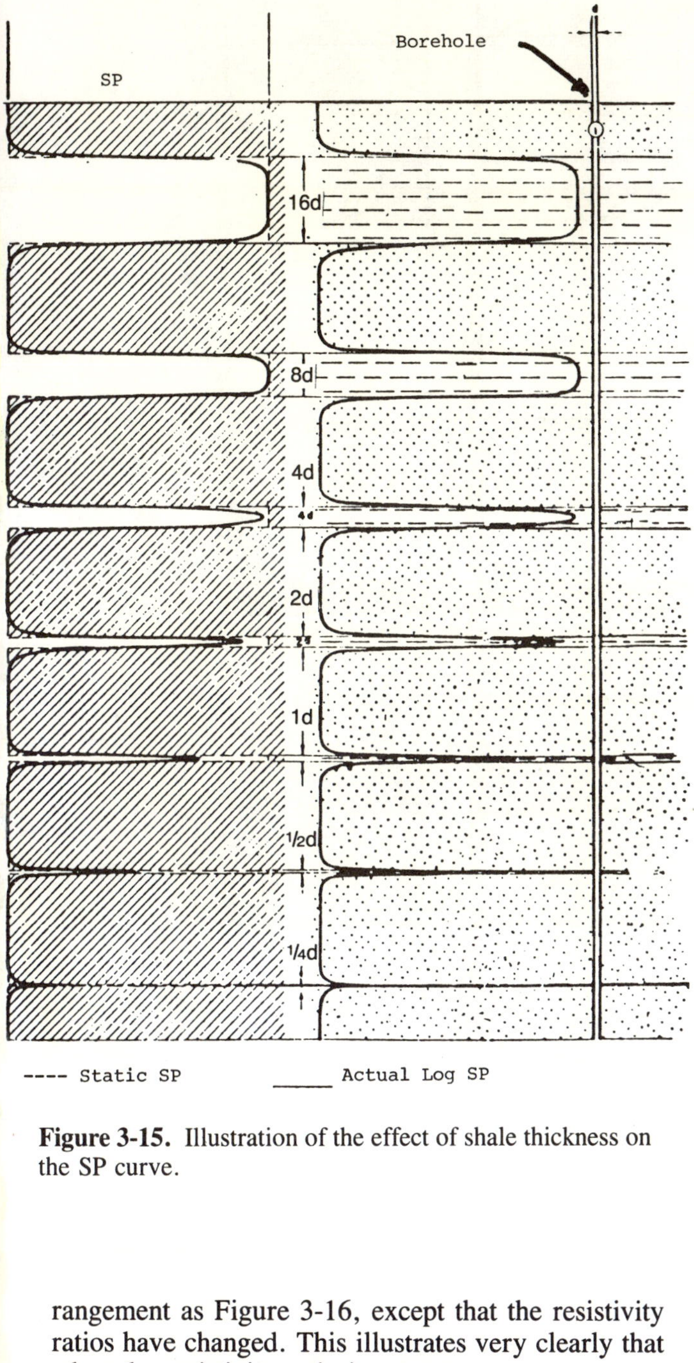

Figure 3-15. Illustration of the effect of shale thickness on the SP curve.

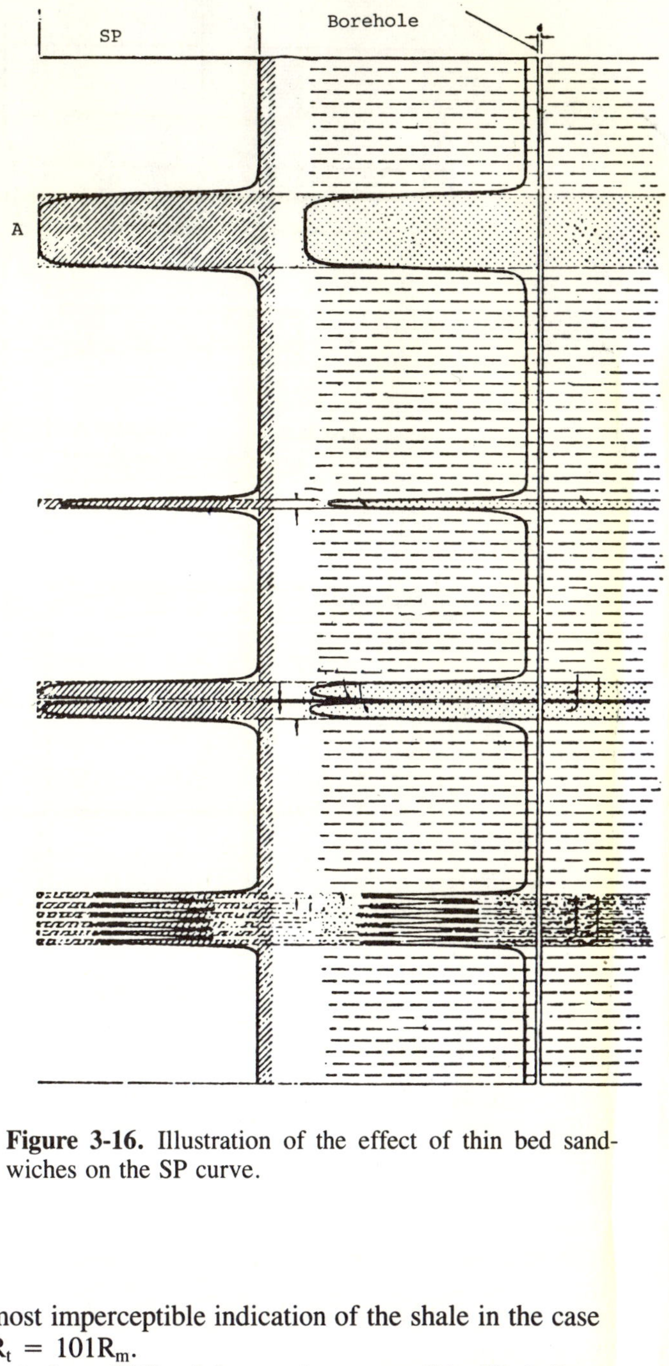

Figure 3-16. Illustration of the effect of thin bed sandwiches on the SP curve.

rangement as Figure 3-16, except that the resistivity ratios have changed. This illustrates very clearly that when the resistivity ratio increases:

1. The SP log becomes more rounded.
2. The peaks for thin beds are reduced in amplitude.
3. More generally, all details become less apparent.

This is particularly noticeable in the progressive decrease of the ripples for the sandwich A, and the almost imperceptible indication of the shale in the case $R_t = 101R_m$.

At least 90% of the maximum possible SP deflection is reached at the center of a permeable stratum where resistivity ratios are low and the thickness of the stratum is over six times the diameter of the borehole. Thus, thick permeable saltwater sands of low resistivity show SP values close to the static SP, and the maximum points on the SP logs do not vary appreciably for changes in thickness. For highly resistive media the SP deflection at the center of a permeable strata tends to increase linearly with their thickness. It should be noted that, except in extreme cases, the

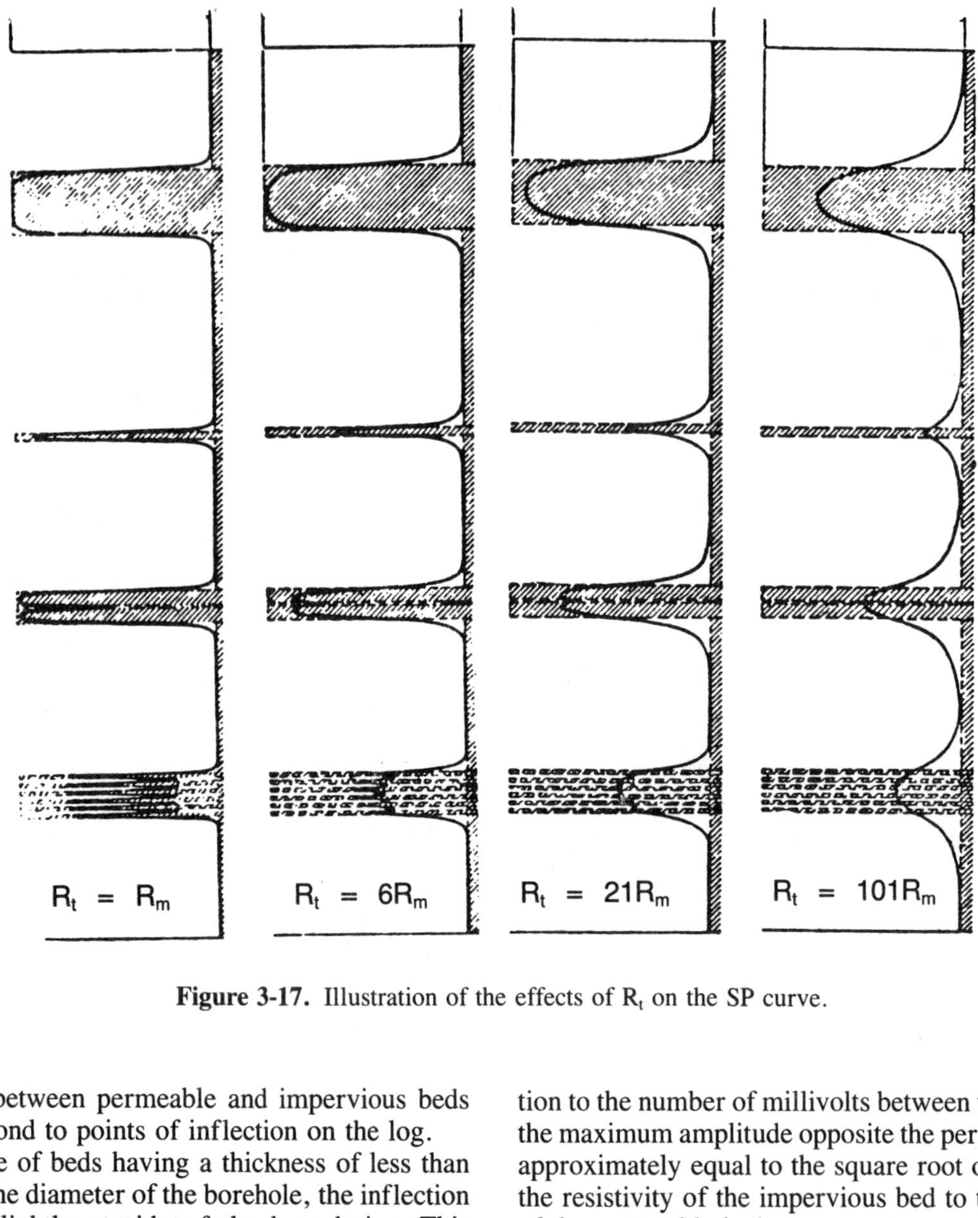

Figure 3-17. Illustration of the effects of R_t on the SP curve.

boundaries between permeable and impervious beds still correspond to points of inflection on the log.

In the case of beds having a thickness of less than one half of the diameter of the borehole, the inflection points are slightly outside of the boundaries. This however, does not appreciably affect the interpretation, since the beds whose boundaries have to be determined in practice are generally substantially larger thicknesses.

When the resistivities are different in a permeable bed and in the adjacent impervious stratum, the shape of the SP log will lack symmetry when it crosses the boundary. It will be more rounded in the more resistive formation. Accordingly, the point of inflection will be displaced toward the top of the peak if the peak corresponds to the less-resistive formation, and vice versa. If both beds are thick, the point of inflection on the SP log may be determined with good approximation. The ratio of the number of millivolts between the base line of the impervious bed and the point of inflection to the number of millivolts between that point and the maximum amplitude opposite the permeable bed is approximately equal to the square root of the ratio of the resistivity of the impervious bed to the resistivity of the permeable bed.

Hole Diameter

In discussing the influence of hole diameter, it is interesting to note that an increase in hole diameter acts approximately like an increase in the resistivity ratio: it tends more to round the deflections on the SP log, and to reduce the amplitude of the peaks opposite thin beds.

Invasion

Invasion by the mud filtrate influences the SP. The deeper the invasion becomes, the smaller the amplitude becomes. For the purpose of analysis we must simply determine if there is some invasion. This determination will allow us to choose the proper correction chart.

The behavior of the SP peak for a given permeable stratum is qualitatively that to be expected for an increased hole diameter with no invasion. Invasion reduces the amplitude of the peaks corresponding to thin permeable beds. It has the same influence on the appearance of inverse peaks corresponding to thin, impervious beds located in invaded permeable formations.

It sometimes happens that a hole is first drilled with one mud and afterwards the salinity of the mud is changed. This does not affect the analysis given, except insofar as the process of filtration may have created a zone of appreciably different resistivity in the permeable formations.

The SP electrofiltration component is practically unaffected by invasion. The only appreciable effect is due to the resistivity changed in the invaded zone, since the electrofiltration emf remains across the mudcake. For that reason, and when the electrofiltration emf is responsible for a substantial part of the SP, the overall invasion effect is generally less pronounced.

In very permeable formations SP anomalies are often observed, which, if not well understood, may cause errors in evaluation. When a very permeable saltwater sand is invaded, fresh filtrate, being lighter than saltwater, will tend to float toward the upper boundary of the sand. An invasion profile such as the one shown in Figure 3-18 will develop. Invasion is very shallow near the lower boundary of each permeable interval and deep near the upper boundary. The SP is affected as follows:

- At the upper boundary the curve is rounded off because of the deep invasion.
- At impervious shale streaks, the SP may have a "sawtooth" profile, as illustrated by Figure 3-18.

Just below the shale streak the SP is less than the SSP (static SP). This anomaly is caused by the accumulation of filtrate below the shale streak; encircling the hole is a horizontal disk-shaped cell, consisting of a shale disk sandwiched between saltwater and mud filtrate. The emf of this cell superimposed on the SSP produces the anomalous profile.

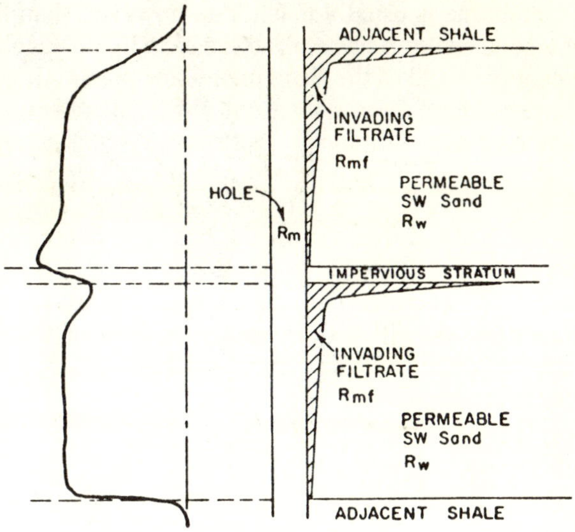

Figure 3-18. Invasion profile and its effect on the SP curve. (Courtesy of Schlumberger.)

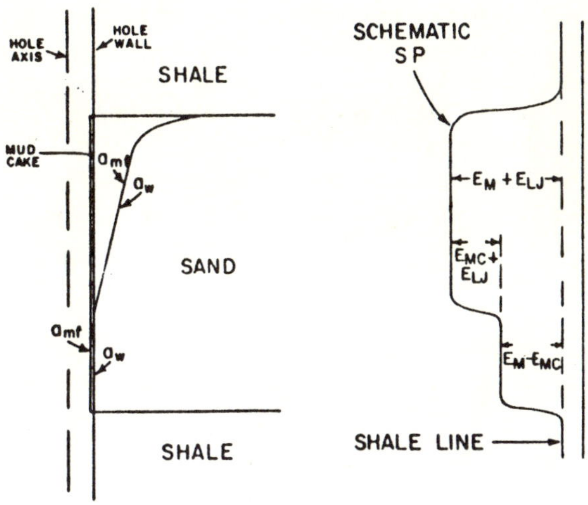

Figure 3-19. Invasion profile with resulting effect of liquid and membrane potential on the SP curve. (Courtesy of Schlumberger.)

The invasion may vanish completely at the lower left part of a very permeable bed, with an invasion profile as shown in Figure 3-19. Where there is no invasion, a reduced SP deflection is observed. Filtrate and the interstitial water are no longer in direct contact, but are separated by the mudcake, which acts as a cationic membrane. The efficiency of the mudcake as a membrane is usually much less, however, than that of a good shale. As a result, there is no liquid-junction potential to add to the shale membrane potential, as is the case where there is invasion. Where there is no invasion, the liquid-junction potential is replaced by a mudcake membrane potential which is of opposite direction. SP deflection is reduced as shown. Since invasion in the lower part of the bed may decrease or increase as mud and hole conditions change, this reduced SP phenomenon may appear or disappear. Sometimes the SP is decreased over most of the bed because invasion exists only in too thin a plane, at the top, to be apparent.

Shaly Sands

The strata under consideration have thus far been bounded usually by comparatively thick beds. It is desirable, however, to examine the effect of interbedded layers of permeable and impervious strata. When there are thin layers of sand in shale, or thin layers of shale in sand, their combination constitutes a "sandwich" which can be considered a shaly sand.

Figure 3-20 shows the following points, with respect to the SP log:

1. On thick sandwiches, the average deflection is approximately proportional to the percentage of permeable beds.
2. The average contour corresponding to a sandwich of finite thickness is the same for a homogeneous permeable bed of the same thickness and resistivity, but for which the total emfs involved would apparently be smaller.
3. The amplitude of the ripples around the average curve decreases very quickly with the decreasing thickness of the individual beds, so that the ripples are hardly noticeable when the individual thickness of both the impervious and permeable beds is less than one half the diameter of the borehole.

The following properties, which are not illustrated, must also be mentioned:

1. The amplitude of the ripples decreases when the permeable beds are invaded by the mud filtrate.
2. The average amplitude of the peaks decreases when the resistivity of the permeable beds increases with respect to that of the impervious beds.

The latter phenomenon is of particular interest when a certain section of a shaly sand is oil bearing, while another section is water bearing.

The expression shaly sand has been applied in a general way to interbedded thin streaks of sand and shale, or in other words, to stratified compounds of sand and shale. Sand beds with unstratified shaly material also come under the category of shaly sand. Whether the different particles enclosed in a shaly sand are stratified or not, the compound behaves substantially in the same way from the point of view of the SP log. By way of exception, shale particles entirely surrounded by mud filtrate or by the original interstitial water generate no SP currents because there is no chain emf. When, however, a shale particle is in contact on one side with the mud filtrate, and on the other side with the original interstitial water, the three media constitute a three-link chain and generate SP currents. SP currents generated around each particle add their effects, and the corresponding SP log is identical to the one obtained where the same amount of shale is present in the form of thin interbedded shale streaks. The amplitude of the SP log does not depend on the type of shaly material repartition in a permeable shaly sand, provided, of course, that the average repartition is uniform. The amplitude, however, is maximum for a clean sand, and it is reduced with an increasing percentage of shaly material.

If there is a progressive change in the nature of the fluid, going from mud filtrate to interstitial water, in the zone of the unstratified shaly sand, the result is still the same as for a stratified shaly sand. In that case, there are, of course, no shale particles simultaneously in contact with the mud filtrate and with the original interstitial water, but there are a large number of shale particles, parts of whose surfaces are in contact with water of varying salinities. Each such particle generates a small emf, but because there are many more particles involved, the effect of the total SP current generated is the same, and the action on the SP log is unchanged. The result is that all the conclusions previously given in connection with stratified shaly sands remain valid in the case of unstratified shaly sands.

When the sand and shale streaks in a shaly sand are very thin, it is almost impossible to represent the detailed variations of the static SP diagram at the scale normally used, that is, 5 inches per 100 ft. On the other hand, the SP log for a shaly sand, in which the average proportion of shaly material is the same at all levels, is identical to the SP log for a clean sand giving an apparently lower emf (as if it were, for example, a clean sand of lower salinity), and may be represented by a uniformly reduced static SP. This lower emf,

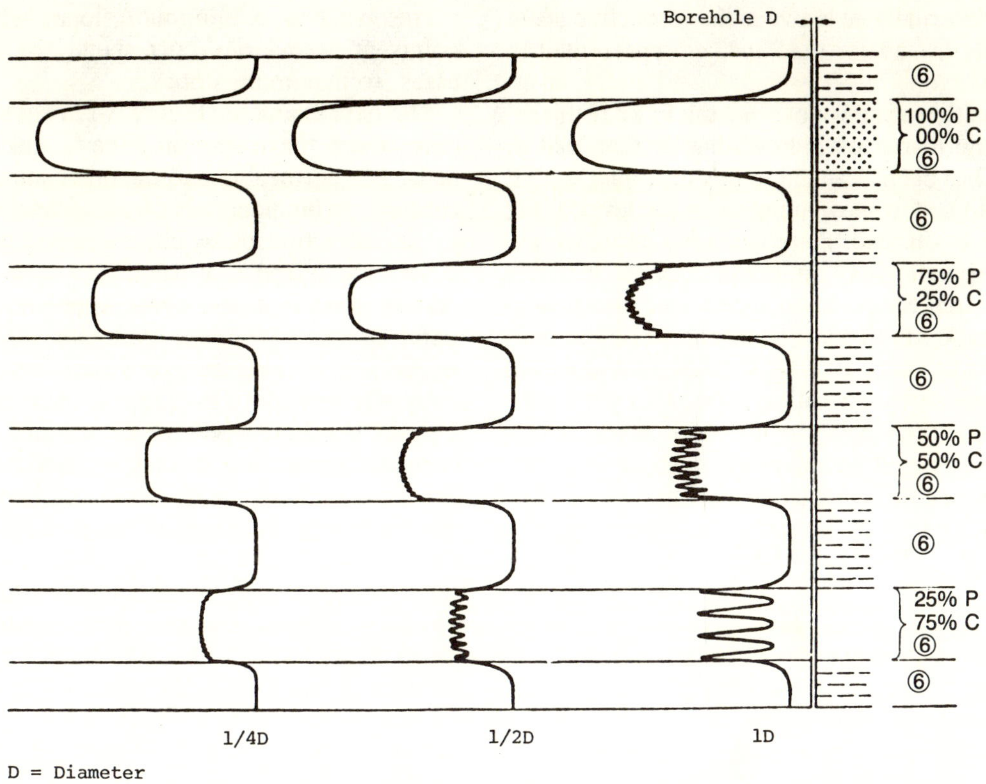

Figure 3-20. Illustration of the effect of shale sandwiches on the SP curve.

which would give the same SP log in the case of a clean sand, is called the *pseudostatic SP*. It is the SP that would be measured in front of the shaly sand if insulating plugs were set at its upper and lower boundaries to interrupt the mud continuity at these levels. The pseudostatic SP represents the maximum possible average deflection for such shaly sand, which is reached only if the shaly sand is thick enough.

The presence of oil in a shaly sand will increase the resistance of the permeable part of the medium. It can be shown that this increase will lower the pseudostatic SP. Accordingly, the amplitude of the deflection on the SP log can be expected to be smaller opposite an oil-bearing section than opposite a water-bearing section. Such a change in the deflections of the log can only be found for shaly sands or for thin beds; it will not occur for thick, clean beds.

As many sands are shaly, it is not surprising that a change in the SP log deflection has been found when passing the oil-water contact in a sand. It is to be noted, however, that the change in the SP log deflection is not a positive diagnostic for the detection of oil, since the same effect would be obtained if the salinity of the interstitial water were reduced, or if the percentage of shale were increased. If there are good reasons to believe that the salinity of the water remains substantially constant in the interval being studied, and that the shale content within the sand is approximately the same, then the level at which SP deflection is less is a good indication of oil content. Such a possibility is at least to be considered if concurrently the resistivity is higher to indicate that an increase in shale percentage is not the probable explanation for the lower deflection.

The presence of gas in shaly sands may affect the SP log in the same manner as the presence of oil. There is a tendency for such shaly sands to show a little less SP and a slightly higher resistivity when they contain gas than when they contain oil. This may be due to less connate water being left in the reservoir where gas is present.

Between a substantially clean sand and a shale, there may be a transition zone of more or less shaly sand. In such a case, the pseudostatic SP for the shaly sand part is intermediate between the respective static SP for the shale and for the sand, as represented in Figure 3-21.

If the transition zone is not very thick, or if the resistivity of the formation is much higher than that of the mud, as has been assumed in this example, the SP log is too rounded to show a plateau at the level of the transition zone. Instead, there is simply an inflection point, corresponding to a minimum slope, at the level where the actual SP potential in the mud is equal to the pseudostatic SP. In addition, there are, of course, inflection points corresponding to maximum slopes at the boundaries of the transition zone. This particular case shows that there could be inflection points which do not characterize a boundary between an impervious and a permeable bed. It is important to note, however, that the special inflection point mentioned here corresponds to a minimum slope, while inflection points at boundaries correspond, in practically all cases, to maximum slopes.

The next example (Figure 3-22) shows a particular case where the transition zone is made of two thin beds, one permeable and the other impervious. Their contact, or boundary, is characterized on the SP log by an inflection point corresponding to a minimum slope. The purpose of this figure is to show that, in certain special cases, there could be an inflection point coinciding with a minimum slope and which, nevertheless, characterizes a boundary.

Briefly now, let's consider a shale base line shift. Usually the shale base line (from which the SP deflections are measured) is easy to determine on the SP log. However, in some wells, base line shift will occur whenever waters of different salinities are sepa-

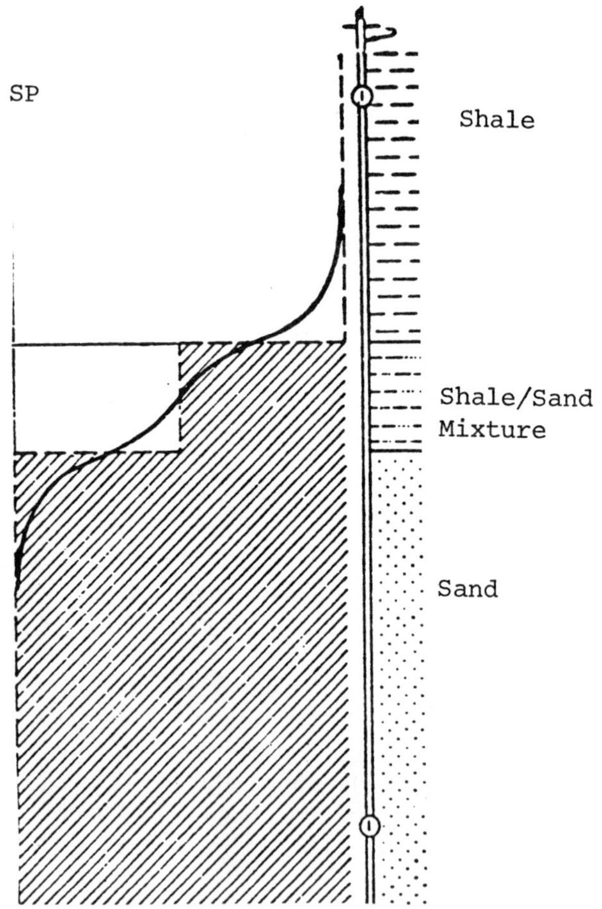

Figure 3-21. Illustration of sand-shale interfaces.

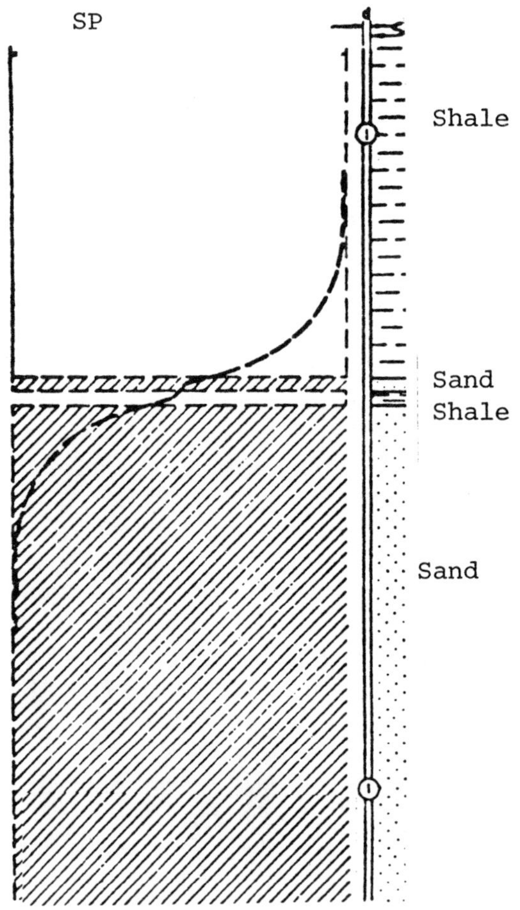

Figure 3-22. Illustration of sand-shale interfaces.

rated by a shale bed which is not a perfect cationic membrane. Large shifts are possible, making determination of the SSP quite difficult.

When there is no shale bed at all to separate waters of different salinities in a permeable bed, there is also an SP base line shift. In such a case, the SP curve shows no variation at the level where the salinity changes, but the SP deflections at the upper and lower boundaries of the permeable bed have different polarities if the salinity of the mud filtrate is between the salinities of the two different interstitial waters. If the permeable bed is not shaly, and if the permeable bed and the surrounding shales are sufficiently thick, SP deflections at the boundaries are the static SP deflections corresponding to the two different waters.

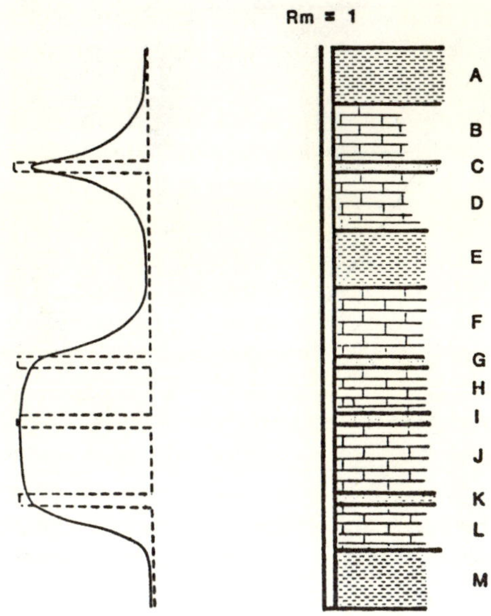

Figure 3-23. Illustration of complex lithologies.

Resistive Beds

The case of limestone fields, and more generally, of permeable beds in compact and highly resistive formations, deserves a special study. Permeable zones, whether oil bearing or water bearing, are somewhat conductive because of the generally high salinity capillary water present in the pores. Other conductive beds, such as shales, are of impervious nature. When the permeable and impervious beds are thick and sufficiently conductive, the SP log approaches the static SP diagram. The number of millivolts recorded in front of a permeable zone differs from the one recorded in front of an impervious bed by an amount approximately equal to the emfs involved, as in the general case.

When, however, the conductive beds are not very thick and are separated by thick, hard formations of high resistivity, the SP log has a shape which is difficult to understand at first glance. Very resistive formations tend to prevent the SP currents from leaving or entering the hole opposite their level. SP currents thus have to flow into the hole almost entirely by way of the permeable beds and of the nearest conductive impervious beds. In so doing, the SP currents produce potential differences by ohmic effect in the mud in front of the hard resistive formation. The result is that the peaks corresponding to the permeable zones spread above and below these zones in an apparently abnormal manner. It will be shown, with the help of an example, that the curious behavior of the SP log is more easily explained, and that the interpretation becomes less difficult, once the basic principles have been established.

Figure 3-23 shows schematically the case of four thin, permeable formations, C, G, I, and K, and three thick shales, A, E, and M. Each of these formations is separated from the others by thick, compact, and highly resistive beds, B, D, F, H, J, and L. In order to characterize the problem more explicitly, it is assumed, with the resistivity of the mud being taken as a unit, that the resistivity of the permeable zones, as well as that of the shales, is approximately ten ohms while the hard formations have a resistivity of five hundred ohms or more. Viewing the SP on the left, the segments which correspond to hard formations are represented by straight lines (more precisely, they should have slight curvatures as will be explained later, but the log would be similar in appearance). The emfs involved are, as usual, represented by the SSP diagram on which the SP is superimposed. As can be seen, the departure of the SP log from the SSP is remarkable in this case, and there is no wonder that this type of log has sometimes been considered abnormal. It is assumed that the SSP in front of the hard formations is the same as that in front of shales. In this schematical example, any other reasonable value could have been assumed for the SSP in front of the hard formations without changing the corresponding SP log. This is because the hard formations, as represented, are much too resistive to allow any appreciable SP current to diverge into the mud, and thereby influence the potential.

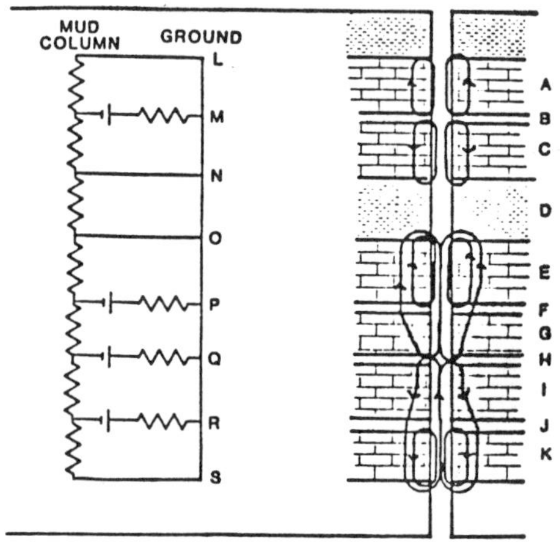

Figure 3-24. Electrical schematic of the SP.

The shape of the SP log is easily understood if the circulation of the SP currents is studied first. This circulation is represented in the schematic of Figure 3-24. The SP currents, which are generated by the different emf, flow into the sands. They cannot traverse the adjacent hard formations through sections located close to the borehole because these sections are too small in area. The hard formations introduce into the circuit large resistances which would practically prevent the current from flowing. In this case, however, the SP current penetrates deeper than usual in the permeable beds and, consequently, enters the hard formations without appreciable reduction in their cross section, as would be required if they were to converge quickly toward the hole. SP currents flow toward conductive beds through which they can return to the mud in the hole close to the circuit. Currents cannot come back to the mud through other permeable beds because they encounter emfs which would oppose the flow of currents in that direction. When the first conductive beds they encounter are of the permeable type, they simply cross them until they reach conductive and impervious beds. This is the case for the SP currents which penetrate the permeable bed H; they have to cross the permeable bed F in order to reach impervious bed D, or they have to cross permeable bed J in order to reach impervious bed K.

The total potential drop along a current path is equal to the total emf involved. The currents divide between the different possible paths and produce in them potential differences, according to the principles of Kirchoff's laws. As it is difficult to visualize the application of Kirchoff's laws to the three-dimensional circuit constituted by the formations and the mud column, the problem can be clarified by referring to the equivalent electrical network on Figure 3-24.

In this network each permeable bed is characterized by an emf, the total of which generates the SP currents and by a resistance which is one of the SP currents encountered in that permeable bed and in the surrounding hard formations, before the current path has expanded to very large cross sections at great distances from the hole. The permeable bed H, for example, is represented by an emf of 100 mv in series with a resistance of ten ohms. This represents the resistance between the cylinder of mud in contact with bed H and infinity. The ground is represented as conductor LMNOPQRS on the schematic, which is supposed to have no resistance at all.

The conductive impervious beds A, D, and K are represented by short circuits between the mud and infinity, as these beds are so thick that they pose very little resistance to the current. No emf is shown between the mud and infinity in the impervious beds, because the total emf has been assumed to be at the boundary between the mud and the permeable beds.

The mud column is represented by resistances which have a constant value per foot of hole, and have been computed on the basis of a hole diameter of eight inches and a mud resistivity of one ohm-meter.

When the bed resistances are small with respect to the resistances of the mud sections, practically all the potential drop is produced in the mud, and the total deflection of the SP log is nearly equal to the total emf involved. This is illustrated for bed B, where the resistance in the formations is ten ohms as compared to two hundred ohms for the resistance in each mud section. When the bed resistances represent a large part of the total resistance, the deflections of the SP log do not reach the maximum emf. All along a borehole opposite a given hard formation, the current in the mud column remains about the same.

At the level of each conductive bed some SP current generally penetrates or leaves the hole; therefore the slopes of the SP log are modified. Returning to Figure 3-23, the SP log changes its slope at the level of permeable bed G, because part of the current leaves the hole and flows into that bed. In the case of permeable beds C and I, the direction of current in the hole is reversed, and so is the slope of the SP.

Where there is only one bed like C between two successive impervious beds such as A and E, that permeable bed is easy to detect on the SP log, even when hard formations like B and D are present, because very definite slope reversal occurs opposite the permeable bed.

Spurious Potentials

Some small number of SP logs display additional fluctuations not due to the actual SP. Extra potentials superimposed on the normal SP are called spurious potentials. Their origins are diverse. In all cases spurious potentials represent unwanted signals that cloud the interpretation. Steps can be taken in most instances to remove the unwanted potentials.

Magnetized Drum

When the cable drum on the logging truck has become magnetized, a regular fluctuation appears on the SP. As the cable unwinds, it crosses the magnetic field emanating from the drum once per revolution. A potential appears each time this occurs, so that the unwanted signal is evenly spaced on the log. The distance between fluctuations corresponds to the circumference of the spooled logging cable. The engineer in charge can remedy the condition by applying an alternating current to the cable thus demagnetizing the drum.

Power Lines

A sixty-hertz sine wave on the SP indicates that the SP ground has been located too near an overhead power line. The remedy in such a case is simply to relocate the ground electrode. Other types of electrical generators cause similar but often irregular spikes on the log. Locating the electrode near truck generators, rig generators, idling vehicles, flowing water and gas lines, mud pumps, and the like may produce fluctuations. Welding gear in use elsewhere on the rig, radio transmitters, air conditioners, coffee pots, and thunder storms may produce uneven spikes.

Bimetallism

A small emf is generated when the logging cable connects dissimilar metals in the borehole. When the cable contacts surface pipe or the rig floor, what is in effect a small battery is set up. The steel tends to go into solution, creating with the lead SP electrode a lead-iron battery with mud as the electrolyte. If the area of exposed steel is small, the extra current may not seriously upset the SP, although pieces as small as a lost rock bit cone are quite large enough. Once the tool passes the cone, it assumes normal appearance.

It should also be noted that all metal contact is to be avoided. Hand rails in the path of the logging cable should be taken down, for example.

Polarized Electrodes

Lead electrodes eventually become coated with a chloride or sulfate film from the muds in which they are immersed. Although normally a desirable feature that improves electrolytic ground contact, this coating can cause the SP ground to act as a reversible electrode. The neutral material of which the SP ground is made will pass certain ions in either direction. There is a limit to the current density that can be passed for a given area. If total electrode area is too small and the limiting current density is exceeded, the electrode is said to become polarized. Over-taping the cable that connects the electrode may reduce the total exposed electrode area. Electrode potential becomes unstable, and almost any sort of log may be obtained. The remedy for over-zealous maintenance is simply to untape the SP ground, exposing more surface area. If this fails to do the trick, increasing the electrical resistance of the logging circuit might be considered.

Telluric Currents

Electromagnetic currents in the atmosphere and earth at high latitudes cause wonderful displays known as the aurora borealis, or northern lights. In boreholes traversing resistive formations, these currents generate large spurious potentials. These natural earth, or telluric, currents "short circuit" through the borehole mud. The log is odd in appearance and will probably not repeat. Commercial logs can be obtained only when northern light activity is at a minimum.

4

Resistivity Logs

A host of devices are employed to measure formation resistivity. Although not superficially apparent, these tools may be classified as either of only two fundamental types. Resistivity tools directly measure the effects of an electromagnetic field on the beds. Resistivity devices require a conductive fluid in the well bore to carry the current, whereas induction tools may be used in virtually any environment.

With any of these tools, different depths of investigation may be achieved by varying the spacings between signal source and sensor. Focusing coils or electrodes are often used to increase the effective depth of investigation and to improve bed boundary response.

A number of tools are of the resistivity type. Laterlogs*, Microlaterlogs*, 18″ normals, 16′ laterals, MICROSFL*, and proximity tools are of this type. All of these are simply electrode arrangements. Each contains a current-emitting electrode, one or more receiving electrodes, and possibly some focusing electrodes. Signal at the receiver is used to directly measure the bed resistivity.

Induction or dual induction devices consist of coils wound on a nonconductive cylindrical base. Some of the coils induce electromagnetic fields in the formations, while other coils focus the fields. Coils are also used as sensors, taking their signals from secondary fields induced by the variable formation currents. Signals at the receiver coils are used to determine formation conductivity. Bed resistivity is found from conductivity through an inverse mathematical relation.

* Registered name of Schlumberger tools.

Early resistivity logs, utilizing electrode-type devices, were called electric logs. Later, logs utilized a combination of induction and electrode devices and were known as induction-electric logs. An SP curve was commonly recorded in conjunction with both types of logs. Modern logs are merely developments of the induction-electric log.

No tool accurately measures resistivity of the intended zone, although some come very close. Between transmission and reception the signals transpire several distinctly different zones. Signals reaching virgin formation must cross the borehole fluid, the mudcake, the flushed zone, and the invaded zone. A portion of the total received signal will come from each of these zones. Corrections for such influences and for the influence of such factors as bed thickness must be applied to the values read from logs.

Normal Devices

The earliest electrical surveys used in well logging were the conventional resistivity logs with an SP. These devices consisted of electrode arrangements and some simple instrumentation. One such electrode arrangement is the *short normal*.

The short normal is a two-electrode measuring device which passes current from an electrode on the tool through the mud and into the formation. Voltage of the returning current is measured between other electrodes. These measured voltages provide resistivity determinations. As depicted in Figure 4-1, in a ho-

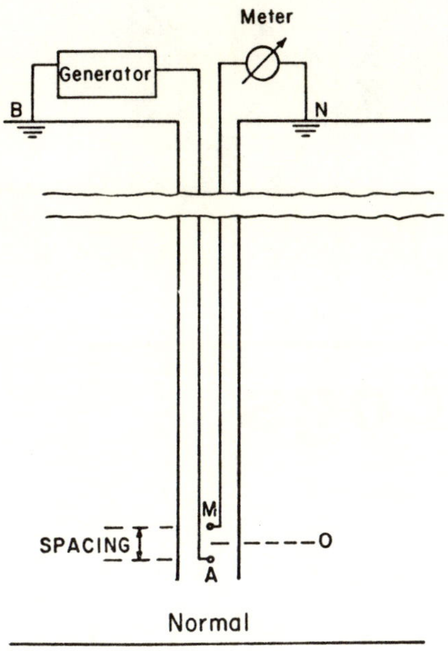

Figure 4-1. Schematic of the normal tool. (Courtesy of Schlumberger.)

mogeneous formation the equipotential surfaces surrounding a single current-emitting electrode, A, are spheres. Voltage between electrode M, situated on one of these spheres, and ground point N is proportional to formation resistivity. Typically, a constant current is passed between electrode A and B. The resulting potential difference is measured between M and N. Electrodes A and N are placed downhole on the logging tool. Electrodes B and M are theoretically located an infinite distance away. In practice, however, virtually the same measurement can be obtained if B is the cable armor and M is a downhole electrode which is electrically insulated from the cable and sufficiently far removed from A and N. The distance between the electrodes A and M is called the spacing. The Schlumberger short normal employs a 16″ spacing. A deeper investigating design used 64″ spacing. The deepest point at which the measurement is being made is midway between A and M. This point is considered the zero point for the device. The radius of investigation is very nearly equal to twice the electrode spacing.

The mathematical relationship for this type of electrode arrangement is depicted in the following:

$$R = K\frac{V}{I}$$

where R = resistivity
K = constant
V = voltage
I = current

Resistivity is equal to K times voltage divided by current. The current, as previously stated, is constant. The proportionality constant K is related to electrode spacing and geometrical factors of the measuring device and formations. The only variable is the voltage. Log trace deflection, responding to voltage changes, can thus be scaled in resistivity units.

Figure 4-2 is a typical example of a normal log curve. You will notice that the distance between the

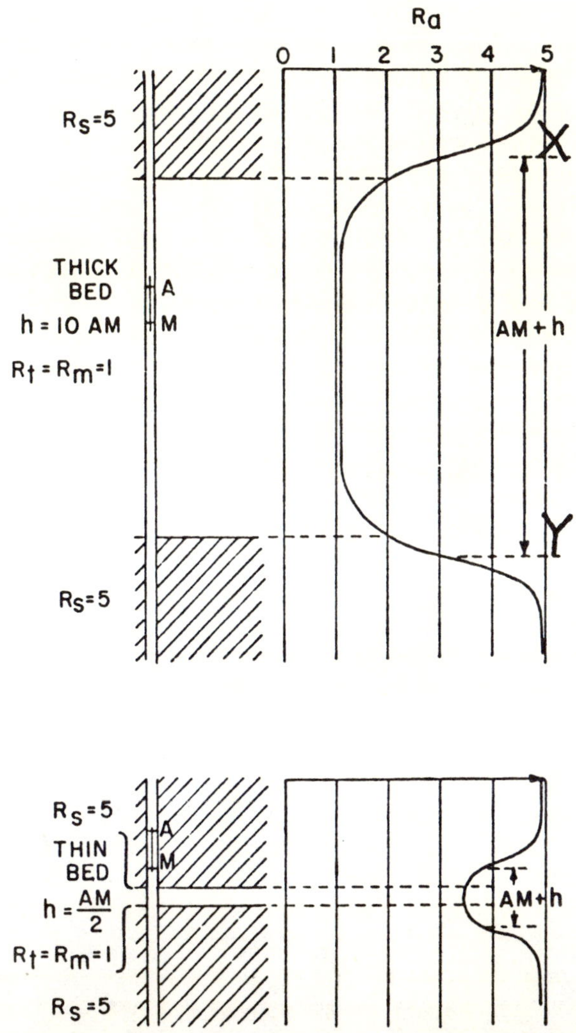

Figure 4-2. Illustration of the normal curve. (Courtesy of Schlumberger.)

inflection points X and Y represents the bed thickness plus the distance between the electrodes A and M.

Looking at Figure 4-3, we see a different electrode arrangement known as a lateral device. A constant current is passed between electrodes A and M and the potential difference is measured between electrodes M and N, which are located on two consequent spherical equipotential surfaces centered on A. The voltage measured is proportional to the potential gradient between M and N. The measuring or zero point for the device is at O, midway between M and N. Tool spacing is the distance from A to O. Generally speaking, the spacing is approximately equal to the radius of investigation.

It is important to note that these devices record an apparent resistivity. Resistivities will be affected by the geometrical dimensions of all the media surrounding the device, such as the borehole, the invaded and uncontaminated zones, and the adjacent beds. If quantitative R_t calculations are to be made, bed thickness must be 3 to 4 times the tool spacing. This means that the once popular 16′ lateral was really useful only where bed thicknesses exceeded 50 feet.

Figure 4-4 is a typical lateral curve. Notice the lack of curve symmetry and the difficulty associated with bed boundary determination. The curve also indicates resistivities in excess of R_t in thick beds. Logs were difficult to interpret, at best, since the log was clouded with shadow zones and reflection peaks that depended on the electrode spacing and bed thicknesses.

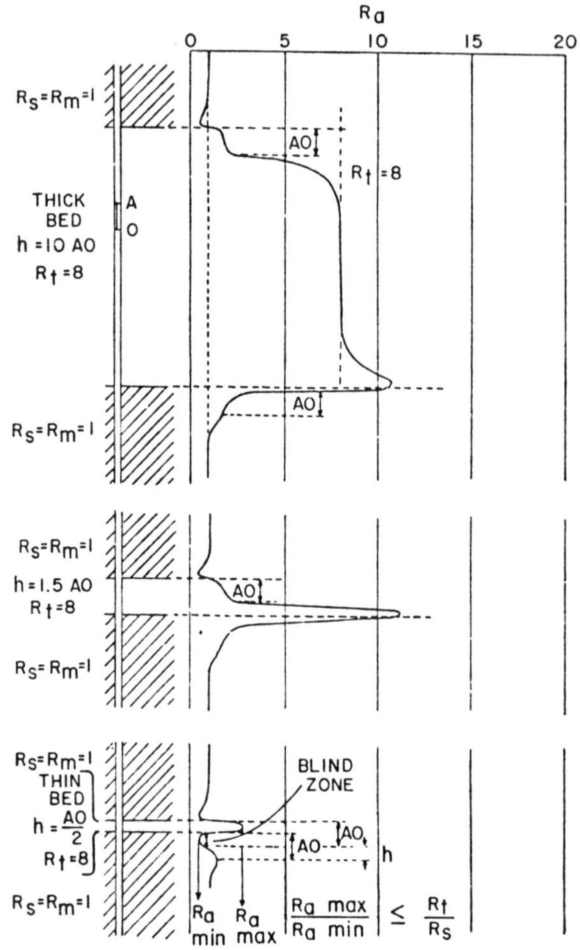

Figure 4-4. Illustration of the lateral curve. (Courtesy of Schlumberger.)

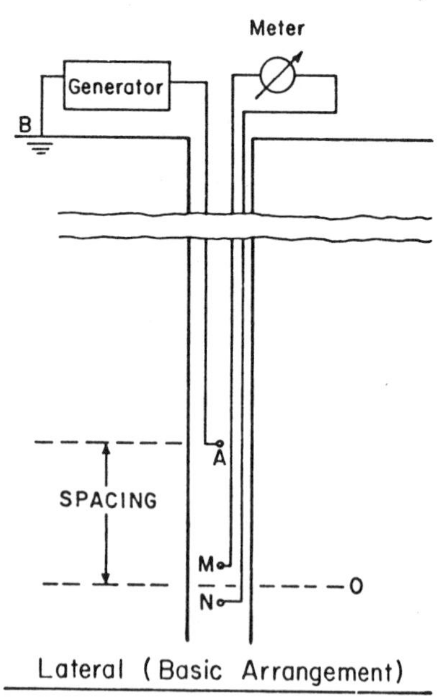

Figure 4-3. Schematic of the lateral tool. (Courtesy of Schlumberger.)

Lateral and normal devices were often used in conjunction. Each device had a different depth of investigation. Curve separation indicated the presence of permeable beds in optimal conditions. Although the longer-spaced conventional logs are no longer common, they continue to exist in a permeability-indicating device known as micrologs. Micrologging devices employ extremely short-spaced normal and lateral electrode arrangements mounted on a pad which is forced against the borehole wall.

In today's logging industry, there are numerous devices available for the purpose of measuring the resistivity of the formation. Accurate resistivity determination is necessary for the interpretation of hydrocarbons from a well. Each of the modern devices is designed to produce a good resistivity measurement for very specific ranges of borehole conditions. The first of these tailored tools we shall discuss is the focused-type devices.

Focused Tools

The influences of the borehole and adjacent formations on electrode devices are minimized by a family of focused current tools. The focused family of tools is designed for accurate R_t determination where R_t/R_m ratios are large, beds are resistive or thin, drilling muds are salty and conductive (where the ratio $R_{mf}/R_w < 4$), and large adjacent bed resistivity contrasts.

Theory of Measurement

The volume of any subsurface formation will influence the recorded resistivity measurements of that formation. A portion of the measured signal is going to be proportional to the voltage drop caused by the measuring current which flows through that element. It is possible to calculate the theoretical distribution and the intensity of the measuring current within the formation for the particular electrode arrangement of most of these resistivity logging tools. This calculation uses one of the basic laws of electrodynamics, which can be stated as follows:

> The total amount of current emitting from an electrode must flow through a medium which encompasses that electrode. If we assume a single electrode placed in an infinite homogeneous medium, the current distribution lines emitted will have a tendency to flow concentrically outwards from the source electrode in all directions.

Figure 4-5 depicts an ideal radial flow with the center of a continuously expanding sphere of current located at the source electrode. Regardless of the size of the sphere, current density is the same at any point on its surface. The total current flowing through it remains relatively constant and identical to the current leaving the source electrode. The surface area of the sphere will naturally increase in proportion to the square of its radius. Thus, current intensity for a given area will decrease as the square of the radial distance increases.

Current density is equal to the total current in amperes emanating from the source electrode divided by the total area of the sphere. Current density is thus defined as amperes per square unit of measurement, which might be in feet or meters. Conventional electric logging devices described earlier employ this basic principle. The problem with this type of device is that when it is placed in a borehole containing salty muds, the low resistivity of the mud provides a conductive path which shorts out the current lines. The device tends to record true resistivity of the mud rather than of the formation.

The focused log was developed in order to overcome the difficulties associated with the electrical logging type devices under these borehole conditions. With a focused log, a large portion of the current is prevented from flowing radially outward in all directions by confining the current to a thin disc. This is accomplished by placing a focusing electrode on either side of the current-emitting electrode. The focusing electrodes prevent the logging current from flowing either up or down within the borehole. Current flows from the tool in a lateral direction, hence the term *laterolog* used in much of the literature to describe this type of tool.

The ideal current distribution pattern for a focused log is depicted in Figure 4-6, which shows a uniform cylindrical disc. The theoretical current density is a function of the radial distance into the formation, and can be calculated as shown in the figure. For a uniform disc, current density varies inversely as the first power of the radial distance. The current density is given in amperes per square unit of measure, which again could be meters or feet.

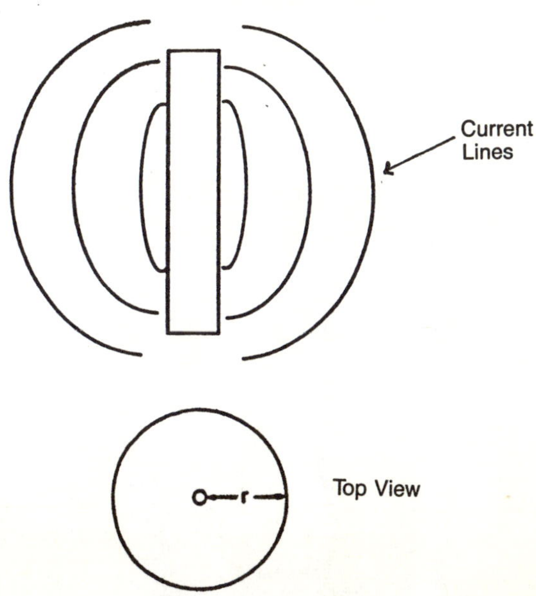

Figure 4-5. Illustration of spherical currents.

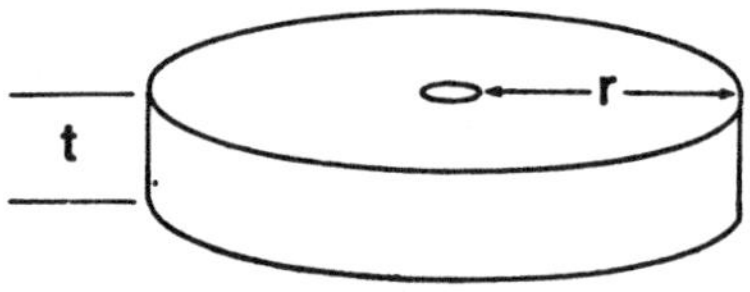

$$\text{Current Density} = \frac{I}{2\pi rt}$$

where t = thickness of the measuring current disc
r = radial distance from the current source
I = total current emitted from the electrode in amperes

Figure 4-6. Illustration of the ideal current distribution of a focused tool.

Figure 4-7 shows a typical focused tool, with bucking electrodes A1 and A2 generating a current field (solid lines on a constant potential V_0) which focuses the current from the alpha ring (dotted lines). Alpha ring current penetrates deeply into the formation before the focusing fields become weak enough to allow it to disperse and return to the pickup electrode.

Figure 4-8 shows an idealized focused log response. As you can see, it has reasonably good bed definition compared to electric logs. Also, the focused log gives us a much better definition of true resistivity. In this example, the shale resistivity is 5 ohm-meters and the resistivity of the sand is 250 ohm-meters. Note that the mud is highly conductive, with a resistivity of .05 ohm-meters.

Bed Thickness

The focused-type device is an excellent tool for delineation of thin beds. It is also good for particularly high-resistivity beds. Adjacent bed effects common to conventional resistivity tools are eliminated when the thickness of the bed is greater than that of the disc of current from the focused tool. If the thickness of the bed is less than the thickness of the measuring disc and is of high resistivity, then the current will be divided between the bed of interest and adjacent formation.

The amount of current flowing through both formations will be a function of their thickness and respective resistivity values. Low resistivity formations adjacent to the bed of interest will decrease the apparent resistivity measurement. The measured resistivity progressively decreases as bed thickness decreases. Assuming a bed thickness less than the measuring disc thickness in a formation that is conductive instead of resistive, the measured resistivity will be greater than the true resistivity of the bed. The smaller the ratio of R_t/R_{sh} and the thinner the bed, the greater the increase will be.

Induction Tools

The induction log was designed to measure formation resistivity in boreholes containing nonconductive fluids. Boreholes containing air, freshwater muds, or

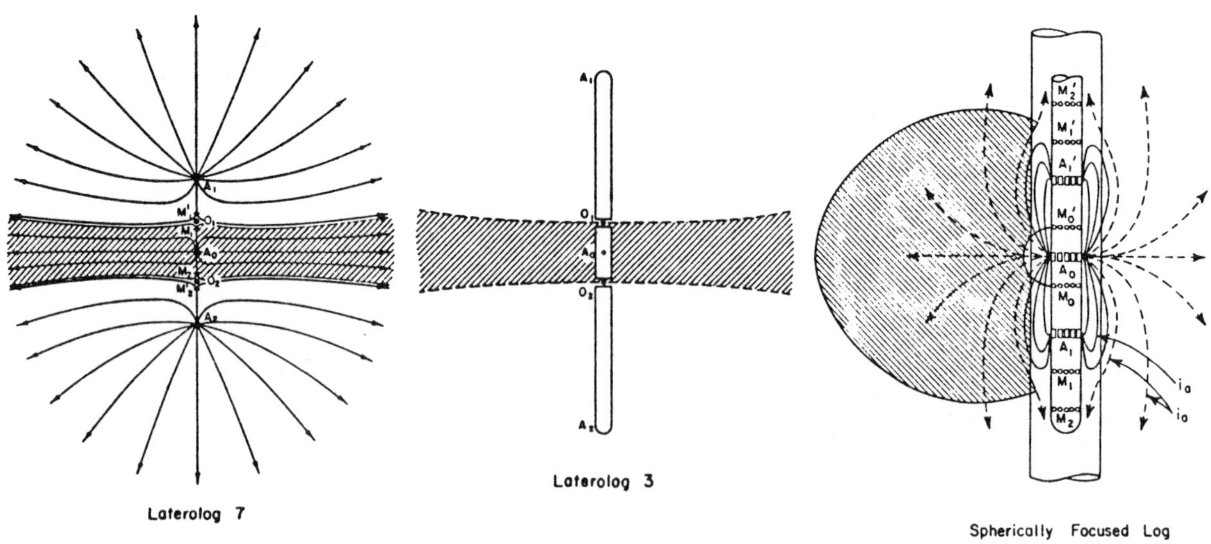

Figure 4-7. Schematic of current flow of focused tool. (Courtesy of Schlumberger.)

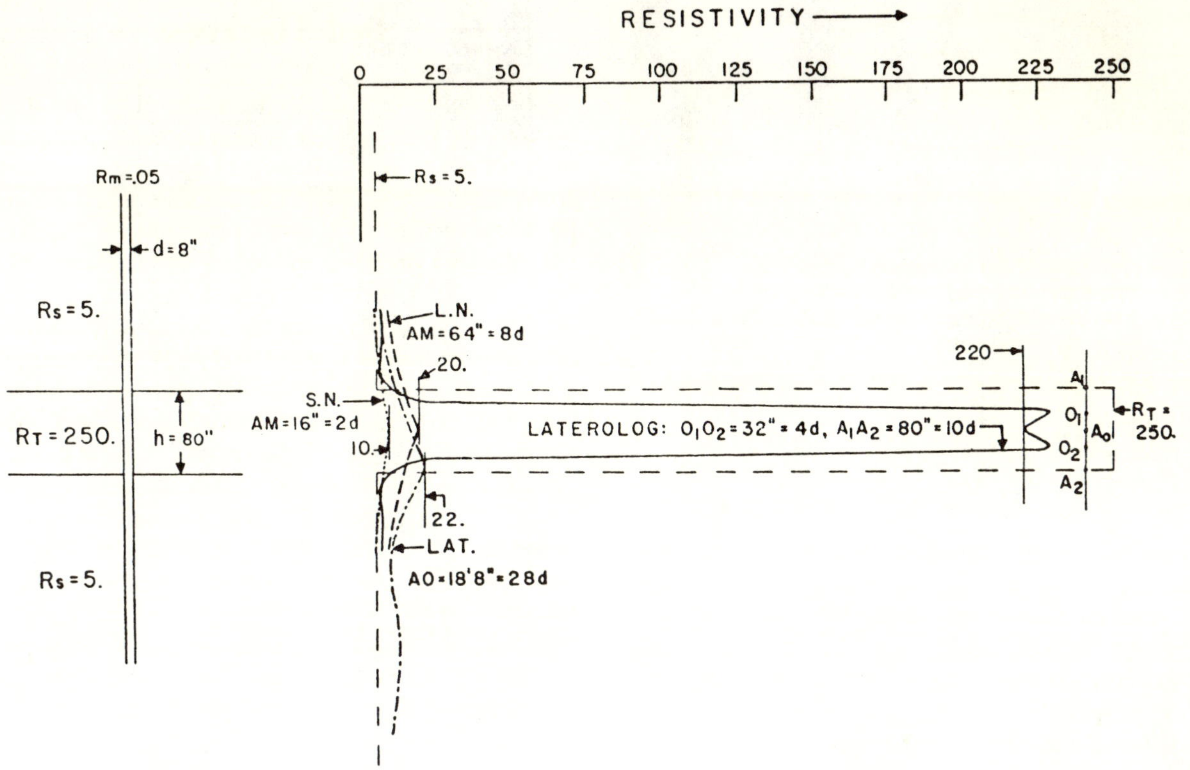

Figure 4-8. Illustration of the focal curve response. (Courtesy of Schlumberger.)

oil-based muds make excellent environments for induction logging. The induction log actually measures the conductivity of a formation rather than its resistivity. Conductivity is the reciprocal of resistivity, and vice versa.

$$C = \frac{1}{R}$$

$$R = \frac{1}{C}$$

Conductivity is expressed in **mhos**, (ohm spelled backwards). For well-logging applications units of one thousandths of a mho or millimhos are more common. Conductivity in millimhos is found by:

$$C = \frac{1{,}000}{R}$$

$$R = \frac{1{,}000}{C}$$

One ohm is then 1,000 millimhos, two ohms are 500 millimhos, and so on.

Figure 4-9 shows the basic configuration of an induction tool. An alternating current generated by the transmitter oscillator is applied to the insulated transmitter coil, to produce an alternating magnetic field. The magnetic field penetrates the formation and induces a current flow in it. Formation current in turn induces a secondary magnetic field around the receiver coil. This field is converted to a current in the coil which is proportional to the conductivity of the formation. The spacing between the receiver and transmitter coils is a compromise between the depth of investigation and the thin bed resolution desired of the tool.

Figure 4-10 shows a typical induction-electric log presentation. Direct output of the induction measuring circuitry is the conductivity curve in track 3. Corresponding resistivity is calculated automatically by an onboard computer, and output as the dotted induction resistivity curve in track 2. Off-scale resistivity values appear in track 2 at a different sensitivity. The solid curves in track 2 are resistivity from a shallow investigating electric device, in this case a 16″ normal. Opti-

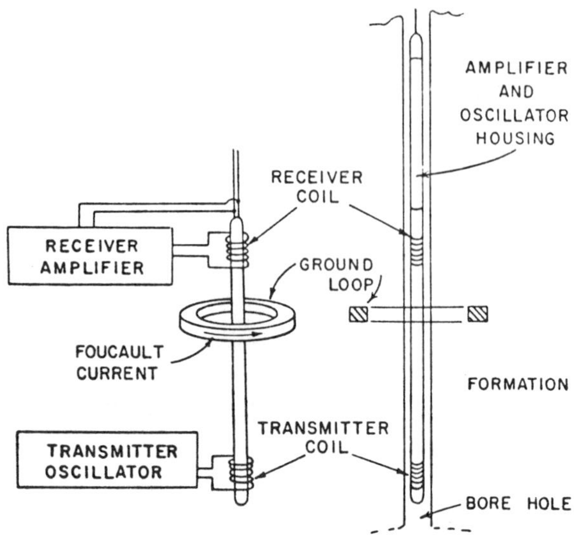

Figure 4-9. Schematic of the induction tool. (Courtesy of Schlumberger.)

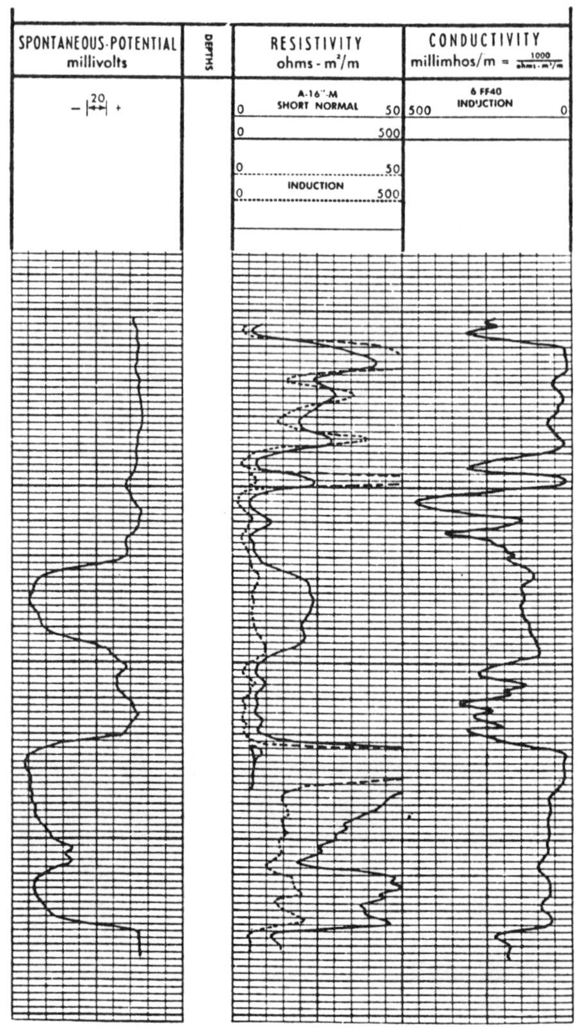

Figure 4-10. Illustration of the conductivity and resistivity curves. (Courtesy of Schlumberger.)

mal conditions for running induction logs also produce excellent SPs.

Under favorable logging conditions, the induction tool yields good values of true resistivities; however, charts may be needed where corrections can be found for bed thickness, large hole diameter, deep invasion, etc.

The dual-induction tool is one of the most advanced resistivity devices available. It is particularly useful where invasion diameters are large. An SP and/or gamma ray curve is also recorded with it. The dual induction tool records three resistivity curves having three different depths of investigation. A shallow curve measures the flushed zone resistivity. A medium curve measures invaded zone resistivity. A deep curve measures primarily the uncontaminated or virgin zone and is very close to R_t. The ratios of shallow to deep and medium to deep are used to determine R_{xo}, R_t, and the diameter of invasion, d_i.

The dual-induction log may be recorded on either linear or logarithmic scales. The logarithmic presentation in Figure 4-11 permits a greater dynamic range of resistivities and is convenient for determining ratios, since the difference of the logarithms of two values is equal to the ratio.

Skin Effects

When a formation is very conductive, the electromagnetically induced secondary currents become large. This makes their electromagnetic fields quite important. The magnetic fields of each ground loop induce an additional magnetic force in adjacent ground loops. This mutual inductance has the effect of reducing the conductivity signal recorded on the log. This signal reduction is termed *skin effect*. Induction logs are usually compensated for skin effects by the surface gear.

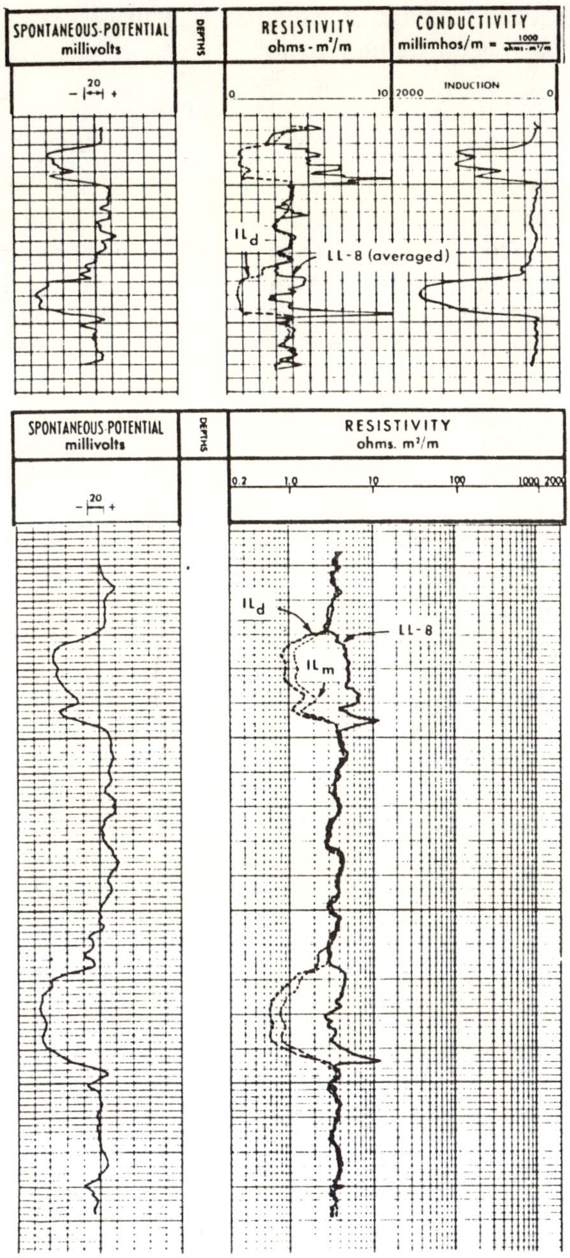

Figure 4-11. Illustration of the dual induction-resistivity curves showing invasion. (Courtesy of Schlumberger.)

Borehole Signal

Modern induction and dual-induction tools use coils located on the sonde to focus the magnetic fields generated by the transmitter coils. The action and purpose of these coils is very similar to that of focusing electrodes. Signal fields are prevented from flowing directly up the borehole. Nevertheless, a certain amount of the total signal at the receiver is due to the borehole fluid. The magnitude of this signal is a function of the hole size and of the standoff device run on the tool. Borehole corrections will be covered later, but for now it is sufficient to mention their influence.

Effect of Bed Thickness

Another purpose of the focusing coils is to increase the effective investigation depth and thin bed resolution of the tool. Bed corrections will also be deferred, but note that where bed thickness is less than tool spacing, adjacent beds will affect induction readings, just as they affect the readings of any other tool. The effect may be especially great where beds are less than five feet thick. Dual-induction tools may resolve accurately to about three feet.

Invasion Effects

As a rule of thumb, R_t values should be less than 2.5 times R_{xo}, and the diameter of investigation no greater than 100 inches for optimum R_t determination from induction logs. This means that the induction log must be used with some caution where salt muds are used in drilling the hole. In such a case, the invaded zone may well be more conductive than the virgin zone. Induction logs respond more to the less resistive zone. A good rule of thumb for determining where the induction log is applicable is to run it in holes where the ratio R_{mf}/R_w is greater than six.

Formation Factor and Water Saturation

Logs are used for many purposes, such as correlation and lithology identification, but the most important reservoir property sought by an analysis of a suite of logs is *water saturation*. The underlying assumption of all analysis to this end is that, within the pores of a rock, all that is not water is probably hydrocarbons. This presumption usually proves to be correct.

Log analysts seek discrepancies. Those which cannot be explained away become candidates for further testing. Wildcats strain the art of analysis to its limits, and logs become only one of the tools necessary to the evaluation.

An important point has been made that warrants further discussion. Well log analysis is an art based on a science. The novice quickly feels the force of this statement when he/she uses this new analytical tool. A good deal remains to be learned, and the numbers computed are not as sophisticated as we might desire.

That is not to say that the actual computations are wrong. This is far from being the case. All parties should simply appreciate them for what they are: indicators. The underlying concepts are quite simple. An idealized value is compared to the measured, observed value. An anomaly, or difference, between the two values indicates the presence of hydrocarbons. How is this difference related to water saturation, or in other words, how much hydrocarbon is present?

The early work of G.E. Archie pointed the way to this relationship. The ratio of idealized value to the observed value is proportional to water saturation. Mathematically,

$$S_w \propto \frac{\text{ideal value}}{\text{observed value}}$$

where S_w = the water saturation

Comparisons of this type may take many forms. For example, we could calculate saturation by comparing the actual resistivity of the salt water in the pore spaces, R_w, to the apparent resistivity of the fluids, R_{wa}, as measured by our logging instruments. Thus:

$$S_w \propto \frac{R_w}{R_{wa}}$$

where R_w is an idealized value denoting a 100% water mix and R_{wa} expresses resistivity of a mix hopefully containing some hydrocarbons with the water.

On the other hand, we could deal with the pore spaces, comparing the volume of water, Φ_w, to the total void volume of the rock, or Φ:

$$S_w \propto \frac{\Phi_w}{\Phi}$$

Finally, we might approach the calculation by considering the rock and its fluids as a system, using curves affected by both. If we express the resistivity of a rock-fluid system idealized to contain only water as R_o, and define R_t as the observed actual resistivity of the system, we may write:

$$S_w \propto \frac{R_o}{R_t}$$

Water saturations can be calculated from a single curve. One means of determining R_w in our first expression is to derive it from the SP curve in a zone known to contain only water. R_{wa} can be calculated in the zone of interest from the same curve. The comparison R_w/R_{wa} is based upon the realization that the presence of hydrocarbons suppresses the SP. Note, however, that two assumptions were made in the process: first we have assumed that the resistivity of the water has not changed between the known water zone and the zone of interest, and secondly, we have assumed that the rock itself is identical to that in the water zone, or in other words, that no other influences have crept in that might similarly suppress the SP. Unfortunately, shale has just such an effect, and a large one at that. Similar assumptions must be made if we are to attempt single-curve analysis using the other two expressions.

How may we then test our assumptions to determine whether they are warranted? It is for this reason that we run suites of logs. Each curve is sensitive or insensitive to various formation parameters and when combined, several curves may enable us to assess the effect of an undesirable property such as shaliness, so that we discount its influence in our calculation.

The principal step of log analysis is to convert all of the various physical properties of a zone, such as porosity or resistivity, corrected with the aid of other curves when appropriate, to a common term so that they may be compared. This means that if we have a porosity and a resistivity value, the resistivity must be expressed in terms of equivalent porosity, or vice versa. There must, therefore, be found a conversion factor that will enable us to do this.

The conversion factor whereby the log analyst relates porosity and resistivity is called the formation resistivity factor (F), or formation factor for short. This formation factor, which we shall call F, is a property of the rock, just as hardness and shape are properties.

Formation factor is a numerical expression of the texture of the rock. To develop a feel for what this means, let us consider a cubic meter of saltwater (Figure 4-12). The resistance of this cube is given by the expression:

$$r = \frac{R \times L}{A}$$

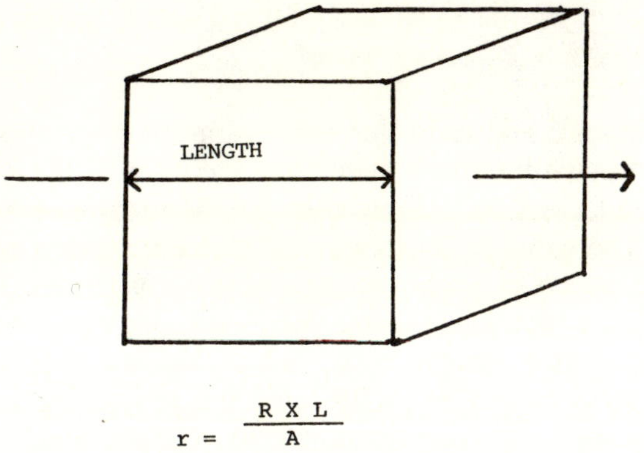

Figure 4-12. Illustration of ionic current through an unimpeded path.

where R = resistivity of the water
r = resistance of the cube
L = length of the cube
A = area of the cube perpendicular to the direction of measurement, L

If R has a value of 0.05 ohm-meters and L and A are 1 meter, then the resistance of the sample is 0.05 ohms.

If we replaced some of the water with a cubic meter of rock composed of straight capillary tubes, the resistivity of the water and the length of the sample would not change. The rock matrix takes no active part in the conduction of ionic current, being for all practical purposes an insulator, so the total conducting area of the cube available for our measurement depends on the area represented by the openings of the capillary tubes. Suppose this area were only half the total area of exposed rock face.

$$r = \frac{.05 \times 1}{.5} = 0.1$$

We see that reducing the conductive area of the sample's face effectively increases its resistance. The exposed conductive area of the sample just discussed is proportional to the porosity of the rock, and for our cubic meter, the equation may be restated as:

$$r = \frac{R \times L}{\Phi}$$

where we use the Greek letter Φ to denote porosity.

Real rocks are not composed of straight capillary tubes. The paths along which the ions of a conductive solution must flow twists and turns, making the length of the sample effectively much greater than one (Figure 4-13). If the length L_e could be found, our equation would become:

$$r = \frac{R \times L_e}{\Phi}$$

We are now ready to define formation factor in terms of resistivity by comparing our idealized cube of saline water to a cube of the same dimensions but composed of rock whose pores are filled with water of

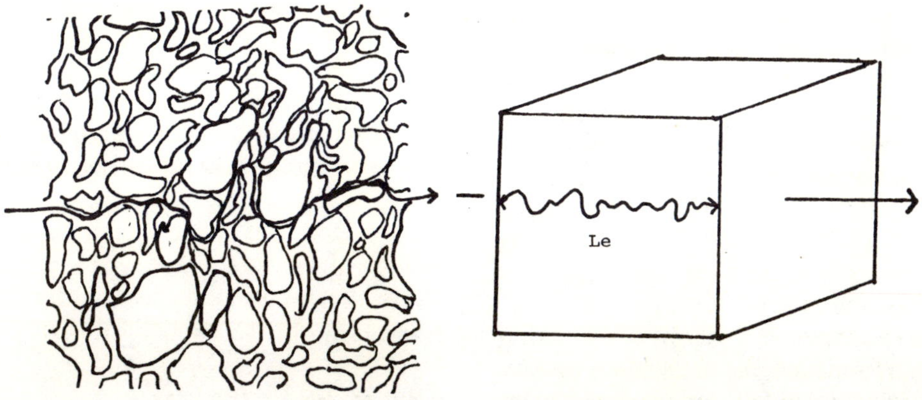

Figure 4-13. Illustration of ionic current through a formation.

the same resistance as that in the cube containing only water.

Formation factor =

$$\frac{\text{resistivity of rock saturated with fluid}}{\text{resistivity of the saturating fluid}}$$

or

$$F = \frac{R_o}{R_w}$$

R_o, the resistivity of the fluid saturated rock, was given by the expression:

$$r = \frac{R \times L_e}{\Phi}$$

and R_w, the resistivity of the saturating fluid, was given by:

$$r = \frac{R \times L}{A}$$

Thus

$$F = r \times L_e/\Phi/r \times L/A = \frac{A \times L_e}{\Phi \times L}$$

The area of the cube is one meter, so our expression for F now becomes:

$$F = \frac{(L_e/L)}{\Phi}$$

Two inferences can be drawn from this expression. The first is that the formation factor of a rock is independent of the fluids in the pores, since the water resistances are cancelled. The second is that F is dependent not only upon porosity, since the ratio (L_e/L) is controlled by the texture of the rock (i.e., the shape, disposition, and interconnectedness of the pores). It is evident that two rocks with identical porosities may have different formation factors.

Porosity can also be expressed in terms of formation factor. By plotting formation factor versus porosity determined from laboratory data for many different rocks and reducing to equations the curves generated on such graphs, a general relationship was found of the form:

$$F = \frac{a}{\Phi^m}$$

The values a and m in this expression relate porosity to F in much the same manner as the ratio (L_e/L) related resistivity to F. They describe the texture of the rock. The value m is often called the cementation exponent, since its value tends to increase as the extent to which a rock is cemented increases, but this is not the only factor controlling its value. Both a and m depend upon properties of the original sediment, and upon whatever processes converted the sediment to rock. The values of a and m will vary from one rock to the next.

Values of these constants can be determined with advanced techniques of log analysis, but several equations are widely accepted for field use:

$$F = \frac{1}{\Phi^2}$$

This is the so-called *Archie Equation,* named after G.E. Archie, an early pioneering analyst. It works well in carbonate reservoirs, whose porosity and permeability are mainly secondary.

$$F = \frac{0.62}{\Phi^{2.15}}$$

The *Humble Equation,* mainly the work of W.O. Winsauer and his colleagues, adequately describes sandstones and other granular structures whose porosity is mainly primary. Because of earlier calculating tools incapable of handling the complex power of the Humble Equation, a simple version of it yields similar values. This is the *Tixier Relationship*:

$$F = \frac{.81}{\Phi^2}$$

To relate these rather idealized models to the resistivities measured by well logs of real rock/fluid systems, let us return to an earlier expression:

$$F = \frac{R_o}{R_w}$$

Transposing it:

$$R_o = FR_w$$

The real resistivites of fluids (which may be mixtures of fluids other than salt water), we shall denote as R_{wa}, or apparent water resistivity. We may now state:

$$R_t = FR_{wa}$$

where R_t = the system resistivity recorded by the log

Hydrocarbon presence in the rock pores serves to increase the measured value of R_t by increasing the length L_e of the ionic path. Oil and gas behave exactly as the rock matrix does in their effect on conductivities, i.e., they are insulators. Normally hydrocarbons exist in the rock pores as suspended droplets, acting like fine grains of rock inserted in the spaces between the larger grains. The apparent resistivity of the water in the pore space increases, as it has been shown that F is independent of fluid resistivities.

Calculating Water Saturation

Earlier we sought to calculate water saturation by comparing the resistivity of a zone to an idealized resistivity of the same rock containing only water. This poses certain difficulties, as logs do not directly measure R_o. Short of actually sampling the rock and its fluids some means must be found to determine R_o from the logs. Recalling two of our earlier statements:

$$S_w \propto \frac{R_o}{R_t} \text{ and } R_o = FR_w$$

By substituting:

$$S_w \propto \frac{FR_w}{R_t}$$

F may be found from a porosity log and the appropriate F-Φ relationship. R_w may be obtained by measuring produced samples or calculating it from the SP. R_t, of course, is directly measured by the logging instruments. We have a workable means of calculating the ratio R_o/R_t. But in what way is it proportional to S_w? Archie showed that the constant of proportionality, for resistivity comparisons, is a power of S_w by reducing graphs of measured laboratory data to equations. The relationship becomes:

$$S_w^n = \frac{FR_w}{R_t}$$

where n, the constant of proportionality, is the resistivity index exponent, commonly referred to as the saturation exponent.

In totally oil wet rocks, n can have a value as high as 3.0, 4.0, or greater. Such rocks are extremely rare. In water-wet rocks, n has a value of 2.0. In shaly uncompacted coastal sands, it has a value of 1.8. These two values will suffice for virtually all situations.

The use of n as a basis of comparison works equally well for other resistivity contrasts. Since we know that $R_o = FR_w$ and $R_t = FR_{wa}$, substituting these expressions for R_o and R_t enables us to say:

$$S_w^n = \frac{R_w}{R_{wa}}$$

The quantity R_{wa}, although not directly measured, is commonly computed by the surface computer and printed directly on the log as an auxiliary curve. It is computed using the equation:

$$R_{wa} = \frac{R_t\Phi^m}{a}$$

Obviously, we must have both a resistivity and a porosity log in addition to some feel for the appropriate values of a and m in order to perform the computation.

The contrasts R_o/R_t and R_w/R_{wa} are useful in the field as a quick-look interpretation method for reducing the number of zones to consider. Clean salt water zones generally exhibit R_t and R_{wa} values lower than those found in shales. These R_t and R_{wa} values, when read in a water-bearing zone near the zone of interest, are arbitrarily assumed to be R_o and R_w. It is easily observed that when R_t and R_{wa} exceed three times the values of R_o and R_w, the calculated water saturations will be less than 60%. This is usually considered to be the upper limit of commercial producibility. Chapter 6 will cover the technique in greater detail.

5

Sonic Logs

No logging device exists that measures formation factors directly; however, this important parameter can be determined if porosity and the appropriate F-Φ relationship is used. Knowledge of porosity is valuable to the reservoir engineer assessing the well's ultimate deliverability. Unfortunately, no logging tool measures porosity. Porosity must be calculated from other parameters. Sonic logs provide one means of obtaining the necessary data for such calculations.

Theory of Propagation

Acoustic energy is mostly thought of as sound waves. However, those frequencies and amplitudes humans are capable of detecting comprise only a small portion of the total sound spectrum. Acoustic energy is simply mechanical energy propagated through matter by a series of "compressions" and "rarefactions" that deform the media through which they pass. Vibrations from a source displace molecules in the surrounding environment, alternately increasing and decreasing the relative density and pressure of the surrounding medium. Continued oscillation of the source object creates a continuous stream of compressions and rarefactions. The acoustic wave in effect causes the medium through which it passes to oscillate. The molecular interaction is similar to that of billiard balls. Each molecule "bumps" into and displaces the next.

Types of Sonic Measurements

A sound wave is influenced by the medium through which it passes. Liquids, for example, have a different effect on the wave than do solids. In a liquid, the molecules are relatively free to "slip" past one another so that the sound wave does not propagate as rapidly as in a solid medium. Gaseous media attenuate the strength of the wave more than do solids or liquids.

Measurements of the various properties of the sound wave can yield useful information about the characteristics and contents of the beds through which the wave passes. Measurable properties of the sound signal include velocity, frequency, amplitude, and attenuation. Velocity expresses the rapidity with which the wave passes through the material. Frequency is an expression of the oscillation rate of the wave. For logging purposes this frequency occurs in the sonic range which is about 20,000 vibrations per second. This log is commonly known as the *sonic log.* Amplitude in an auditory context means how "loud" or strong the signal is. Attenuation is an expression of the loss of signal strength with its passage through a given substance.

Generating the Sonic Signal

The sonic tool generates sound waves by means of a piezoelectric or magnetostrictive electromechanical device. Piezoelectric crystals expand and contract as a varying voltage is applied across them. Magnetostrictive devices consist of a metallic cylinder which expands and contracts under the influence of a variable magnetic field. Repeatedly pulsing either device produces sonic vibrations.

The transmitter is normally enclosed in a housing that protects it from the borehole environment. This housing is often filled with a fluid to enhance acoustic coupling between the transmitter and the mud column.

Signal Path in the Borehole

The emitted sonic wave radiates in all directions from the transmitter. This signal travels along a multitude of paths, shown schematically in Figure 5-1. These are:

- Along the sonic tool itself
- Through the mud column
- Along the borehole wall
- Through the formation

The deepest investigating waves are traveling into the formation at depths equal to or greater than one wavelength. Wavelength is a function of sound velocity and frequency. The depth of investigation is given by the expression:

$$\text{Depth of investigation} = \frac{v}{f}$$

where v is the sound velocity in feet per second, and f is the center frequency of the transmitter crystal in cycles per second.

If the formation velocity is 15,000 ft/sec and the center frequency of the sonic tool is 20,000 vibrations per second, sound traveling into the formation will be investigating an area .75 feet behind the borehole wall. Sonic waves also travel slowly along the mud column surrounding the tool. Liquids have inherently low sonic velocities. Eventually, all of the signals will arrive at the receiver, which converts the sonic energy into electrical pulses for transmission to surface computers.

The Sonic Wavetrain

The receiver signal seen at the surface is quite complex. It is a composite of arrivals from all possible pathways, and consists of more than one wave type. Figure 5-2 illustrates the sonic signal and each of its components. Sonic tools are designed so that normally the first distinct signals detected are *compressive waves* that have traversed the body of the rock itself.

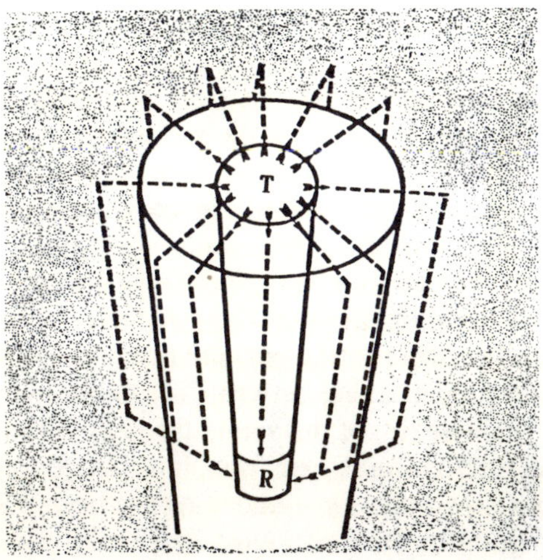

Figure 5-1. Illustration of the sonic signal path.

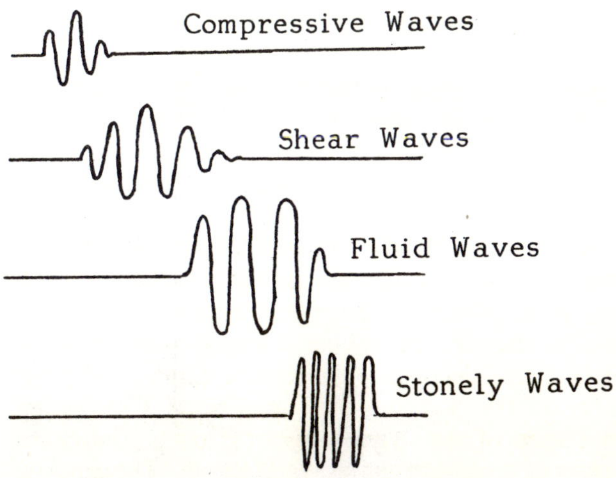

Figure 5-2. Schematic of the sonic signal.

Compressive waves are so called because they are the part of the sonic impulse that propagates by compressing the matter through which it passes. Compressive waves cause the rock to vibrate "back and forth" along the direction of the sonic propagation. Immediately after the passage of the compressive wave, the rock vibrates from "side to side" as the compressed material elongates. Lateral vibration of the rock produces a slow shear wave, which appears next, slightly over-lapping the late compressive waves. Figure 5-3 illustrates graphically the mechanics of the material deformations that give rise to these waves. Sonic wave energy first compresses the material, shortening it in length and causing it to expand horizontally. As soon as a compressive pulse passes, the material relaxes, returning to its former dimensions. The lateral relaxation produces a secondary compression in the direction of the main compressive wave propagation. The velocities of the waves depend upon the material's bulk elasticity or resistance to shortening and expanding. It is easy to visualize why shear waves arrive later then compressive waves.

Still later, and overlapping the last shears, another group of compressives arrive. These are the fluid waves, which travel very slowly in the liquid path afforded by the mud column.

The last arrivals are the stonely waves. These travel to the receiver through the mud column/borehole wall interface. Waves traveling along the tool itself are not normally observed, having been effectively attenuated by tool design.

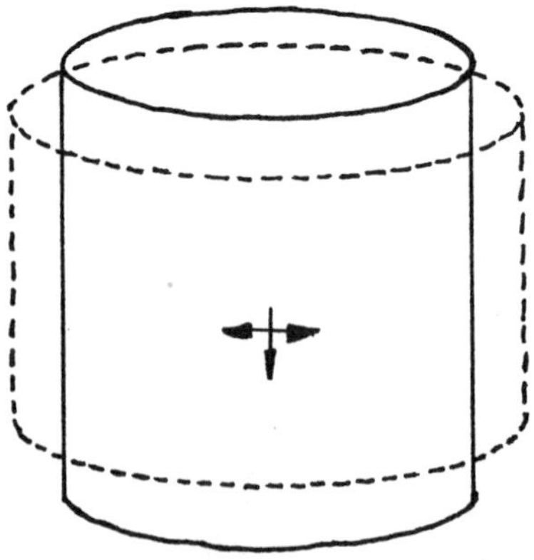

Figure 5-3. Illustration of material deformations caused by the sonic energy.

Sonic Logs

The sonic log is a recording of the time required for a sonic sound wave to traverse a given distance through the formation. This time increment is the interval transit time (Δt). Interval transit time is commonly measured in microseconds per foot (μsec/ft.). Δt is the reciprocal of the sound velocity expressed in feet per second. The relation between these two expressions of compressive and propagation rate is given by:

$$\Delta t = \frac{10^6}{\text{velocity}}$$

Figure 5-4 shows a typical sonic log presentation. The heading defines the scale of the solid transit time curve in track four. Note that the transit time increases from right to left, so that the sonic response follows that of the resistivity log. The dashed curve represents the calculated porosity. To complete the log, a caliper and gamma ray or SP curve often appears in track one.

An alternate sonic presentation is shown on Figure 5-5. Sonic logs in combination with other logs are becoming increasingly popular. This presentation omits the porosity curve and places the solid Δ curve on a linear grid in track three, often with a dashed caliper curve. Resistivity is presented on a logarithmic scale in track two. Track one may contain the SP, gamma ray, and/or a computed R_{wa} curve.

Tool Design

A typical sonic tool consists of two transmitters and two receivers. Figure 5-6 shows the configuration of the sonic tool.

First assume that the tool has only one transmitter, that is, TR_l on Figure 5-6, and one receiver, R_l. TR_l fires a sonic sound pulse. The sound travels through the formation and arrives at the receiver. The time it took for the sonic sound pulse to travel from TR_l to R_l divided by the spacing between TR_l and R_l would represent the velocity of the sound through the formation as Δ in microseconds per foot. In order for a counter to count how long the sound traveled from TR_l to R_l, the exact moment of fire must be known. This task is rather difficult since the electrical characteristics of the wireline change from time to time.

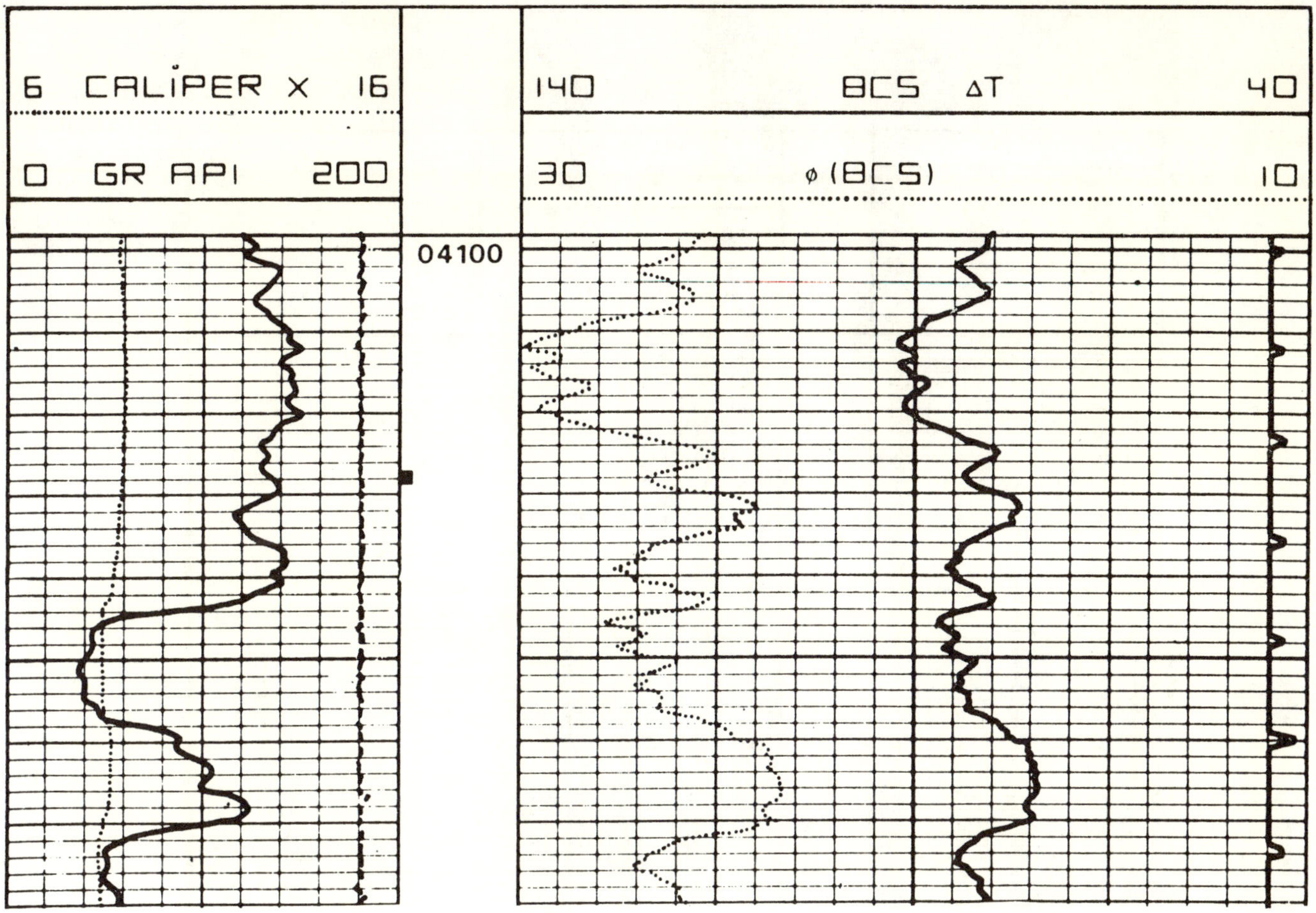

Figure 5-4. Illustration of the sonic and porosity curves. (Courtesy of Gearhart.)

In order to overcome the difficulty, a second receiver, R_2 is incorporated into the tool.

Surface computer gear issues a command for TR_l to fire. Sonic sound is emitted and travels along the path TR_l to R_l. R_l, upon receiving the sonic impulse, starts the time counter. R_2 stops the time counter as soon as it receives the impulse. R_l starts time; R_2 stops it. Δ is the time it took the sound to travel from R_l to R_2 divided by the spacing between R_l and R_2.

All is well with this tool design until a washout is encountered. Referring to Figure 5-7, a washout causes the sonic impulse to travel through more borehole mud at R_l than it does at R_2. Since sound travels slower through fluids, it adds an error to the Δ measurement. Tool tilt has the same effect as a washout, placing the R_l receiver closer to the borehole wall and R_2 farther away.

Again, referring to Figure 5-7, a sonic tool with two transmitters eliminates errors due to tilt, washouts, and rugose boreholes. t_1 sees an error due to the washout. t_4 from the lower transmitter is also subjected to the same error. Mathematically, all of the error can be subtracted as follows:

$$\Delta t = \frac{(t_3 + t_4\ (\text{error})) - (t_1\ (\text{error}) + t_2)}{4}$$

As can be seen, in the process of averaging, the error is subtracted or canceled. This system is called *bore hole compensated sonic.* This technique can also be implemented with one transmitter and two receivers. This system works the transmitter twice as hard (or one half the data rate) and memorizes the position of the transmitter as if it were both a lower and upper transmitter. While some wireline companies still use a single-transmitter tool, it is the author's opinion that a two-transmitter tool is far superior.

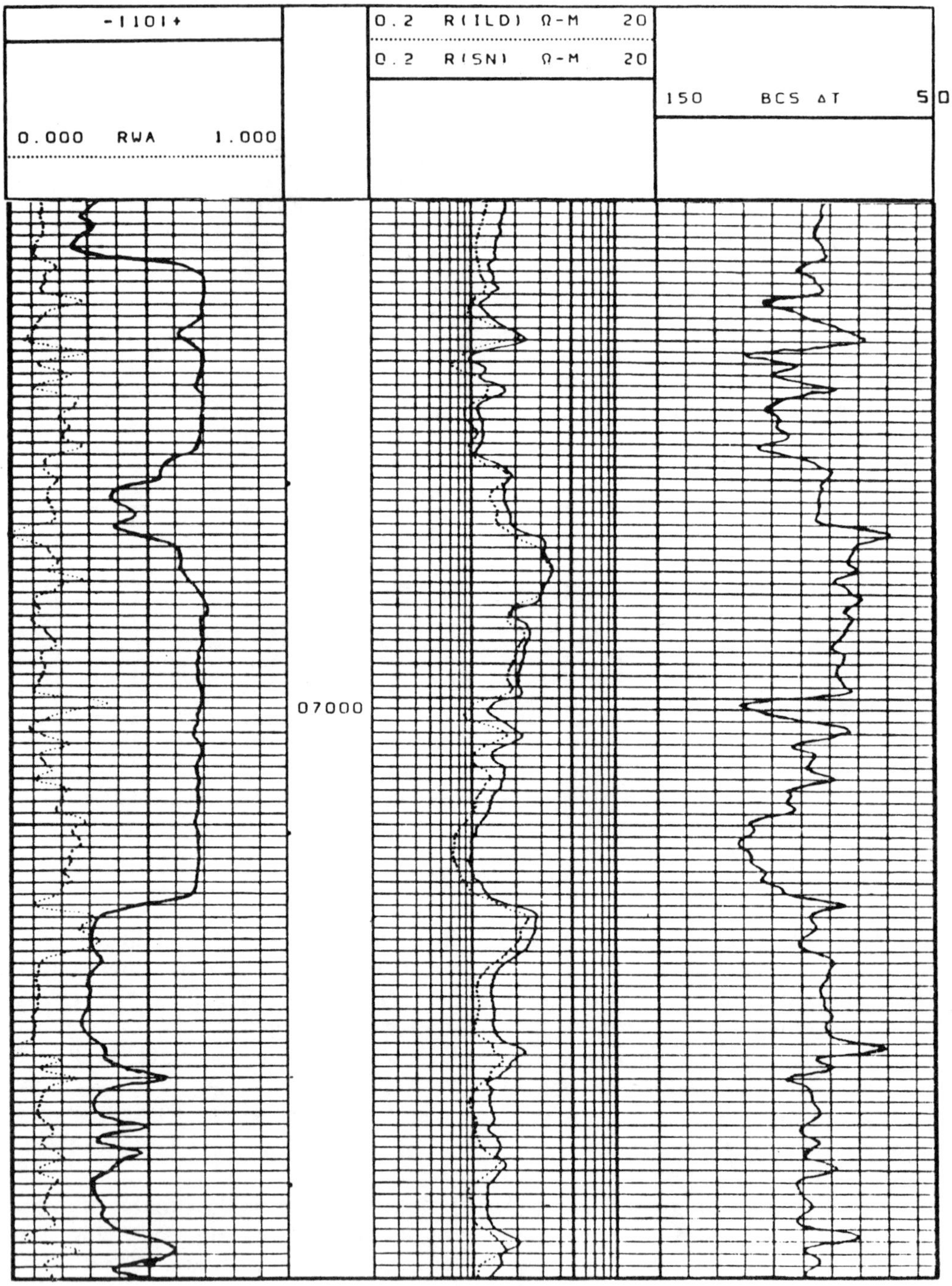

Figure 5-5. Illustration of the resistivity-sonic combined presentation. (Courtesy of Gearhart.)

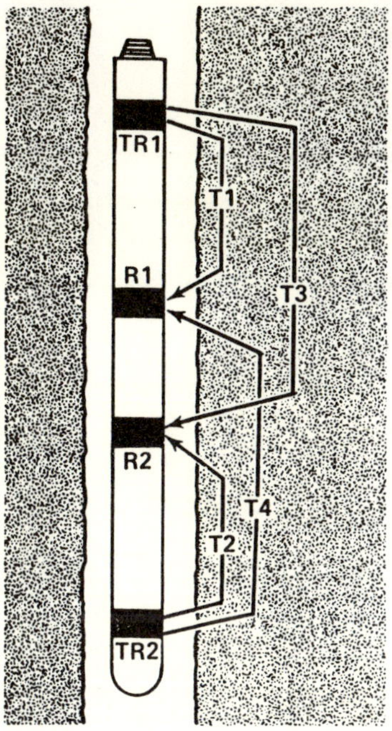

Figure 5-6. Illustration of multiple transmitter receiver signal path. (Courtesy of Gearhart.)

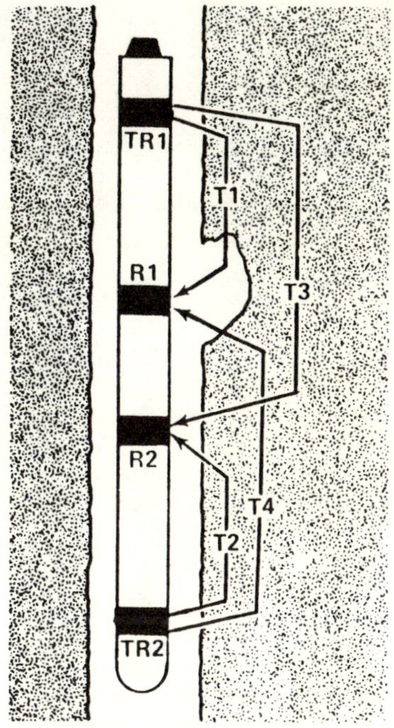

Figure 5-7. Illustration of the signal through a washout. (Courtesy of Gearhart.)

The Wyllie Time-Average Equation

Porosity is a derived quantity. Since the sonic tool is only capable of measuring travel time, some way of relating it to porosity must be employed. While many relationships between porosity and travel time have been proposed, the most commonly accepted is the *Wyllie Time-Average Equation.*

$$\Delta t = \Delta tf\,\Phi + \Delta t_{ma}\,(1 - \Phi)$$

where Δtf = fluid travel time ($189\mu s$)
Δt_{ma} = formation matrix travel time
Φ = porosity
Δt = travel time from log

The equation basically states that the total travel time recorded on the log is the sum of the time the sonic wave spends traveling in the hard part of the rock, called the *rock matrix*, and the time spent traveling through the fluids in the pores, as if the wave encountered first the solid rock and then a layer of pure fluid. Although the sonic signal does not actually encounter the different materials in this manner, the model is not a bad one. Sonic wave movement is a bulk phenomenon that tends to average the contribution of each material when they are homogeneously mixed.

Porosity can be solved for directly, by transposing the Wyllie Equation:

$$\Phi = \frac{\Delta t_{log} - \Delta t_{ma}}{\Delta t_f - \Delta t_{ma}}$$

Pore fluid velocities do not vary appreciably when the fluid is a liquid. A Δt_f of 189 μsec/ft is commonly used when the actual value is unknown. Δt_{ma} varies considerably among different rocks, so lithology must be known to obtain even a rough estimate of the porosity. Average values of Δt_{ma} for various rocks are commonly used as a starting place for analysis (See Table 5-1).

Matrix Δt may vary considerably for a given lithology. Obviously, a poor choice of matrix will result in poor porosity calculations. The effects of temperature and pressure on Δt_{ma} are negligible over the range of practical interest. There are rare instances where the matrix bears the entire overburden pressure because the zone is underpressured.

Figure 5-8 shows a chart that solves the Wyllie Time-Average Equation. Travel time is read from the log and entered on the bottom of the chart. A line is

Table 5-1
Commonly Used Δt_{ma} Values*

	Average	Range
Sandstone	57	(53–100)
Limestone	47	(44–53)
Dolomite	43	(40–45)
Chalk	47.5	(47.5)
Anhydrite	50	(50)
Shale	—	(60–170)

** A more extensive list is given in the appendix.*

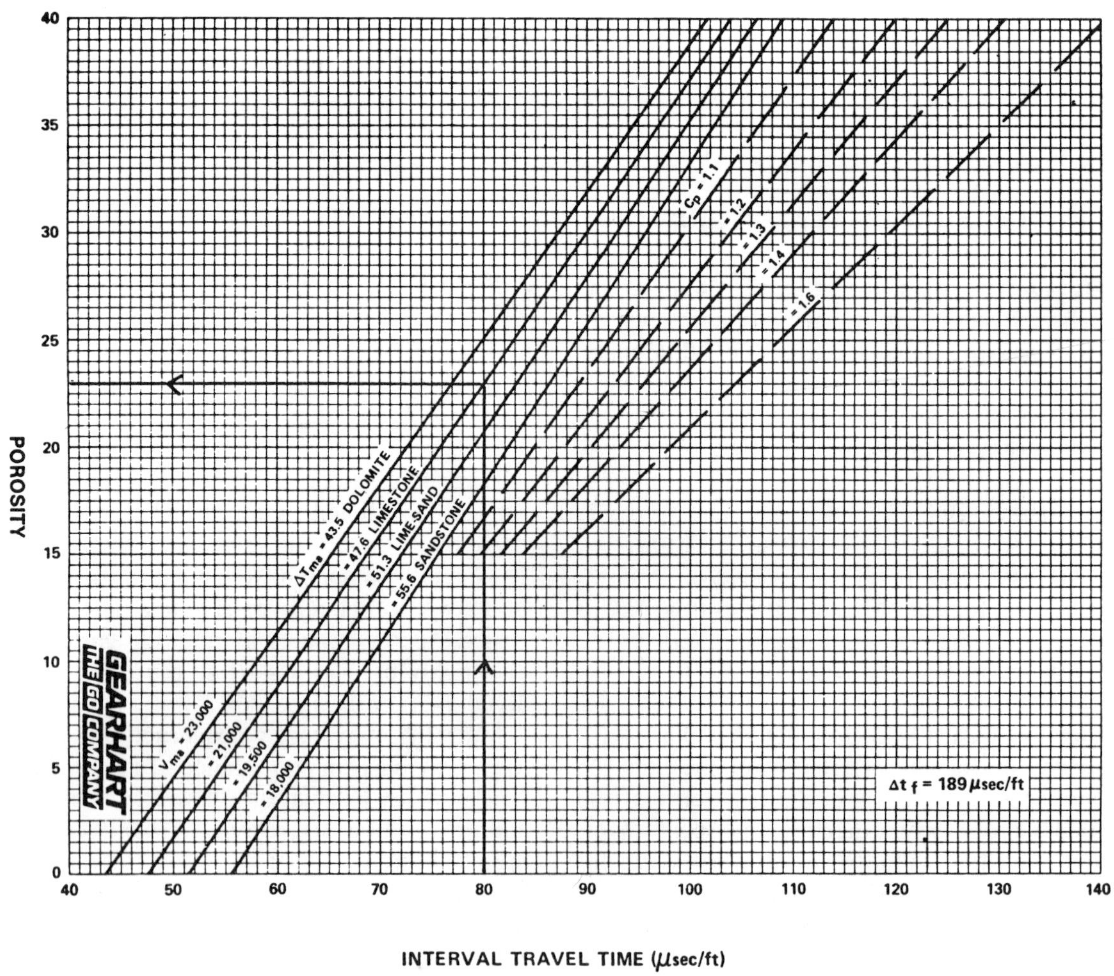

Figure 5-8. Borehole-compensated sonic porosity versus travel time. (Courtesy of Gearhart.)

extended vertically until it intersects the desired matrix line. The line is then extended at a right angle to left, and porosity is read.

Shortcomings of the Wyllie Time-Average

The model tends to fall down where pore space is not uniformly distributed. Vugular rocks may have large individual pores, but if the pores are not well connected and dispersed, the fastest arrivals will be those waves that have "gone around" the pores. Transit times will be shorter than the overall porosity indicates they should be. Porosities calculated from such transit times by the Wyllie Equation will be lower than the porosities derived from cores of the same rock. For this reason, it is often said that sonic tools ignore secondary porosity.

Several qualifications about sonic log and core porosity comparisons are in order at this point. First, the sonic wavelength is so much greater than the size of the vugs or fissures that the sonic tool inherently averages a much greater volume of rock than is available to the core analyst. Secondly, cores have been somewhat shattered by the coring process. Downhole pressures have been relieved, and the rock has expanded and cooled. Whole core analysis makes use of samples extracted from the wellbore during drilling and specifically not measured by the log. Sidewall core analysis inspects samples from the near vicinity of the borehole, whereas the sonic tool measures the properties of the zone at a deeper point. It is sometimes argued that if log porosities do not agree with core porosities, the log must be incorrect. This must be regarded as an absurdity, except where the tool has obviously failed.

Sand compaction affects the measured travel time. The porosities calculated from logs of uncompacted sands are too high. A compaction correction term must be incorporated into the Wyllie Time-Average Equation. The problem is that Δt is too long. The increased Δt is due to the mobility of the grains in the formation and to energy losses in the water rings at grain contacts. Since the grains are movable, the sonic energy has no solid path through the formation, and the sound is slowed down when a part of its energy is transferred to the grains.

The degree of sand compaction is usually determined on logs from the travel times in nearby shales. If Δt_{log} in the shale is less than 100 μsec, the nearby sands can be treated as compacted. Otherwise, a correction is usually applied to the porosity determined from the Wyllie Equation. This correction can take several forms. The most common method is to use an equation of the form:

$$\Phi = \frac{\Delta t_{log} - \Delta t_{ma}}{\Delta t_f - \Delta t_{ma}} \times \frac{100}{\Delta t_{sh}}$$

where Δt_{sh} is the transit time of a nearby shale read from the log.

The chart on Figure 5-9 can also be used to apply the compaction correction. Once porosity has been obtained, a Δt reading is obtained from a nearby shale. Porosity is entered on the left-hand scale. Δt in a nearby shale is entered on the bottom scale and a correction is plotted.

Example

The following parameters apply:

Porosity = 22%
Shale = 96 microseconds/ft
B_{cp} from the chart is 1.4

$$\frac{1}{B_{cp}} = \frac{1}{1.4} = .71$$

Corrected porosity = .22 × .71 = .157

or 15.7% porosity

Other methods of compaction correction use an equation of the form:

$$\Phi = \frac{\Delta t_{log} - \Delta t_{ma}}{\Delta t_f - \Delta t_{ma}} \times \frac{1}{C}$$

The value of the constant "C" is determined by several popular methods. In clean (i.e., shale-free) water-bearing sands near the zone of interest, porosity derived with the unmodified Wyllie transform can be compared with density or induction-derived porosity. The relevant equations are:

$C = \Phi_v(aR_w/R_t)^{-1/m}$ for the induction

$C = \Phi_v/\Phi_d$ for density

Both of these methods suffer the serious drawback of requiring clean sands, a condition none too common in precisely those areas where a compaction correction is most likely to be required.

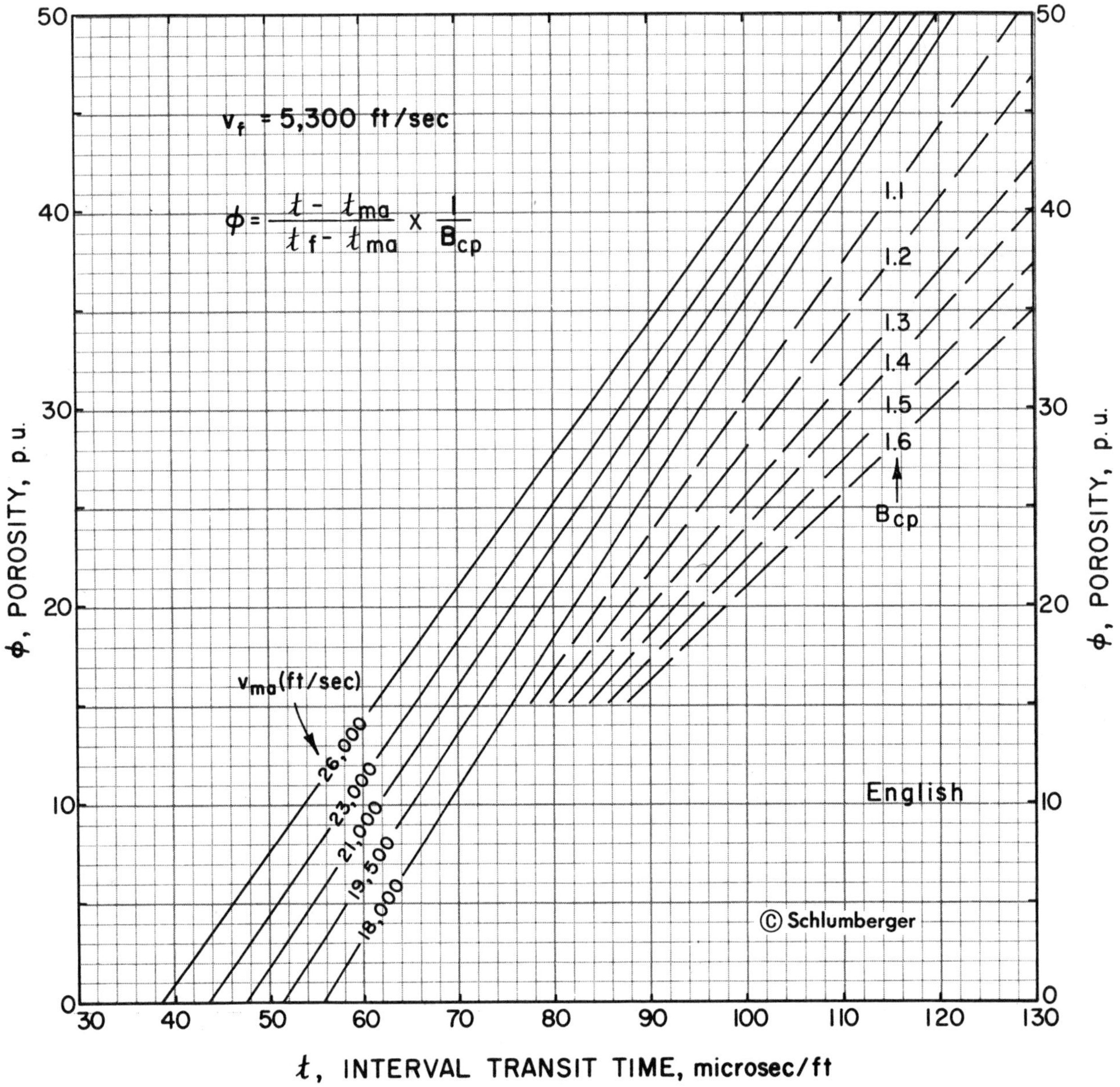

Figure 5-9. Porosity evaluation from sonic. (Courtesy of Schlumberger.)

A better method for calculating C in a shaly sand would be to compare the sonic and neutron log responses, using an equation of the form:

$$C = \Phi_v / \Phi_n$$

Another approach to determining the value for C is to express it as a function of depth. The relation

$$C = 6.966 - 1.414 \log D$$

where Log D is the base 10 logarithm of the depth, has been found to work well on the Louisiana/Texas Gulf Coast.

Other correction equations too numerous to list here exist. Choice of any of the foregoing compaction-correction techniques is largely a matter of preference and convenience. Compaction corrections are made to compensate for the weakness of the Wyllie model, which is not general enough to describe every situation that may be encountered.

If any technique is to be recommended, one that seeks to correct the source of error at the outset is perhaps the best. Recall that the real problem in uncompacted sands is that Δt is too long. Correcting it before entering the Wyllie Equation may yield porosities that better agree with the porosities from other logs and cores.

Three types of fluids can be found in a given formation. The Wyllie Time-Average Equation accounts for only one type of fluid. This, of course, is formation water, which is accounted for by 189 microseconds in Δt fluid.

The second fluid commonly found in pore spaces is oil. Since oil is slightly less dense than water, sonic velocity in oil will be slightly slower than in water, but not appreciably so. The same Δt_f may be normally used. However, it should be noted that some amount of dissolved gas is often found in oil at reservoir conditions and does have an effect on the sound velocity. An empirical approach to porosity determinations when oil is known to be present in the pores is to multiply the porosity by .9 to effect an *oil correction.*

The third fluid that must be considered is gas. Gas, being much less dense than either water or oil, has very slow sonic velocities. Gas compressibility also has a large effect on travel time. Even small amounts of gas dissolved in water or oil will increase Δt. A fluid must be almost all liquid to be considered incompressible. The usual approach is simply to multiply the apparent porosity calculated with a Δt_f of 189 by .7 to effect a *gas correction.* A much better technique is to crossplot the sonic log response with another porosity curve.

An interesting anomaly occurs when an uncompacted sand contains gas in the pores. The gas is relatively uncompressed and the sonic wave loses some energy and slows down when it temporarily compresses the gas further. Travel times tend to be too high and porosities are also too high. This is the basis for detection of gas from the sonic log. Density and neutron porosities will not agree with the sonic. The phenomenon is not as readily observable in deeper compacted formations.

Shale Effects

Shale presence, like unconsolidated rock, will also cause porosity to be too high. The real reason for this is that Δt_{ma} is wrong, but just as with compaction corrections, techniques exist that modify porosity after calculation.

Shale, of course, exists in several forms in the reservoir, so that it is not easy to "correct" its influence. Generally, the sonic tool tends to see shale as pore space when the sand is relatively unconsolidated, and as matrix when large percentages of older shales are present.

Shales display two interesting properties. The first is anisotropy. Individual shale particles may behave much like a gel, which becomes liquid or goes into solution, when shaken vigorously. Passage of the sonic wave supplies the "shaking" energy.

Large accretions of shale, on the other hand, tend to exist as laminae, or thin sheets of argillaceous material. Between each sheet may be bound a layer of water one molecule thick. Sound waves passing through the stack attenuate and slow as if crossing fractures.

No single matrix value may be assigned to shale. Several clay minerals compose even the purest shales. The most common of these is montmorillonite, kaolinite, and illite. Each clay mineral has its own characteristic grain size, density, and affinity for water. Each additionally requires a different energy environment for deposition, and comes from different source rock. The travel times of even bedded shales vary considerably from one region to the next.

Two methods of determining porosity of a shaly sand are possible. One method is to arrive at a Δt_{ma} more representative of the sand/shale mixture before calculating porosity, which then need not be corrected further. Δt_{ma} of the mixture may be approximated by summing the contribution of each component to the overall matrix velocity.

$$\Delta t_{ma}\ (\text{mix}) = (\Delta t_{ma} \times V_{ma}) + (\Delta t_{sh} \times V_{sh})$$

In this expression, Δt_{ma} is the matrix Δt of the sandstone, and V_{ma} is the fractional volume of total rock matrix composed of sand. Δt_{sh} is assumed to be the same as Δt_{log} in nearby thick shales. V_{sh} is the fraction of total rock volume occupied by shale. Of course, Δt_{ma} of even more complex mixtures, involving several different minerals, may be determined in this way.

The other method involves correcting the apparent porosity calculated by assuming a clean matrix. Correcting an apparent porosity requires knowledge of the manner in which the shale exists in the sand. For sands containing laminated shale, apparent porosity is corrected by subtracting from it a correction term equal to the product of shale volume in the sand times the apparent shale porosity of nearby thick shale beds, or:

$$\Phi_{effective} = \Phi_{apparent} - (\Phi_{sh} \times V_{sh})$$

The apparent porosity of nearby shales is calculated using the same Δt_{ma} used in calculating porosity in the sand.

For dispersed shaly sands, the effective, or shale corrected, porosity is calculated by subtracting the decimal fraction of shale present from the apparent porosity, or:

$$\Phi_e = \Phi_a - V_{sh}$$

$$\Phi_{effective} = \Phi_{apparent} - V_{sh}$$

Structural shale cannot easily be corrected for when calculating porosity from a sonic log. Generally, structural shale is indistinguishable from matrix. Effective porosity is assumed equal to the apparent porosity.

Porosity determination in a shaly formation is at its best when the sonic tool response is compared to the responses of other "porosity" tools. Such crossplot comparisons are not appropriate to the intended scope of this present text. The determination of shale volume will be covered in a later chapter.

Figure 5-10 summarizes in flowchart form, the many steps to porosity determination from sonic logs. It must first be determined whether or not the rock is compacted. Apparent porosity is calculated first, using an appropriate equation. If the formation is shaly or contains gas, apparent porosity must be corrected to obtain the effective porosity. This intermediate porosity may be compared to effective porosities from other types of tools to obtain the true formation porosity.

Note that it is true porosity and not apparent porosity that should properly be used in F-Φ relationships when determining the water saturation of the zone. In the interest of time and to obtain quick decisions for setting pipe, wellsite analysis often omits fully accurate determination of all input parameters. It should be realized that such calculations are only qualitative in nature and that field porosities and water saturations may bear little relation to reality.

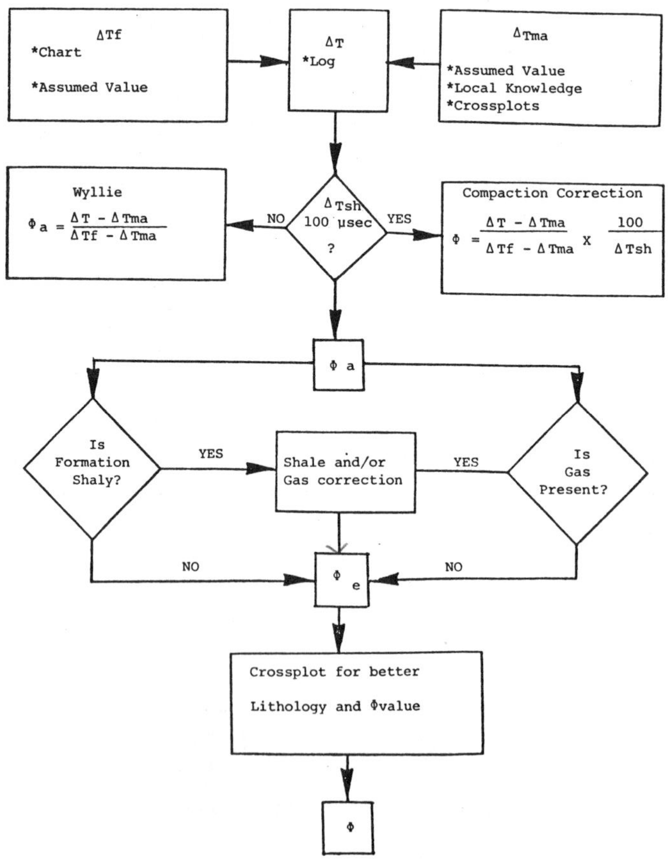

Figure 5-10. Flow Diagram of the correction and computation of sonic porosity.

6

Analyzing the Log

Each curve thus far studied has given us just a small element of the total picture. Just as a story has individual paragraphs, so does a log. The best advice given to any potential log analyst is simply to never look at just one curve. If you see something you like, look for confirmation again.

While looking at the sonic curve, train your eyes to look at the caliper. Remember that the depth of investigation is usually no more than a foot. You can usually tell if a formation is invaded by observing the caliper curve. Invasion builds up mudcake, and thus it can be seen on the caliper curve. Invasion implies porosity and permeability.

If the resistivity curve implies a hydrocarbon, the sonic wave would be slower passing through. This suspicion could be indicated on the log by a longer Δt. While it is by no means guaranteed to be so, it happens frequently enough to warrant inspection. This is particularly true in gas-bearing formations.

If the analyst confirms the analysis through the use of more than one curve, he may proceed with greater confidence, knowing that each element is building a picture.

Up to this point we have discussed each individual curve on a log. We now have sufficient information to do an analysis on a dual-induction, sonic combination log.

Our analysis will proceed by stages, as outlined below:

1. Define the bed boundary
2. Use pattern recognition
3. Use a quick-look technique
4. Determine R_w from the SP
5. Determine R_t
6. Determine porosity
7. Calculate formation factor F
8. Calculate S_w

Throughout this chapter, two methods will be employed. The first will be a graphical solution and the second, mathematical. For the reader not interested in either one, simply pass over it to the next step.

Defining the Bed Boundary

As previously mentioned, the bed boundary occurs at the point of inflection on the SP. This point is where the curve of the SP changes slope, contrary to the popular notion that it is midway on the curve.

Starting our analysis, Figure 6-1 shows a portion of log with the zone of interest. The lithology is limestone. We begin by drawing two lines on the SP curve.

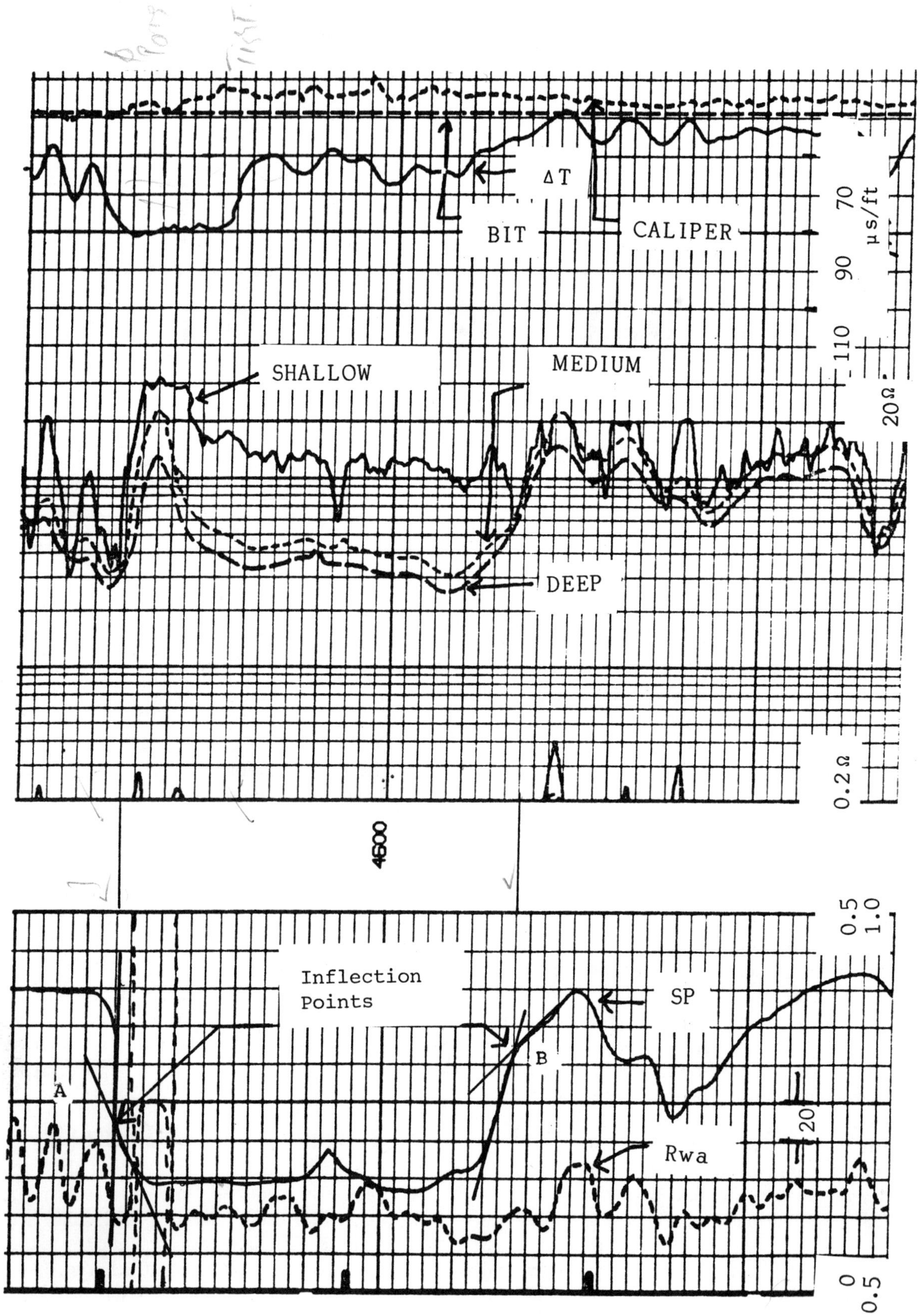

Figure 6-1. Locating bed boundaries using the inflection points of the SP.

Note that one line cannot cover the curve because there is a change in slope. In order to do so, another line must be drawn. Where the two lines cross, an inflection point occurs, corresponding to the bed boundary. Point B on Figure 6-1 depicts this more pronouncedly than point A.

As a matter of good practice, draw the bed boundaries in the depth track to make the formation of interest well defined.

Pattern Recognition

Now that we have defined the zone, we shall next attempt to determine the overall characteristics of the bed. The technique we shall use is only qualitative. By simply looking at the log, seeking characteristic curve patterns, we shall separate the more obvious water-bearing and hydrocarbon zones.

Consider Figure 6-2. Represented here is a schematic induction log with an SP. The dotted curve is induction resistivity. The solid curve near it is from a shallow reading device. The discussion which follows applies equally to any other deep and shallow investigating resistivity device.

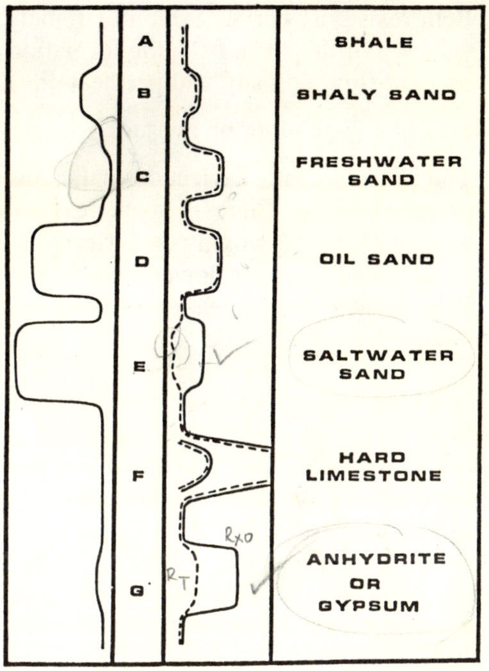

Figure 6-2. Illustration of resistivity SP patterns.

Pattern A is a shale because:

1. The SP does not deflect from the shale base line.
2. Both resistivity curves have a low value.

This value defines a resistivity base line, similar to the shale base line on an SP. Note also that both resistivity curves read approximately the same. This implies that no invasion has taken place, just as we might expect in a nonpermeable shale.

Pattern B is a shaly sand because:

1. The SP does not deflect to maximum potential as it does in pattern E.
2. Both resistivity curves record a value only slightly higher than shale.

This means that the zone must contain some rock more resistive than shale. Sand is possibly indicated because it does not contain the bound water associated with shale.

Pattern C is a freshwater sand because:

1. The SP deflects to the right of the shale base line. For this to happen, the drilling mud must be saltier than the formation waters; hence, formation water is fresher.
2. Both resistivity curves show a value higher than that of the saltwater sand in pattern E.

The high resistivity is due to the freshwater.

Pattern D is an oil sand because:

1. The SP does not attain maximum deflection as in pattern E. Two conclusions may be drawn. First, formation water must be more saline than the mud filtrate. Secondly, the sand must contain some shale. No hydrocarbon suppression would occur if this were not the case.
2. The shallow curve is higher because fresh filtrate and residual oil are present.
3. The deep resistivity is high because the virgin rock is saturated with oil.

Pattern E is saltwater because:

1. The SP shows maximum negative deflection.
2. The shallow resistivity is high because the invaded zone contains fresh filtrate. The separation between the resistivity curves indicates good permeability.
3. The deep resistivity is low because the virgin waters are salty.

Pattern F is an impervious low porosity limestone because:

1. The SP does not deflect from the shale base line, indicating an impervious stratum.

2. Both resistivity curves exhibit extremely high values. For this to be true, the formation must be tight (low porosity and permeability).

Pattern G is an anhydrite or gypsum because:

1. The SP shows little deflection, indicating little or no porosity. There is some permeability, otherwise the SP would not deflect.
2. Both resistivities are lower than in the tight lime because the formation contains metallic ions.

A porous lime behaves much in the same manner as a sandstone.

Looking at Figure 6-3, we see the deep, medium, and shallow curves all following each other. This is similar to Pattern D of Figure 6-2. Further down the hole, Pattern E indicates the formation to be water wet. At 4,592 ft the SP notches, indicating a shale streak. The resistivity curves appear to be coming together, indicating no permeability. If the shale bed were significantly thicker, the SP would indeed return to the shale base line and the three resistivity curves would be as Pattern A of Figure 6-2 indicates.

The formation thus indicates a hydrocarbon at the top, followed by water. The zone is worth further investigation.

Quick-Look Techniques

Let us now attempt a rough estimate of the water saturation at this point. Let's assume that the formation is uniformly porous and permeable throughout. Making these assumptions, we can also assume that the lowest resistivity in the bed represents R_o. This technique can only be applied where the zone is obviously 100% saturated with water.

The value of R_t is chosen from the same curve except it is chosen at the point where hydrocarbons are suspected.

Simply stated, R_o is taken directly from the deep-reading resistivity curve where it's obviously 100% water wet. R_t is read from the same curve where the hydrocarbon is suspected. Figure 6-4 illustrates the point.

In this case, R_t reads 10 ohms and R_o reads 3.2 ohms.

The resistivity index is the ratio of R_t to R_o.

$$\text{R.I.} = \frac{R_t}{R_o} = \frac{12}{3.2} = 3.75$$

The water saturation is:

$$S_w = \sqrt{\frac{R_o}{R_t}} = \sqrt{\frac{3.2}{12}} = 0.516 \text{ or } 51.6\%$$

For limestones, generally the cut-off water saturation point is 50%. This case is so close to cut-off that it is worth further investigation. A key point for quick-look analysis is that if the resistivity ratio is equal to or greater than 3, the saturation will be less than 60%.

R_{wa} Technique

Looking at Figure 6-4, adjacent to the SP is the R_{wa} curve. In theory, this curve should yield values for R_w. The R_{wa} curve is computed by the surface gear given porosity, resistivity, and the parameters a and m.

$$R_{wa} = \frac{R_t \Phi^m}{a}$$

The values of a and m are chosen by the engineer at the time of logging. While this computation should yield R_w, it is subject to many errors. A general value for a and m is chosen. Porosities used are not corrected. In thin beds R_t must also be corrected. For these and several other reasons, this curve cannot be trusted to yield correct values of R_w, and therefore it is labeled R_{wa}, or apparent. While this curve cannot be used quantitatively, it is of value qualitatively. As with the deep reading curve, we are going to use just the R_{wa} curve.

At the point where the formation is 100% water wet, we will label it R_w. Where we suspect hydrocarbons, we label it R_{wa}, as in Figure 6-4.

R_w reads 0.1 ohms, and R_{wa} reads 0.8 ohms. Note that for R_{wa}, the curve goes off scale, so care must be taken to read the backup scale. A quick-look water saturation can now be calculated:

$$S_w = \sqrt{\frac{R_w}{R_{wa}}} = \frac{.1}{.8} = .354 \text{ or } 35.4\%$$

The R_w/R_{wa} technique also suggests further investigation.

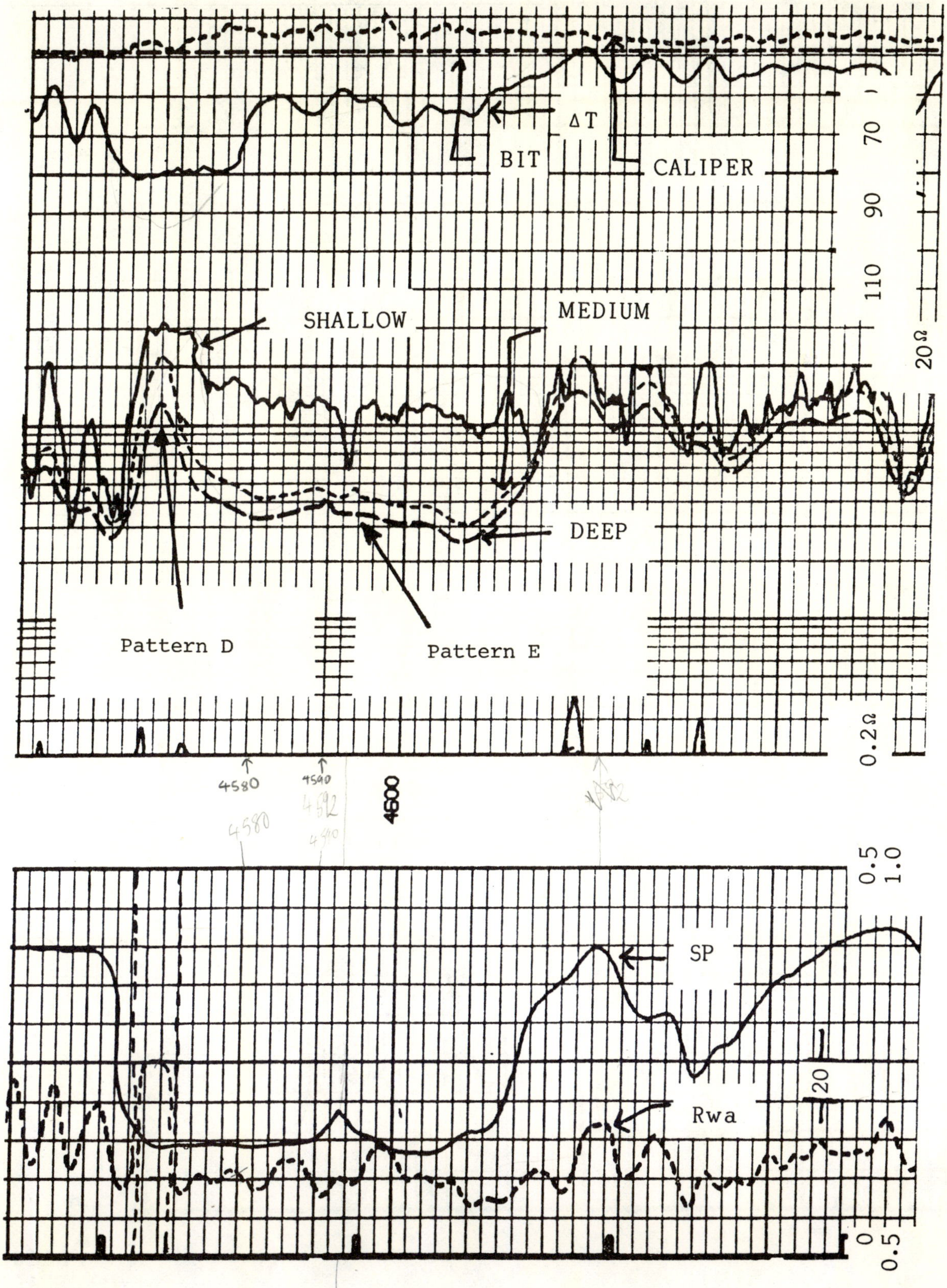

Figure 6-3. Illustration of patterns on the resistivity curves.

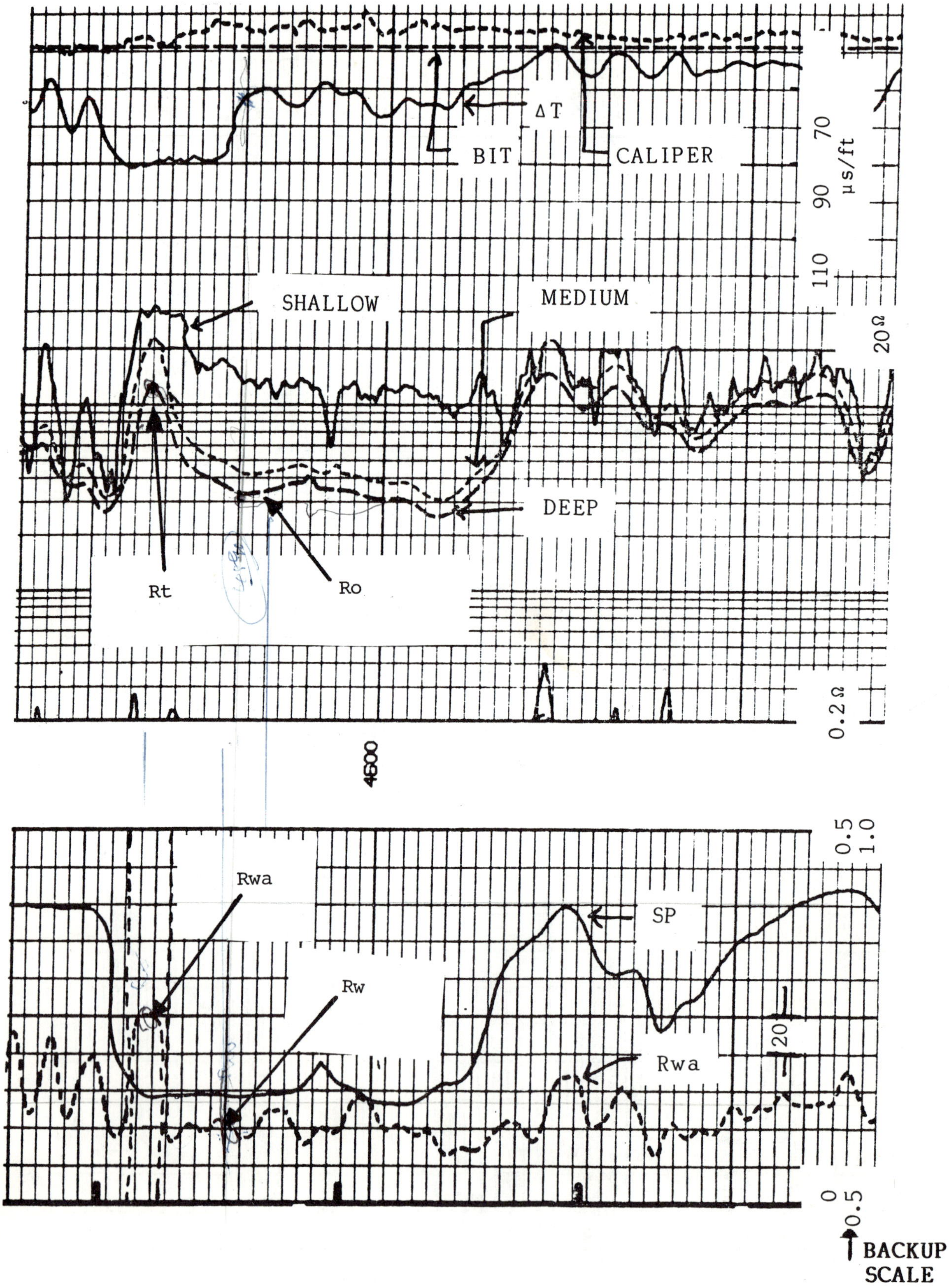

Figure 6-4. Illustration of R_o and R_t.

Determining R_w from the SP

R_w can now be considered as the cornerstone of analysis. A small error in the R_w will geometrically progress itself to a larger error in the final analysis. I cannot stress enough the value of R_w being as accurate as possible.

R_w is ideally derived from chemical analysis of virgin formation fluid. As an alternate, R_w may be obtained through local knowledge of a field that has been well developed. If none of these sources are available, the best alternative is deriving an R_w from the SP, as outlined below.

Step one is to establish the sand shale lines (Figure 6-5). Between the lines there are 5.2 divisions. Taking the millivolt scale and multiplying it by the number of divisions gives us the total SP deflection.

$$20 \times 5.2 = 104 \text{ mv}$$

Since the deflection is to the left, the SP is negative.

$$-104 \text{ mv}$$

The next step is to determine the formation temperature. Figure 6-6 is the heading on the log. Total depth is 7,002 ft (depth logger). Bottom hole temperature is 159 degrees (maximum record temperature). Zone of interest is at 4,590 ft (from the log, Figure 6-5).

The only thing needed now is the mean average temperature. This is the temperature of each day of the year averaged, or the temperature that would be measured a few hundred feet below the surface. Such information can be obtained from weather services, geological societies, the United States Geological Survey, or the U.S. Oceanography and Atmospheric Service.

For this example, 80°F will be used.

Calculating Formation Temperature Determination

Chart Method. Figure 6-7 represents the steps necessary to obtain information temperatures.

1. Bottom hole temperature is entered on the bottom of the chart: 159° on the 80° mean temperature scale.
2. Total depth is entered on the left side: 7,000 ft.
3. The two points are plotted and fall on the geothermal gradient line marked 1.2° per hundred ft.
4. Enter the depth of the zone of interest on the left side. Here it is 4,590 ft.
5. Extend the line horizontally until it intersects the 1.2 diagonal line.
6. Drop the line vertically and read the formation temperature: 132° is indicated.

Note: Circled numbers on the chart match each step above.

Computation Method. Using the equation:

$$T_f = (BHT - T_{avg}) \times (Df/TD) + T_{avg}$$

where
T_f = formation temperature
BHT = 159° (from heading)
T_{avg} = 80° (average mean temperature)
Df = 4,590 ft (zone of interest)
TD = 7,002 ft (total depth from heading)

Thus:

$$T_f = (159 - 80) \times (4{,}590/7{,}002) + 80$$
$$= 131.8°$$

Comparison of the Two Methods. With the chart method, we obtained 132°. Mathematically, we obtained 131.8°. As you can see, there is no difference in the two. Either method can be used with confidence. As far as the charts that can be used, any of the wireline companies' chart books can be employed.

A mud resistivity must now be determined at formation temperature. Again, from the log heading (Figure 6-6) we read R_m = 2.7 ohms @ 75 degrees.

Determining Mud Resistivity

Chart Method.

1. On the left-hand scale of Figure 6-8, find and plot 75°.
2. Find and plot 2.7 ohms on the right-hand scale. Draw a line between the two points.
3. Read the right center scale indicating the K PPM or thousands of parts per million sodium chloride equivalent. In this case, it is just under 2 thousand.
4. Next, plot the formation temperature previously found on the left hand scale (132°).
5. Extend a line from 132° so that it passes through the 2 K PPM point on the right center scale. This line should be extended all the way to the resistivity scale on the right.
6. Read the new mud resistivity at formation temperature. In this case, it is 1.6 ohms at 132°.

(text continued on page 68)

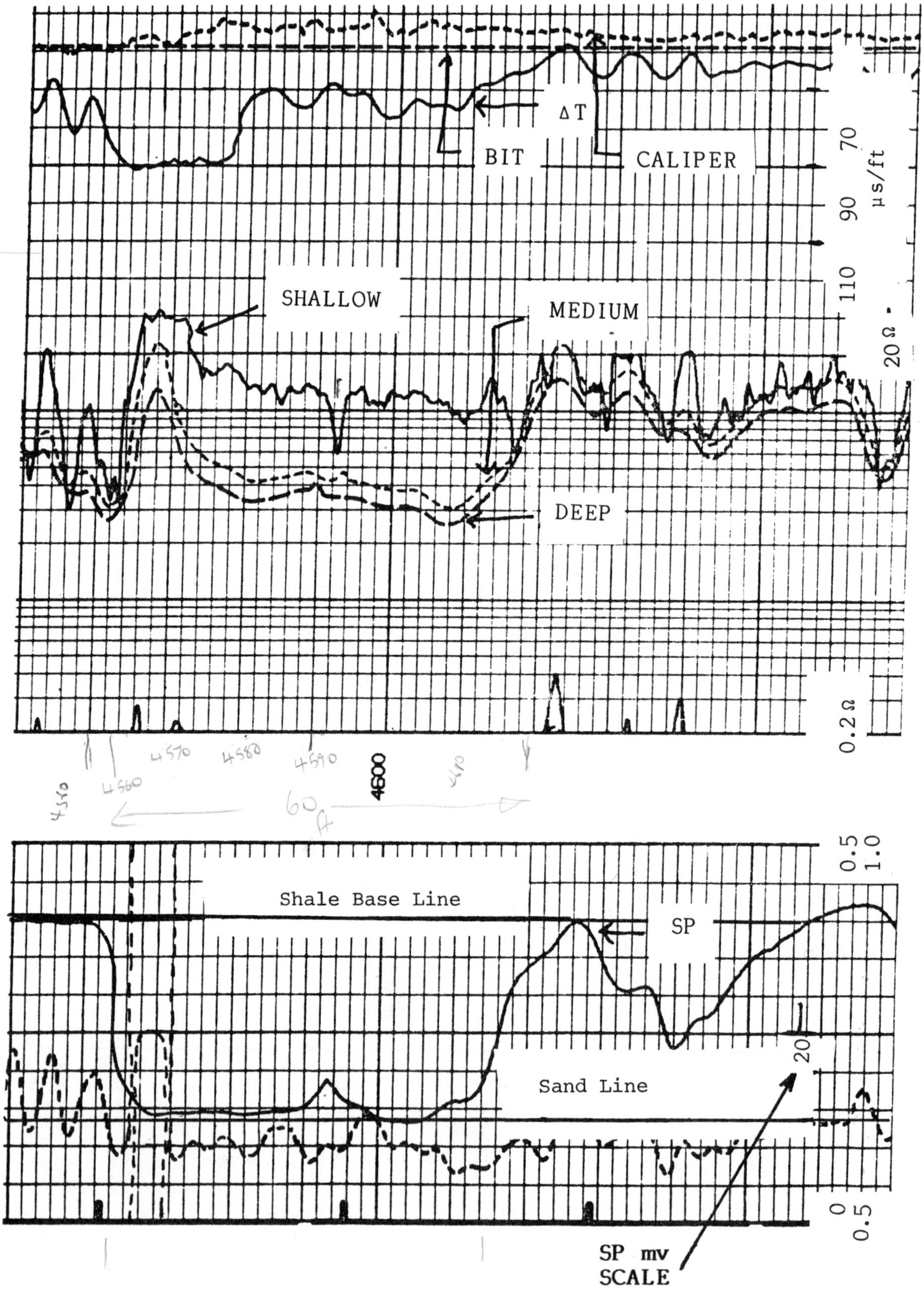

Figure 6-5. Illustration of the SP shale and sand lines.

COMPANY ·

WELL

FIELD

COUNTY STATE

Location

Sec. Twp. Rge.

Other Services:

COMPANY WELL FIELD County State

Permanent Datum Elev. Elev.: K.B.

Log Measured From Ft. Above Perm. Datum D.F.

Drilling Measured From G.L.

Date	8-21-84			
Run No.	ONE			
Depth—Driller	6998			
Depth— Logger	7002			
Btm. Log Inter	6998			
Top Log Inter.	568			
Casing—Driller	8 5/8@ 572	@	@	@
Casing—	568			
Bit Size	7 7/8			
Type Fluid in Hole	water Base Mud			
Dens. I Visc.	10.3 I 70	I	I	I
pH I Fluid Loss	10 I 8 ml	I ml	I ml	I ml
Source of Sample	PIT			
R_m @ Meas. Temp.	2.7 @ 75 °F	@ °F	@ °F	@ °F
R_{mf} @ Meas. Temp.	3.2 @ 75 °F	@ °F	@ °F	@ °F
R_{mc} @ Meas. Temp.	1.6 @ 75 °F	@ °F	@ °F	@ °F
Source R_{mf} R_{mc}	Mea. I Mea.	I	I	I
R_m @ BHT	1.33@ 159 °F	@ °F	@ °F	@ °F
Time Since Circ.	3 Hrs.			
Max. Rec. Temp.	159 °F@ TD	°F @	°F @	°F @
Equip. I Location	UCC1 I Tyler	I	I	I
Recorded By	Jones			
Witnessed By	Fische			

Figure 6-6. Typical example of a log heading.

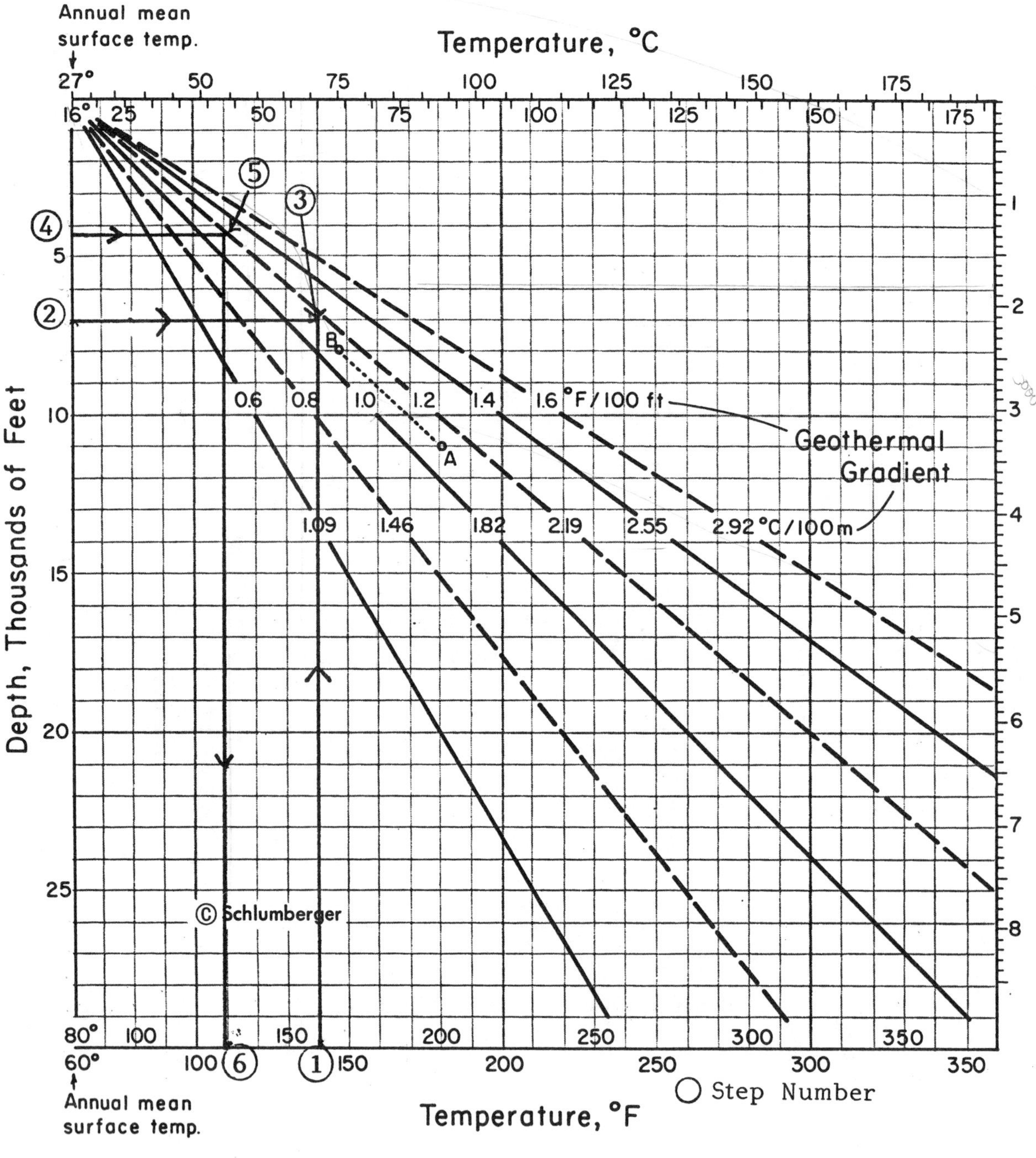

Figure 6-7. Finding formation temperature. (Courtesy of Schlumberger.)

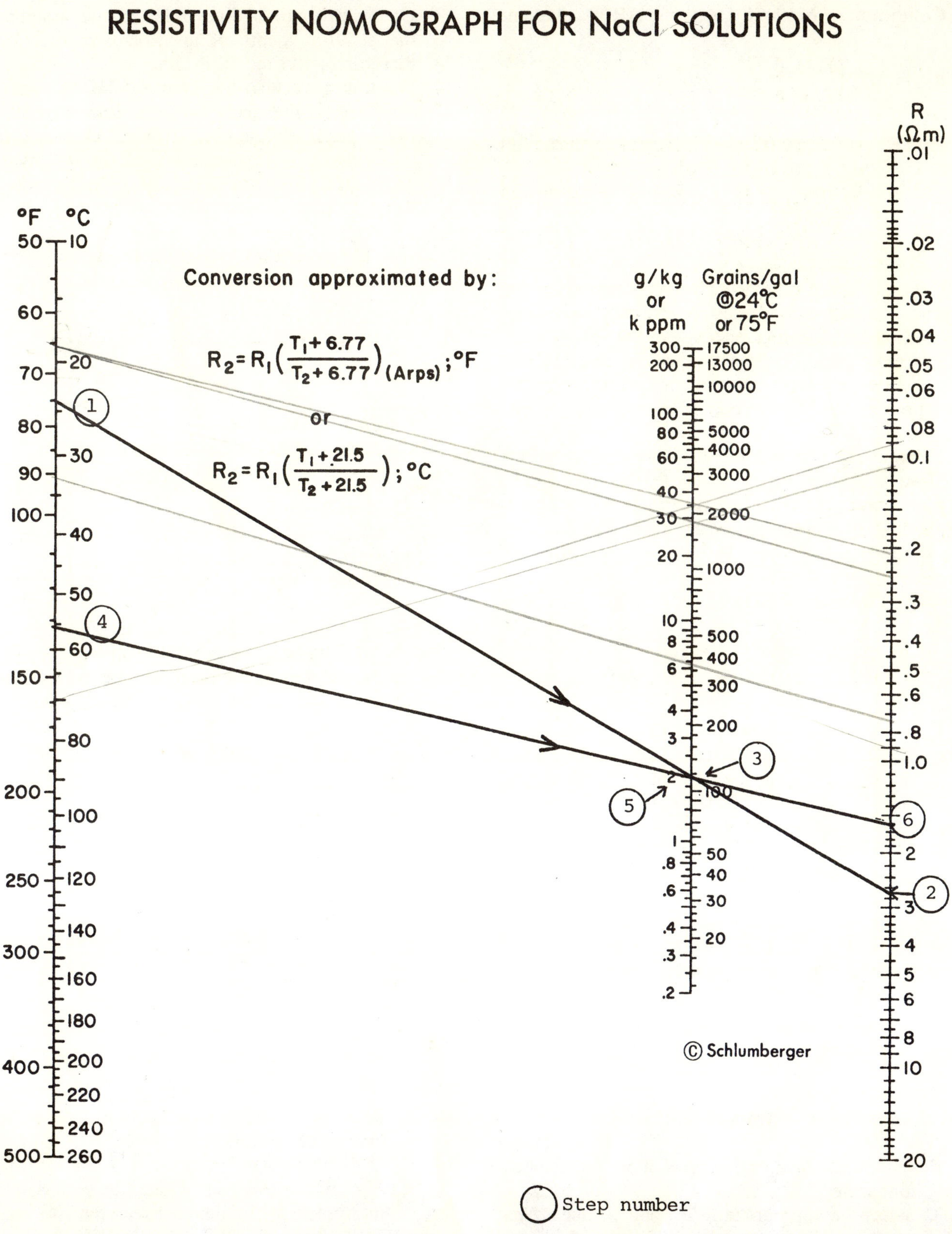

Figure 6-8. Finding R_m at formation temperature. (Courtesy of Schlumberger.)

Computation Method. Using the ARPS equation:

$$R_2 = R_1 \left(\frac{T_1 + 6.77}{T_2 + 6.77}\right)$$

where R_1 = 2.7 @ T_1 = 75° (from log heading) (R_m from log heading)
R_2 = new R_m @ T_2 = 132° (formation temp.)
R_2 = 2.7 (75 + 6.77/132 + 6.77) = 1.59

Comparison of the Two Methods. Again, as you can see, either method yields good results. A resistivity of 1.6 ohms at formation temperature will be used.

There are other charts and equations available for this conversion. The author has found that the ARPS equation and the "Schlumberger Nomograph" are the easiest and most accurate means of conversion.

Looking again at the zone of interest, we read the shallow resistivity in the zone that is the wettest (100% water saturated) and determine the thickness of the bed. Figure 6-9 shows the measurements. Formation thickness is about 56 feet, and R_i is about 10 ohms.

Taking the ratio of R_i to R_m:

$$\frac{10}{1.6} = 6.25$$

Up to this point, we have gathered all of the parameters needed to see if an SP correction is necessary.

R_w for the SP cannot be calculated unless we are sure that the deflection of the SP on the log has the same amplitude or potential as the static SP, or SSP. If it is not, it must first be corrected before proceeding.

Figure 6-10 is a chart that can accomplish our purpose. Note that this chart requires our knowledge of how deeply the formation has been invaded. For purposes of this example, it will be assumed to be 30 inches. A detailed discussion on obtaining this information is also presented later in this chapter.

SP Correction (Figure 6-10)

1. Enter the formation thickness of 56 ft on the bottom scale.
2. Select the appropriate R_i/R_m pair of lines. This example calls for a ratio of 6.25. Since this value does not exist on the chart, we select the ratio of 5.
3. Of the two pair of lines selected, we assumed that a 30-in. depth of invasion occurred. We therefore select the 30-in. line.
4. We draw a vertical line from the bed thickness scale until it intersects the 30-in. line. Note that in this case, the line goes off of the chart before it can intersect. This indicates that the 104 mv of SP we read from the log is at static condition.

   SP = SSP

   If in fact it were not, we would extend a horizontal line to the right-hand scale and read a correction factor where:

   SSP = SP × factor

So far, we have:

1. Selected and read SP
2. Corrected SP for SSP
   a. Determined formation temperature
   b. Determined R_m at T_f
   c. Estimated depth of invasion
   d. Read formation thickness
   e. Used SP correction chart

The final step is extracting the R_w from the SP.

Before we continue, we need to keep in mind that mud fluids and formation fluids frequently contain salts other than sodium chloride. For this reason, we will be converting all of the fluids to an equivalent of NaCl. The terms R_{mfeq} and R_{weq} refer to this equivalent.

Chart Method.
1. Read from the log heading (Figure 6-6) the mud filtrate resistivity or R_{mf}: R_{mf} is 3.2 ohms at 75°F.
2. Convert R_{mf} to 132° (formation temperature) in the same manner that R_m was converted. See Figure 6-11. R_{mf} becomes 1.9 ohms at 132°F.
3. Convert R_{mf} to R_{mfeq} or R_{mf} equivalent.
   a. Enter an R_{mf} of 1.9 on the bottom scale of the chart on Figure 6-12.
   b. Draw a vertical line from 1.9 ohms to the 132° line on the chart. Note: you will have to interpolate a 132° line between the 100° and 150° lines.
   c. Extend the line horizontally to the left and read R_{mfeq}. In this case it is 0.7 ohms at 132°.

(text continued on page 73)

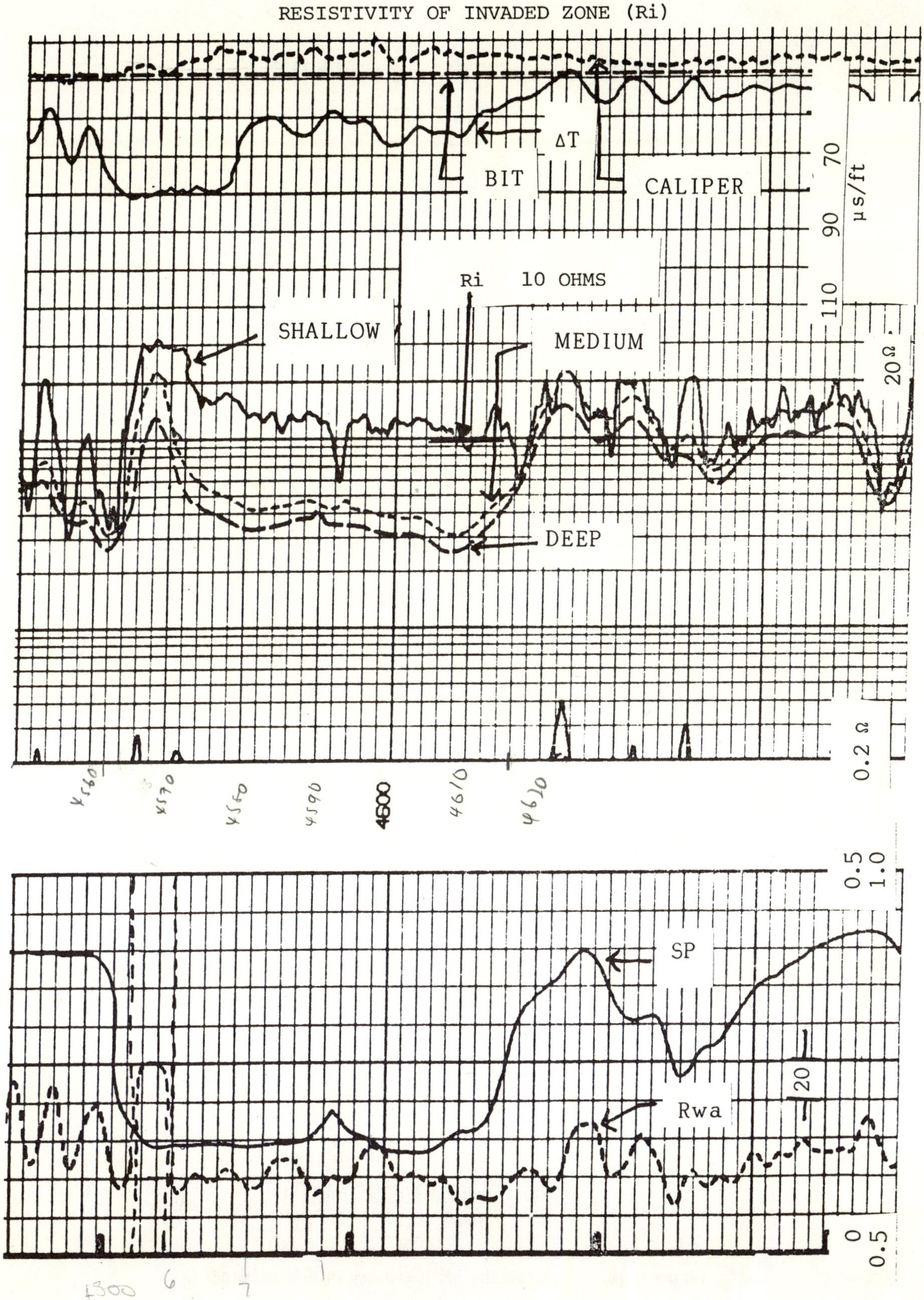

Figure 6-9. Locating R_i

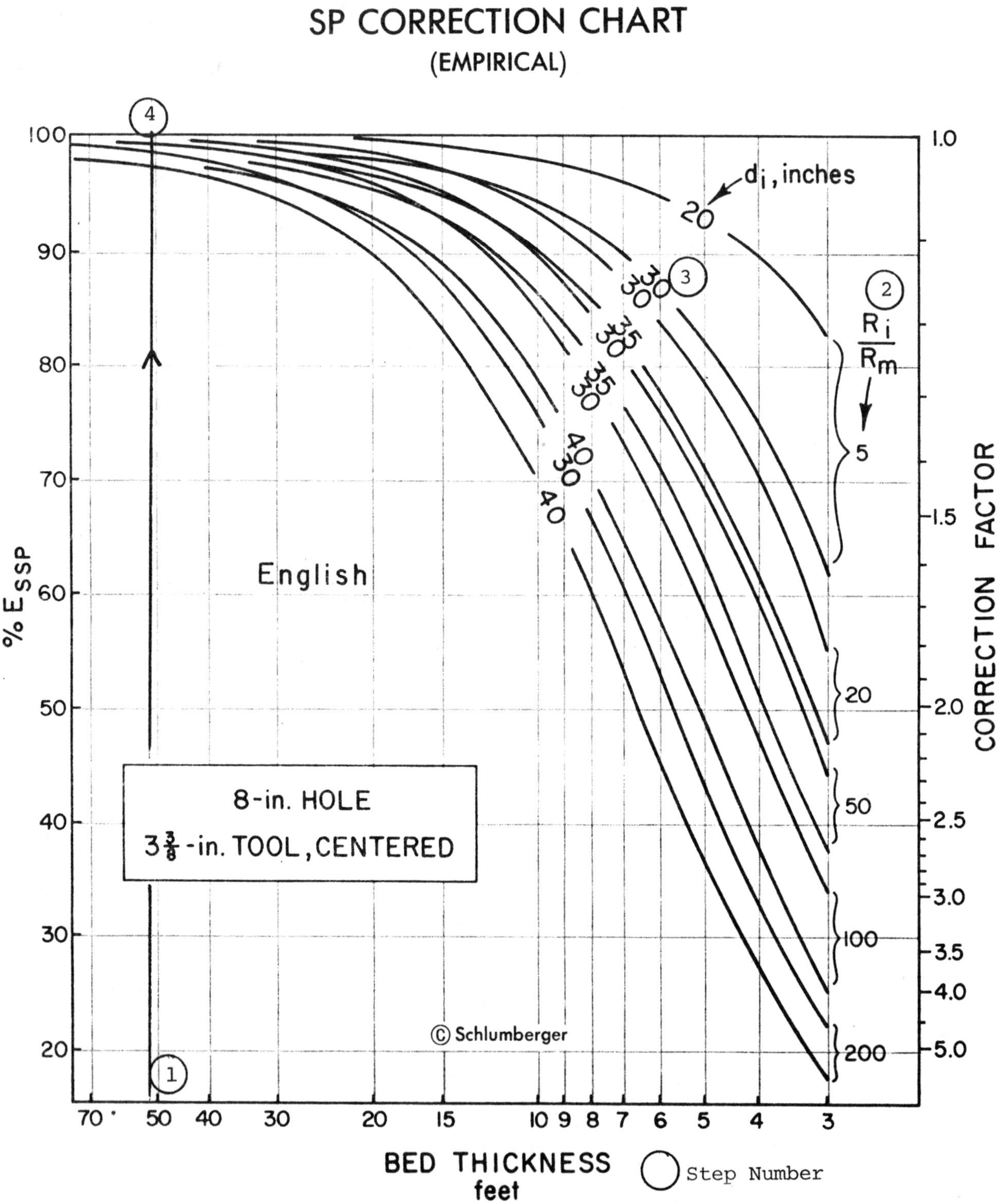

Figure 6-10. Correcting the SP. (Courtesy of Schlumberger.)

RESISTIVITY NOMOGRAPH FOR NaCl SOLUTIONS

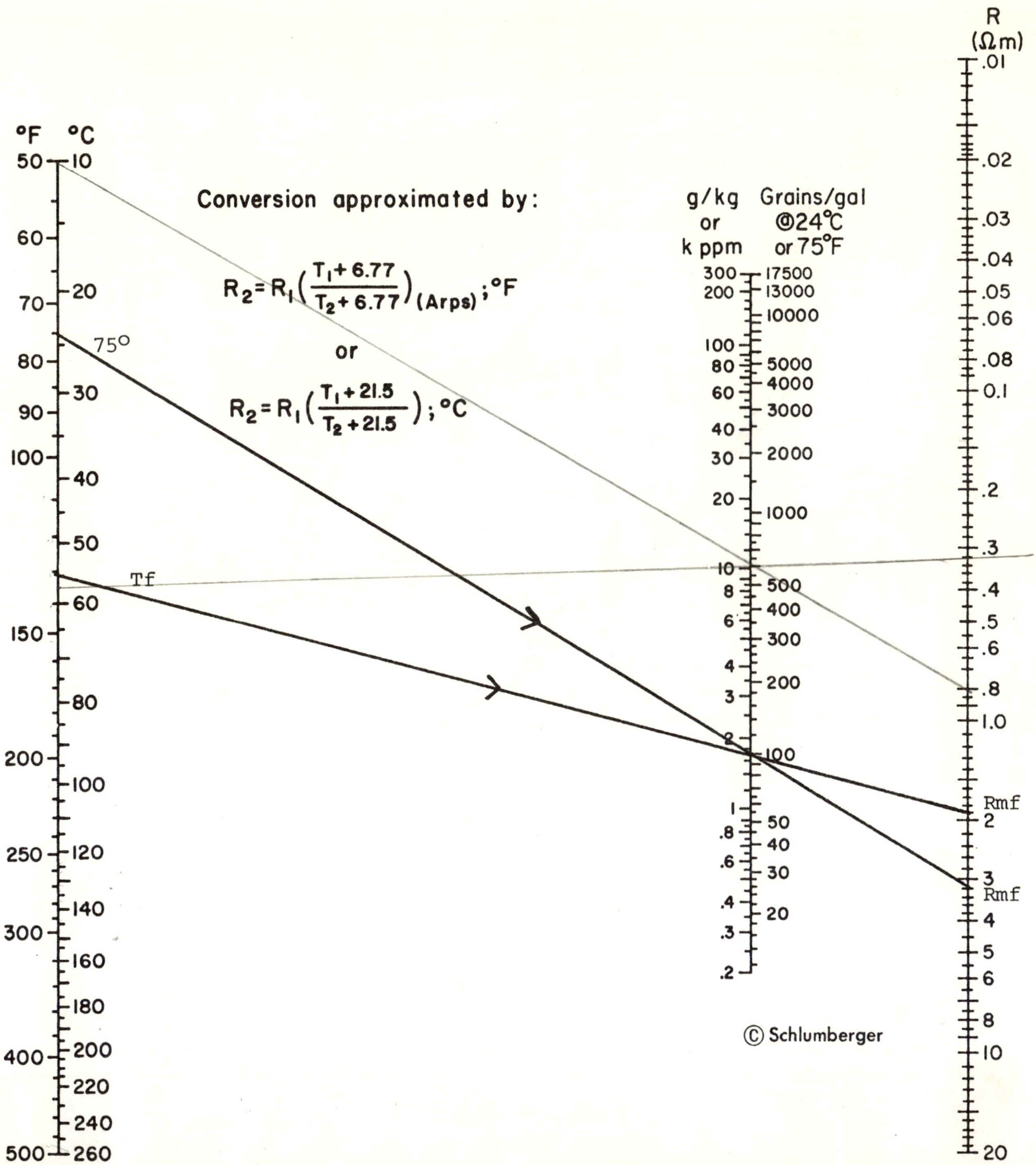

Figure 6-11. Converting R_{mf} to formation temperature. (Courtesy of Schlumberger.)

R_w VERSUS R_{weq} AND FORMATION TEMPERATURE[3]

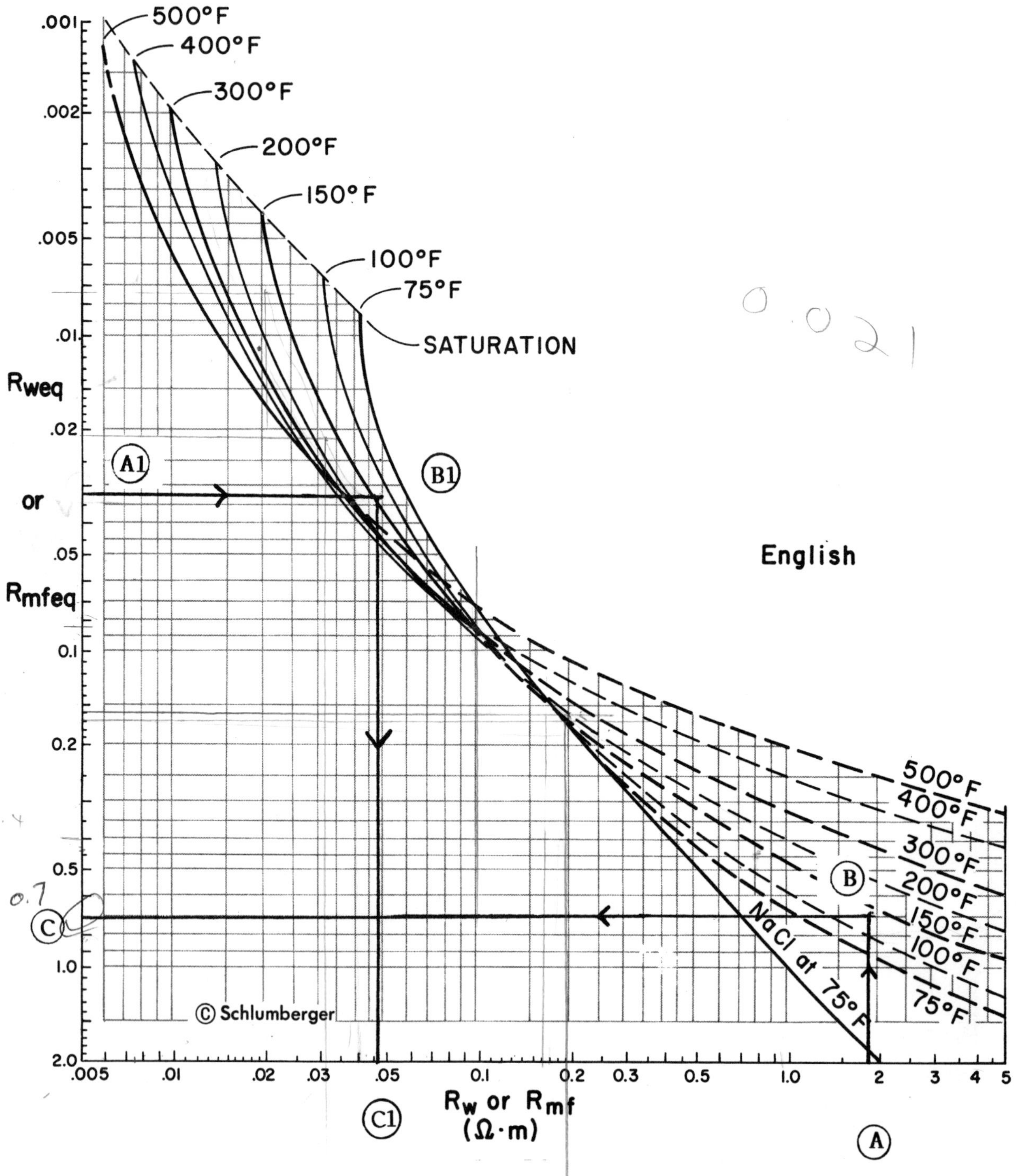

Figure 6-12. Converting R_{mf} to R_{mfeq}. (Courtesy of Schlumberger.)

4. Using the chart of Figure 6-13, enter the −104 mv of SP previously read on the left-hand scale;
   a. Extend a line from the SP scale through the temperature scale (using 132°F) to the R_{mfeq}/R_{weq} scale and stop at this scale.
   b. Extend a line from this point to the R_{weq} scale through the R_{mfeq} scale so that the line passes through an R_{mfeq} of 0.7 (previously calculated).
   c. Read the R_{weq} of 0.032 ohms at 132°.
5. Using Figure 6-12, convert R_{weq} to R_w. This is the reverse process used to obtain R_{mfeq}. Follow steps A1, B1, and C1. Read an R_w of 0.046 ohms at formation temperature.

Computation Method.

1. From the log heading (Figure 6-6) read the mud filtrate resistivity. R_{mf} is 3.2 at 75°F. If R_{mf} is not at 75°, then convert it using the ARPS equation.
2. Choose the proper equation:
   If R_{mf} @ 75°F > 0.1 then use:

$$R_{mfeq} = R_{mf} \times 0.85$$

If R_{mf} @ 75°F < 0.1 then use:

$$R_{mfeq} = \frac{(146 \times R_{mf}) - 5}{(337 \times R_{mf}) + 77}$$

Clearly, R_{mf} is greater than 0.1 therefore:

$$R_{mfeq} = 3.2 \times 0.85 = 2.72 \text{ @ } 75°\text{F.}$$

3. Compute the constant K

$$K = (60 + 0.133T)$$

Where T is 132° (formation temperature)

$$K = (60 + 0.133 \times 132) = 77.56$$

4. Solve for R_{weq}:

$$R_{weq} = \frac{R_{mfe}}{10^{-(SP/K)}}$$

where $R_{mfeq} = 2.72$
$K = 77.56$
$SP = -104$

$$R_{weq} = \frac{2.72}{10^{-(-104/77.56)}} = 0.124 \text{ @ } 75°\text{F.}$$

5. Convert R_{weq} to R_w. If $R_{weq} < 0.12$ then use:

$$R_w = \frac{(77 \times R_{weq}) + 5}{146 - (377 \times R_{weq})}$$

If $R_{weq} > 0.12$ then use:

$$R_w = -0.58 + 10^{(0.69R_{weq} - 0.24)}$$

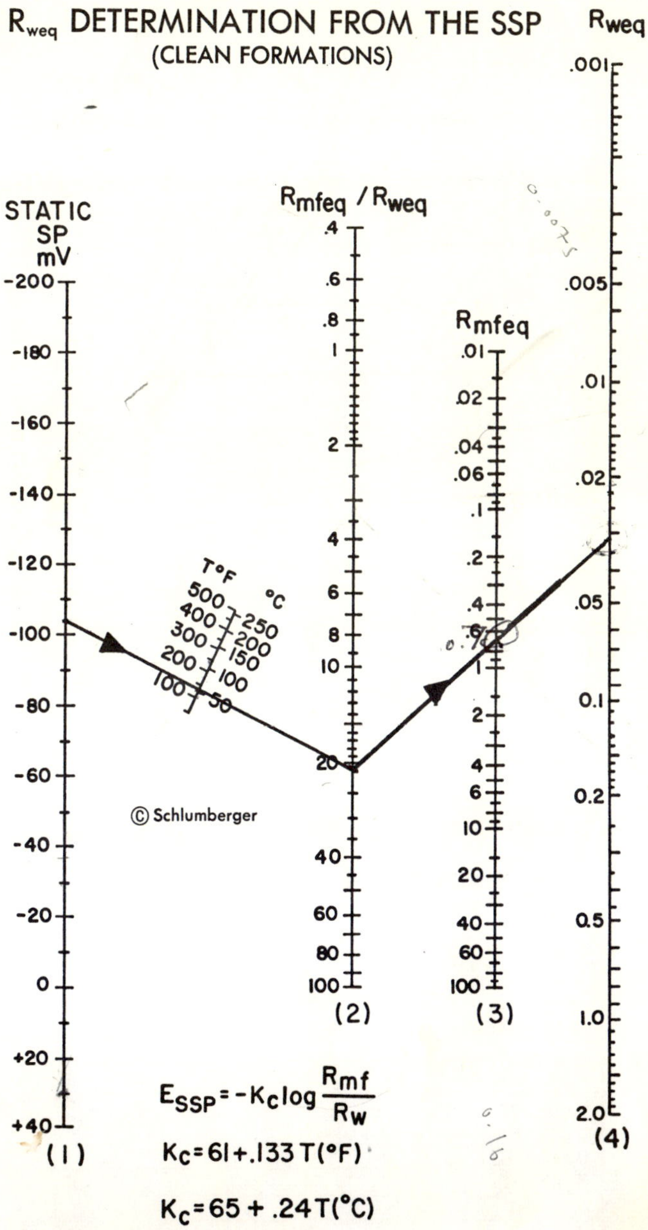

Figure 6-13. Finding R_{weq}. (Courtesy of Schlumberger.)

In this case the latter is applicable, therefore:

$$R_w = -0.58 + 10^{(0.69 \times 0.124 - 0.24)} = 0.121$$

R_w is 0.12 ohms at 75°.

6. The final step is to convert R_w to formation temperature. Using the ARPS equation:

$$R_2 = R_1 \left(\frac{T_1 + 6.77}{T_2 + 6.77}\right)$$

$$R_2 = 0.12 \times \left(\frac{75 + 6.77}{132 + 6.77}\right)$$

$$= 0.07 \text{ @ } 132°F.$$

Note: Using this procedure requires the resistivity values to be at 75°. All of the numerical constants have been adjusted for this temperature. In order to ensure proper results care must be used to maintain all parameters at 75°F.

Comparison of the Two Methods. The chart method yields an R_w of .046 ohms, while the calculation method yielded 0.07 ohms, both at formation temperature. While the error is small, one is well advised to eliminate as much error as possible. As an illustration, the difference between the two values is only 2.4 hundredths of an ohm. Calculating two water saturations each with identical parameters except for R_w results in a saturation difference of 10.6% between the two. We will use 0.07 for further computations.

Determining the Value of R_t

Generally, the deep reading resistivity curve should read the value of R_t. Since this measurement is made deeply into the formation, it should only see virgin rock. In some cases, this is true, while in others, the curve must be corrected.

The first correction to be made will be a borehole correction. The induction tool was originally intended for a nonconductive borehole fluid. Since we are using it in a conductive-type mud, it is possible for some of the induction signal to "short out" through the mud. We therefore need to find out if any of the signal on the log is caused by the mud.

The first parameter needed is the size of the stand-off used on the tool. The stand-off is a rubber device placed around the induction tool to keep it as centralized as possible in the hole, thus preventing much of the signal from entering the mud. This information is usually placed on the log heading and labeled as S.O. in inches. Figure 6-14 shows a 1.5-in. stand-off for this log.

The next parameter is hole size. Looking at the caliper on Figure 6-15, it can be seen that it is about 7.5 in. Note that bit size is 7⁷/₈ in., yet the hole is smaller. This indicates mudcake build-up, which confirms porosity and permeability. Excessive mudcake build-up can indicate a depleted reservoir—certainly not the case in this instance.

The third and last parameter needed is mud resistivity at formation temperature. This value is 1.6, having calculated it previously in this chapter. Using the chart of Figure 6-16, we will use the solid lines for an ILD (Induction Lateralog Deep) device.

Fold Here

Service Ticket No. 10937 Remarks:

Change in Mud Type or Additional Samples		
Date | Sample No.	|	|
Depth—Driller		
Type Fluid in Hole		
Dens. | Visc.	|	|
ph | Fluid Loss	| ml	|
Source of Sample		
R_m @ Meas. Temp.	@ °F	@ °F
R_{mf} @ Meas. Temp.	@ °F	@ °F
R_{mc} @ Meas. Temp.	@ °F	@ °F
Source: R_{mf} | R_{mc}	|	|
R_m @ BHT	@ °F	@ °F
R_{mf} BHT	@ °F	@ °F
R_{mc} BHT	@ °F	@ °F

SCALE CHANGES

Type Log	Depth	Scale Up Hole	Scale Down Hole
SONIC	1825	130	100

EQUIPMENT DATA

Run No.	Tool Type and No.	Pad Type	Tool Position	Other
one	Dual Ind.	12A	**1.5" S.O.**	
	SONIC		CENT.	
				STAND OFF

Figure 6-14. Locating the stand-off parameter.

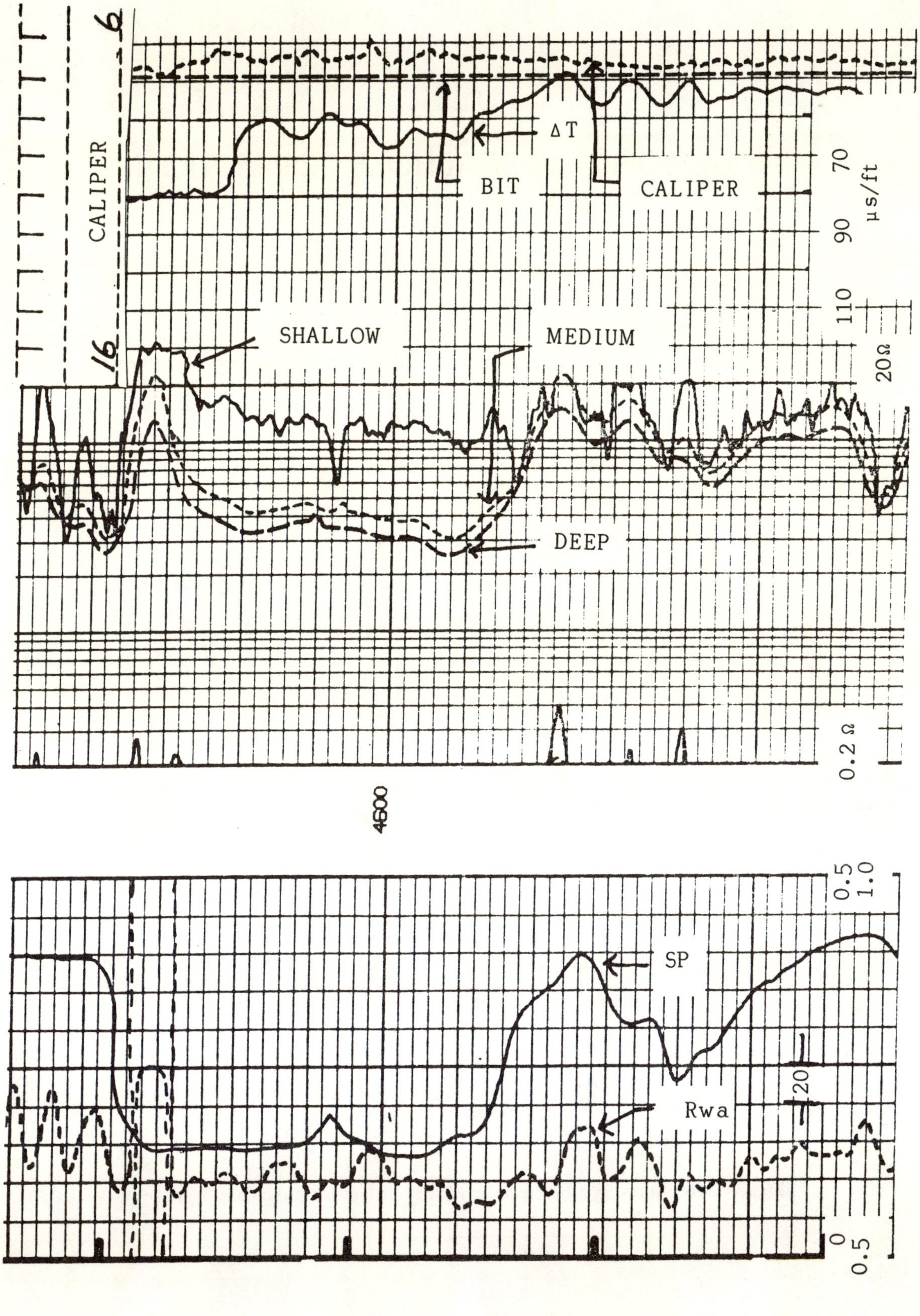

Figure 6-15. Using the caliper curve to judge borehole integrity.

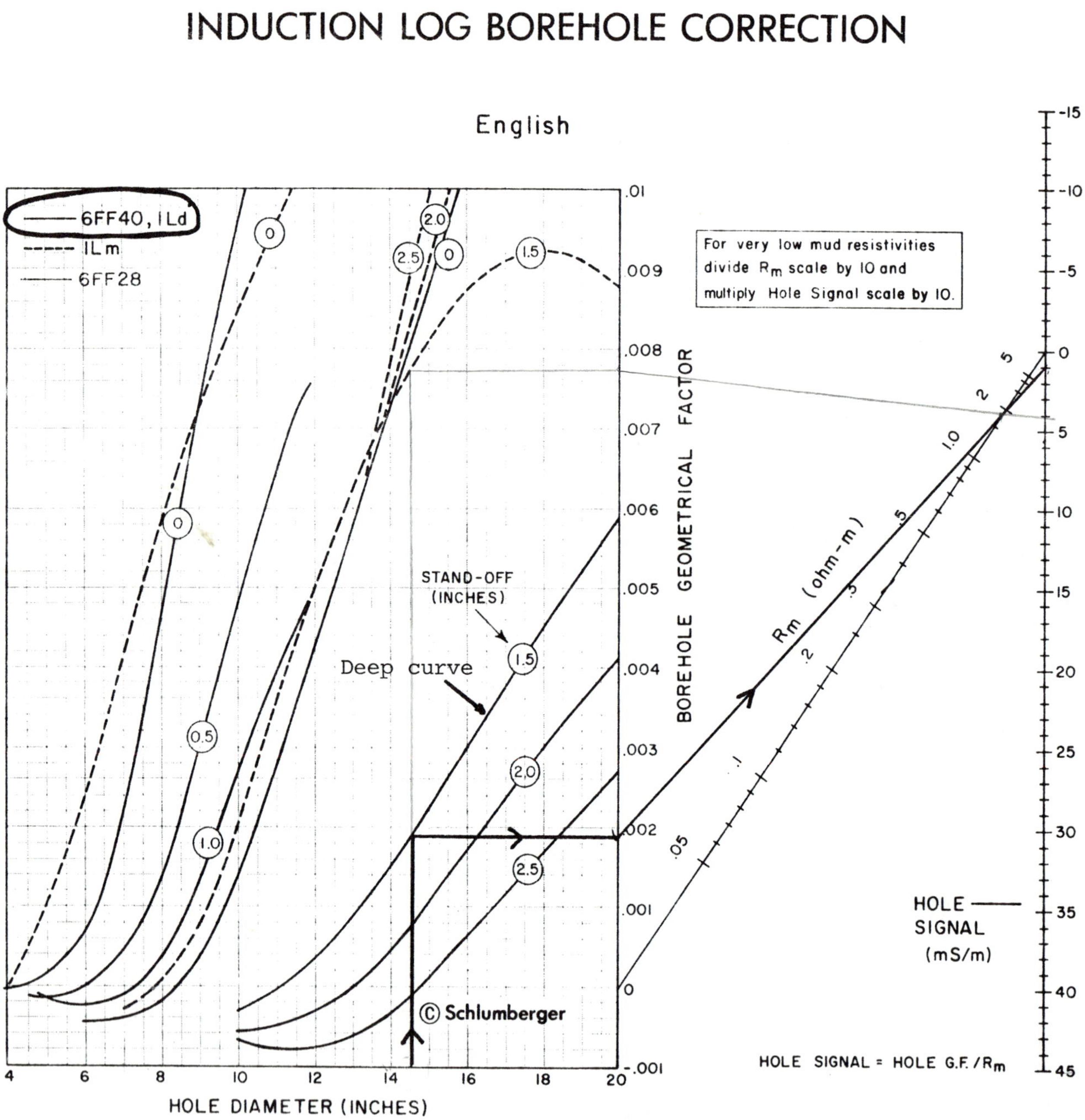

Figure 6-16. Applying a borehole correction. (Courtesy of Schlumberger.)

Entering a hole size of 7.5 in. on the bottom scale, and locating the solid line for a 1.5-in. stand-off, it becomes obvious that no correction is necessary.

For the purpose of instruction, let's assume that a washout occurred at the zone of interest, and we had a 14.25-in. hole.

1. Enter 14.25 in. on the bottom scale.
2. Extend a vertical line to the 1.5-in. solid stand-off line.
3. Extend a line, at a right angle, to the edge of the graph.
4. Locate on the diagonal R_m line a mud value of 1.6. Extend a line from the right edge of the graph, passing through 1.6 to the *hole signal* line.
5. Read the amount of signal that entered the mud. This is about 1 mmho (millimho).

$$\text{mmho} = \frac{1{,}000}{R}$$

and

$$R = \frac{1{,}000}{\text{mmho}}$$

If the 1 mmho of signal entered the hole, that must be subtracted from the deep reading resistivity.

a. Convert R_{deep} to conductivity

$$\frac{1{,}000}{13 \text{ ohms}} = 76.92 \text{ mmhos}$$

b. Subtract 1 mmho

$$76.92 - 1 = 75.92$$

c. Convert back to resistivity

$$\frac{1{,}000}{75.92} = 13.2 \text{ ohms}$$

If a positive hole signal occurs, then subtract it from the log reading. If a negative hole signal occurs, then add it to the log reading.

Negative conductivities are possible because each tool has a built-in pre-correction. When the signals are encountered, it simply means that the tool over-corrected.

As a matter of interest, if the tool had no stand-off, a 5 mmho correction would be indicated. While this does not seem to be alarmingly high, some wireline companies' tools could require 10 to 20 times this amount. It is imperative that a stand-off is used on this tool to protect the integrity of the log.

Again, no correction was indicated here, so R_{deep} remains 13 ohms.

Having this behind us, we correct for bed thickness. Figure 6-17 contains four charts for this purpose. Each of these charts was constructed with varying shale resistivities, as can be seen on the lower left-hand side of each chart.

Recall that our formation of interest is over 50 ft thick. These charts are constructed for beds 28 ft and less, thus indicating no correction. Had we needed a correction, we would have selected the appropriate chart by reading the resistivity of a nearby shale. After the closest selection is made, the bed thickness would be entered on the bottom scale and a vertical line drawn to intersect one of the lines indicating the deep reading resistivity from the log. The line is then drawn horizontally to the left and the corrected resistivity is read.

We still use 13 ohms since no correction was indicated. And finally the last and perhaps most difficult correction to make is the invasion correction. The difficulty lies in the reading of the Tornado Chart, as you can see from Figure 6-19. With a little time and patience, this chart can be used with full confidence.

1. From the log (Figure 6-18) read*:

   $R_d = 13$

   $R_m = 18$

   $R_s = 40$

   where R_d is deep, R_m is medium, R_s is shallow. No correction was indicated on this log.
2. Calculate the ratios:

   $R_m/R_d = 18/13 = 1.4$

   $R_s/R_d = 40/13 = 3.0$
3. Enter these values on the chart of Figure 6-19. Plot until they intersect each other.
4. The vertical dashed lines indicate the diameter of invasion by the mud filtrate. In this case it is about 70 inches.

**Note:* These readings should be corrected prior to this step.

(text continued on page 81)

INDUCTION LOG BED THICKNESS CORRECTION

6FF40 or ILd and 6FF28

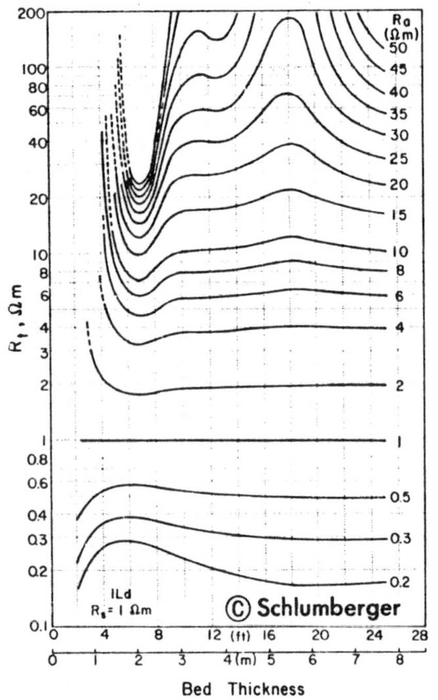

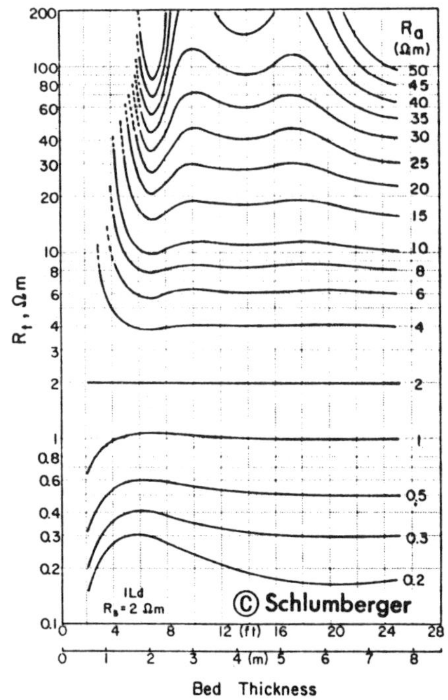

NOTE: These corrections are computed for a shoulder-bed resistivity (SBR) setting of 1 Ω·m. Refer to log heading.

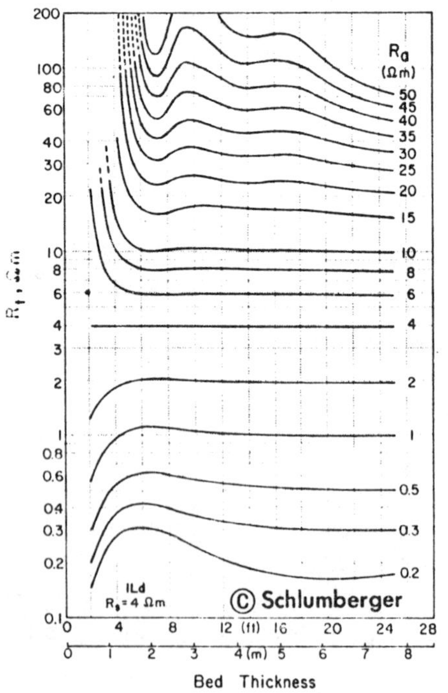

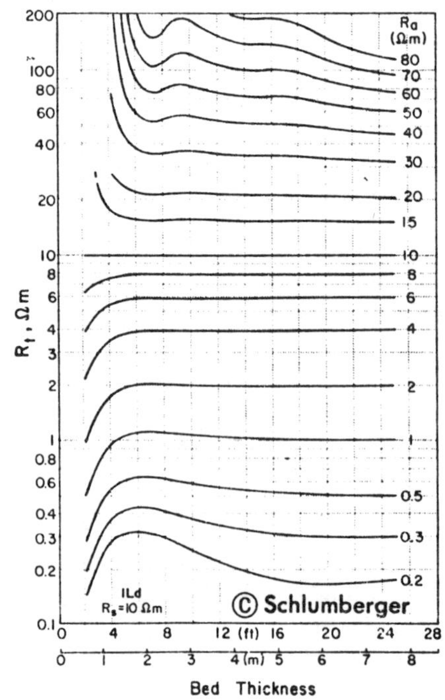

Figure 6-17. Bed thickness correction. (Courtesy of Schlumberger.)

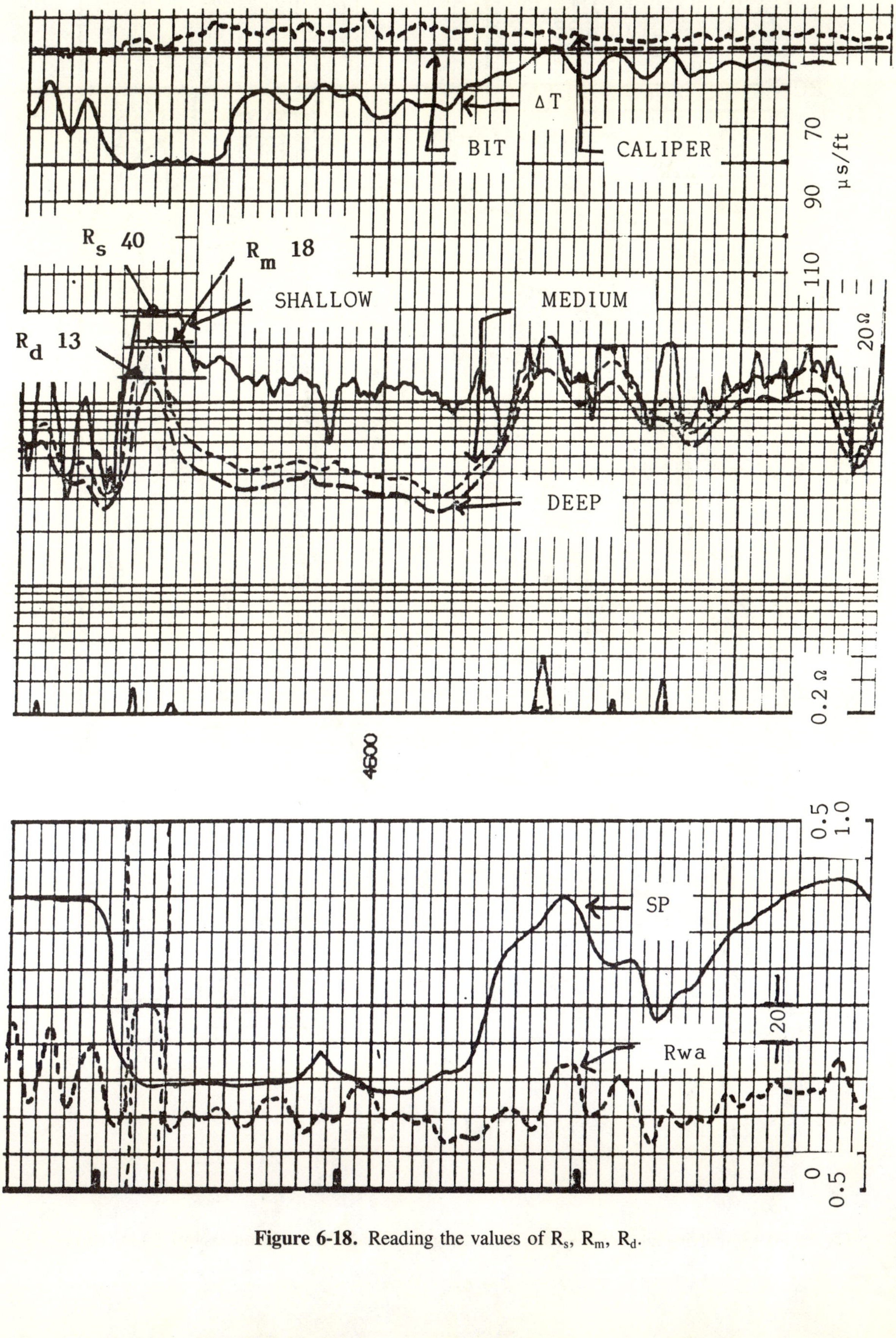

Figure 6-18. Reading the values of R_s, R_m, R_d.

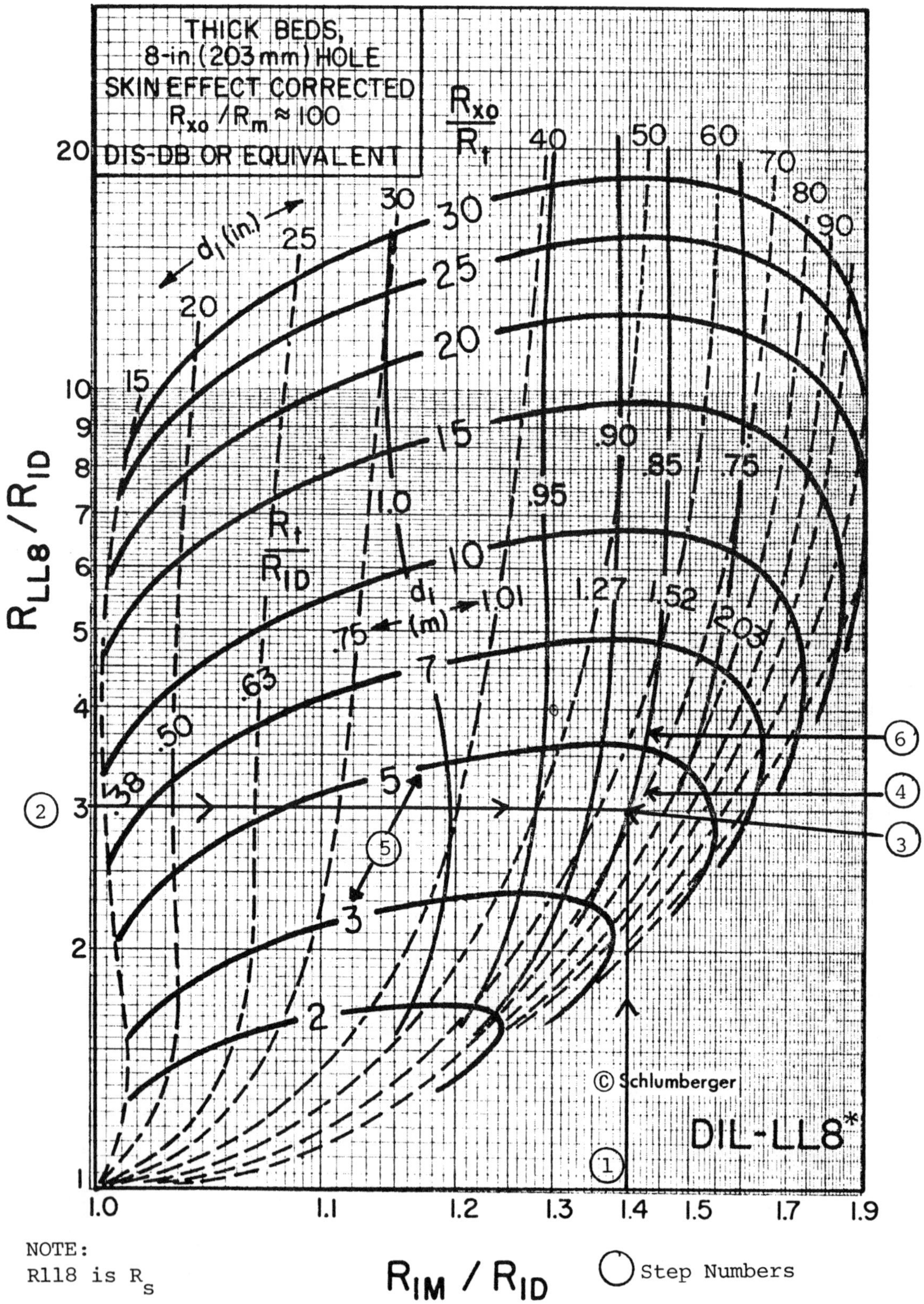

Figure 6-19. Applying an invasion correction using the tornado chart. (Courtesy of Schlumberger.)

5. The solid horizontal lines read the ratio R_{xo}/R_d. Our point plots almost halfway between 3 and 5. Interpolating, we obtain 4.2.
6. The solid horizontal lines read the ratio R_t/R_d where R_t is correct for invasion. Our point plots on the 0.85 line.
7. All that is left is simply to solve for R_{xo} (resistivity of the flushed zone) and R_t.

$$R_t = R_d \times (R_t/R_d) = 13 \times .85 = 11 \text{ ohms}$$

and

$$R_{xo} = R_t \times (R_{xo}/R_t) = 11 \times 4.2 = 46.2 \text{ ohms}$$

R_t is now known to be 11 ohms. Had we used 13 ohms of uncorrected resistivity, another error would have been introduced in the final analysis.

For those readers interested in solving this chart mathematically, it is presented here:

If $R_s/R_d < 2.5$

or

if $R_m/R_d < 1.0$

or

if $R_t/R_d < 0.4$

then

$$R_t = R_d$$

or else:

$$\frac{R_t}{R_d} = 1 - \left(\frac{(R_m/R_d - 1.1)^2\,(0.4 \times R_s/R_d)\,(R_s/R_d + 3) - 1}{(R_s/R_d - 1)^2}\right)$$

then

$$R_t = R_d \times \left(\frac{R_t}{R_d}\right)$$

For this case:

$$R_m/R_d = 1.4$$

$$R_s/R_d = 3.0$$

$$\frac{R_t}{R_d} = 1 - \frac{(1.4 - 1.1)^2\,(0.4 \times 3)\,(3 + 3) - 1}{(3 - 1)^2}$$

$$= 1 - \frac{0.09 \times (1.2 \times 6 - 1)}{4}$$

$$= 1 - .1395 = 0.86$$

$$R_t = 13 \times 0.86 = 11.2 \text{ ohms}$$

Determining Porosity

Referring back to the log (Figure 6-20) we next concentrate on the sonic porosity. The scale for the sonic log is 40–140 microseconds therefore each scale division is 10 microseconds.

In the zone of interest (not just the formation, but in particular, where the hydrocarbon is suspected) Δt reads 80 μsec. Knowing this, we can now proceed with a porosity determination.

Chart Method

1. Using the chart on Figure 6-21, plot a Δt of 80 μsec.
2. Extend a vertical line until it intersects with the limestone lithology line.
3. Extend the line horizontally to the left and read porosity on the left scale. This is 23% porosity.

Computation Method

$$\Phi = \frac{\Delta t - \Delta t_{ma}}{\Delta t_f - \Delta t_{ma}}$$

where $\Delta t = 80\ \mu\text{sec.}$
$\Delta t_f = 189$
$\Delta t_{ma} = 47.6$ (limestone)

Therefore:

47.6

$$\Phi = \frac{80 - 47.6}{189 - 47.6} = .229 \text{ or } 22.9\%$$

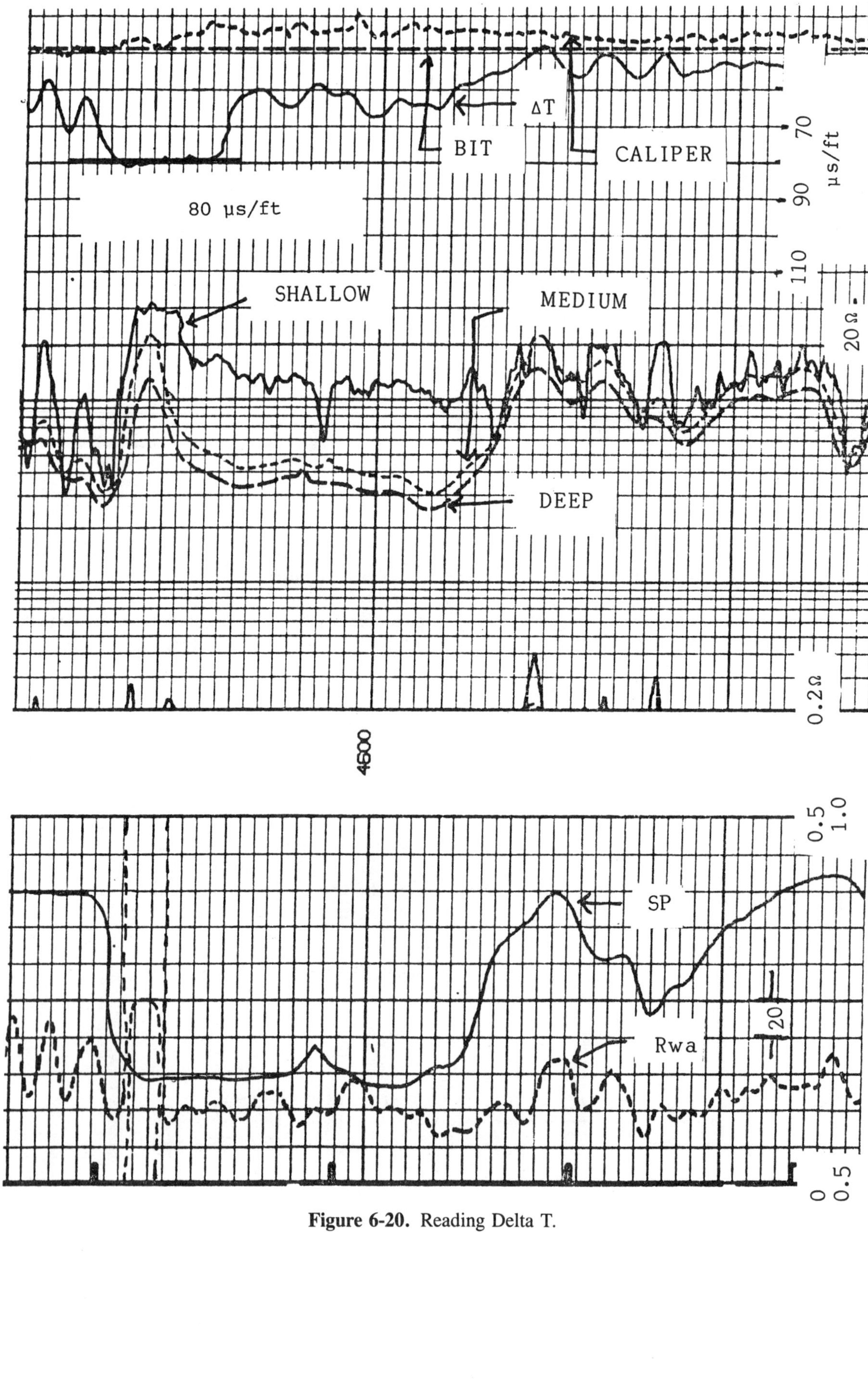

Figure 6-20. Reading Delta T.

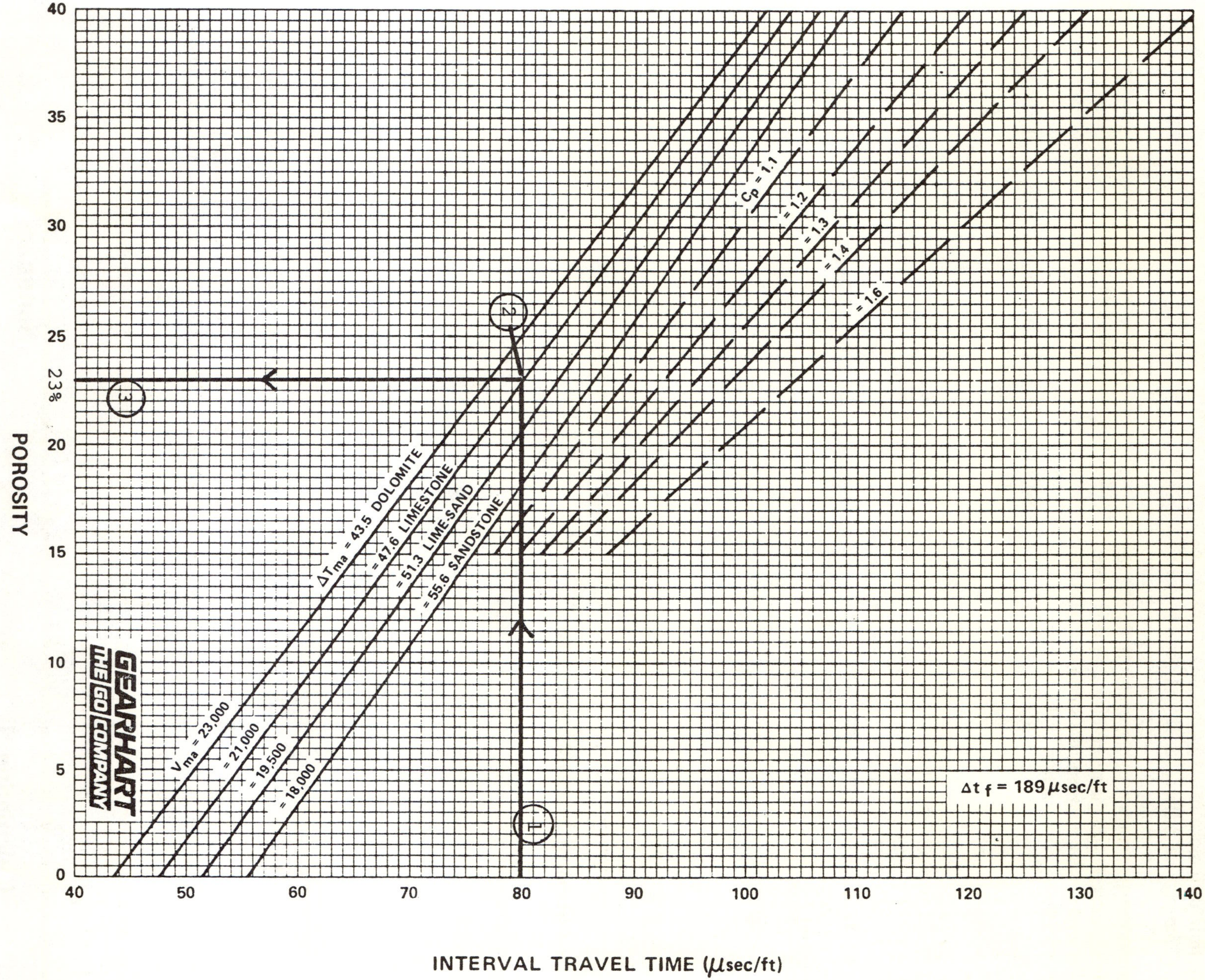

Figure 6-21. Finding porosity from Delta T. (Courtesy of Gearhart.)

Comparison of the Two Methods

Very little difference exists in this case—only one tenth of one percent difference. If you observe carefully, over the entire zone, Δt is largest only in the hydrocarbon-bearing portion of the zone. It should be suspected that gas is in the reservoir. A simple check with the mud logger confirms the suspicion and now we must apply a gas correction to the seemingly high porosity.

$$\Phi_{cor} = \Phi_a \times 0.7$$

$$\Phi_{cor} = 22.9 \times 0.7 = 16\%$$

Without the use of one or two more porosity logs, this is as far as porosity can be corrected.

Now we have at hand the basic parameters needed for the final analysis:

$$R_w = 0.07 \text{ ohms @ } 132°F$$

$$R_t = 11 \text{ ohms}$$

$$\Phi_{cor} = 16\%$$

Calculating Formation Factor

The calculation begins with two primary equations.

$$F = \frac{a}{\Phi^m}$$

$$R_o = F \times R_w$$

where $a = 1$
$m = 2$

thus:

$$F = \frac{1}{.16^2} = 39$$

$$R_o = 39 \times 0.07 = 2.73$$

Calculating S_w

And at long last, we solve for S_w:

$$S_w = \sqrt{\frac{R_o}{R_t}}$$

$$S_w = \sqrt{\frac{2.73}{11}} = .498 \text{ or } 49.8$$

Note that the final calculation contains all of the parameters used in the analysis.

$$S_w = \sqrt{\frac{R_o}{R_t}}$$

Since $R_o = FR_w$, substituting:

$$S_w = \sqrt{\frac{FR_w}{R_t}}$$

and

$$F = \frac{a}{\Phi^m}$$

Substituting:

$$S_w = \sqrt{\frac{aR_w}{\Phi^m R_t}}$$

Conclusion

A water saturation of 49.8 is extremely marginal for a limestone. The use of a neutron/density tool indicated a secondary porosity existed, totaling 19.5%, bringing the water saturation to 41%, which is more optimistic.

The quality of the analysis is strictly based on the quality of the data that goes into it.

Summary

1. Define bed boundary—using two lines to choose the point of inflection on the SP.

2. Pattern recognition—differentiate an oil sand from a water sand.
3. Use a quick-look technique.

   a. Using the deep recording curve, choose R_t in the oil-bearing zone and R_o in the water-wet zone.

   b. $S_w = \sqrt{\frac{R_o}{R_t}}$ or if $\frac{R_t}{R_o}$ is greater than 3, S_w will be less than 60%.

   c. Using the R_{wa} curve, choose R_{wa} in the oil-bearing zone and R_w in the water-bearing zone.

   d. $S_w = \sqrt{\frac{R_w}{R_{wa}}}$ or if $\frac{R_{wa}}{R_w}$ is greater than 3, S_w will be less than 60%.

4. R_w from the SP.

   a. Draw the shale base line and the sand line.
   b. Count the divisions and multiply the scale.
   c. Calculate formation temperature.
   d. Bring R_{mf} to formation temperature.
   e. Calculate or estimate invasion.
   f. Calculate R_{mfeq} (equivalent).
   g. Solve for R_{weq} (equivalent).
   h. Convert R_{weq} to R_w.

5. Find the value of R_t.

   a. Read the deep reading resistivity curve at the zone of interest.
   b. Get the stand-off used from the log heading.
   c. Bring R_m to formation temperature.
   d. Correct R_a (apparent) for borehole effect.
   e. Read nearby shale. Select appropriate chart and correct R_a for bed thickness.
   f. Read the medium curve and the shallow curve. Calculate the ratios of $R_{shallow}/R_{deep}$ and R_{med}/R_{deep}.
   g. Using the Tornado Chart, determine the corrected value of R_t.

6. Porosity determination.

   1. Read Δt in the zone of interest.
   2. Calculate the porosity using the appropriate matrix.
   3. Correct for compaction.
   4. Correct for gas.
   5. Correct for oil.

7. Formation factor.

   a. Choose the correct F-Φ relationship and calculate F
   b. $R_o = F \times R_w$

8. Calculate S_w.

$$S_w = \sqrt{\frac{R_o}{R_t}} = \sqrt{\frac{F \times R_w}{R_t}}$$

Repeat the process for each additional zone.

7

Nuclear Logs

The use of a single porosity-type device, such as a sonic log, while of itself is of tremendous value, cannot account for other anomalies which may exist. With the use of more than one porosity log, we can determine such things as lithology, shale volume, gas zone, gas saturations, mineral identification, and so on.

In the quest for more accurate porosity measurements, the industry developed the nuclear porosity tools. Again, while no one of these tools can supply us with all of the answers we need, their combination can indeed illuminate the path.

Nuclear Logging

There are three ways in which gamma rays interact with matter. The first, μphoto, is the gamma ray attenuation due to photoelectric absorption. Simply, the gamma ray transfers its energy to an electron orbiting one of the formation's atoms. The second, μpair, is attenuation due to pair production. This is where the gamma ray is converted to an electron-positron pair in accordance with Einstein's equation of $E = MC^2$. In both the first and second gamma ray interactions, the gamma ray is lost or absorbed by the formation. The third interaction is μcomp, or compton scattering. This process occurs when the gamma ray collides with an atom and dislodges an electron. The gamma ray transfers, or loses, part of its energy and now travels in a different direction in a lower energy level. At lower energy levels, the gamma ray may undergo photoelectric absorption. Of the three, only compton scatter is used for density measurements since the other two processes absorb the gamma ray.

Electron density is related to bulk density through the equation (for compound materials):

$$\rho E = 2\rho B \frac{\Sigma Z_l}{M}$$

where ΣZ_l is the sum of the atomic numbers of all atoms making up the molecules in the compound, and M is the molecular weight of the compound.

For most materials found in earth formations:

$$2\frac{\Sigma Z_l}{M} \simeq 1$$

Therefore, the count rate in most formations is related to the bulk density of the formation.

Dual-Spaced Density Tools

The modern density tool employs one radioactive source and two detectors. One of the detectors is physically located close to the source (short-spaced detector) and the other is farther away (long-spaced detector).

By necessity, the tool is skid mounted so that ideally the source and detectors are flat against the borehole wall as depicted in Figure 7-1.

For the ideal case, the tool sees nothing but formation. Mud cake, however, is most often present in porous and permeable formations. This mudcake is also seen by the density tool. With the use of a two-detector tool, we can construct spine and rib plots that correct for varying mudcake thicknesses.

Figure 7-2 shows a plot of the two detectors. If no mudcake were present, all points would plot on the straight line called the spine.

If the bulk density of the formation and the mudcake density are held constant, varying the mud thickness results in the plot of Figure 7-2. The curved line is referred to as a rib.

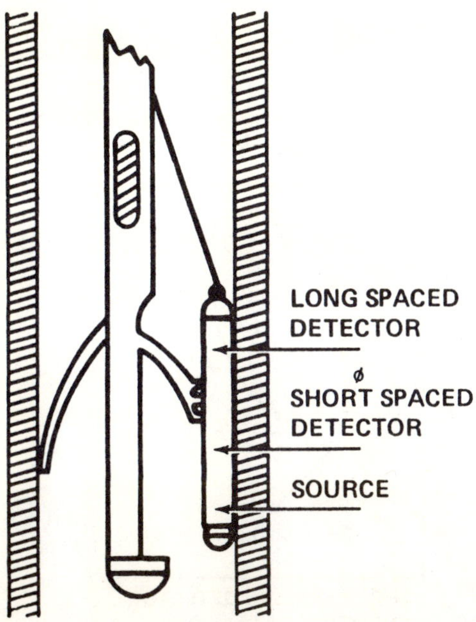

Figure 7-1. Illustration of the density tool. (Courtesy of Gearhart.)

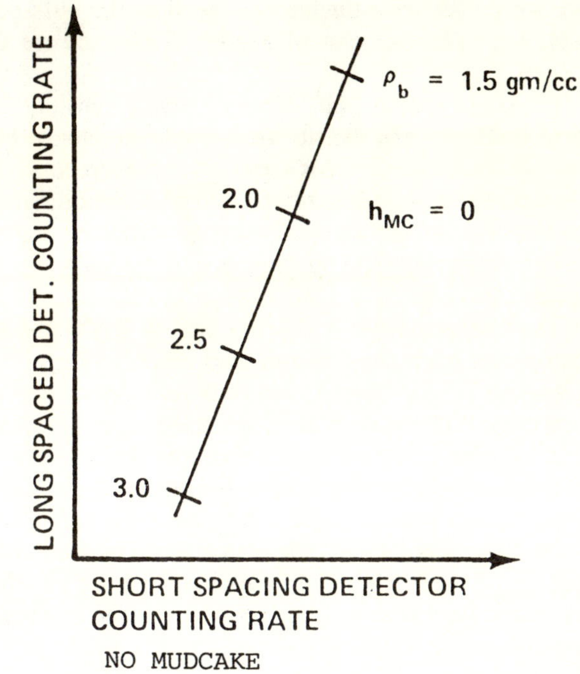

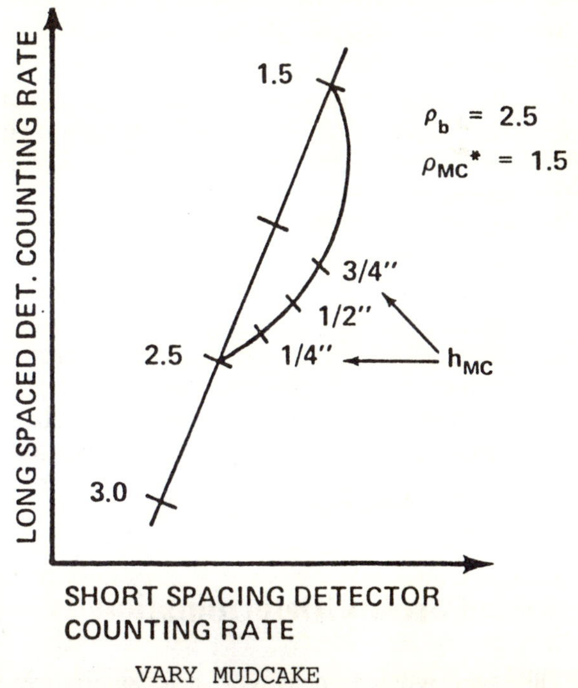

Figure 7-2. Spine and rib plot used to correct for mudcake thickness. (Courtesy of Gearhart.)

If the density of the mudcake was held constant and we varied the mud thickness as well as the bulk density, the resulting plot of Figure 7-3 would be obtained.

If the formation bulk density is held constant and mud thickness and density are varied, the plot of Figure 7-4 is obtained. This means that a correction can be applied for mudcake effect, and thus combine mudcake thickness and density into one variable, leaving only bulk density and the presence of mudcake as variables.

This information is then used to compute the amount of correction needed and is labeled $\Delta\rho$.

Figure 7-5 is a density log with the applied correction curve. The scale of the correction is plus or minus 0.25. As long as the dashed curve is within the bounds of the scale, the density curve is correct. If the curve exceeds the boundary then the density could not be properly corrected, and the data at that point is meaningless. This is often the case in a washout or when mudcake thickness is large or when barite muds are used.

Lost Pad Contact

It is not always possible for the density pad to make good contact with the formation due to the rough nature of the borehole. Any loss of contact will cause the pad to read borehole fluids. Since the amount of fluid is not constant along the face of the pad, no accurate correction is possible.

Prior to using the density curve for calculations, the log must be scrutinized for lost pad contact. If lost pad contact is severe, density and correction curves will exhibit large excursions from the average readings in a zone. In addition to these obvious losses of pad contact, smaller, less noticeable areas of lost pad contact can occur. These are usually accompanied by changes in the caliper reading greater than 0.3 in. per foot.

Porosity Determination

The corrected bulk density of a formation can be converted to a porosity in the same manner as the sonic Δt. Just as before, the fluid density and matrix density must be known. With this information we solve for porosity.

$$\Phi = \frac{\rho_{ma} - \rho_b}{\rho_{ma} - \rho_f}$$

Hence, as ρ_{ma} and ρ_f are known, Φ can be obtained directly from a measurement of ρ_b. Common values of ρ_{ma} are:

2.65 for sands and limestones
2.68 for "limey sands" and "sandy limes"
2.71 for limestones
2.87 for dolomites

When high salinity fluids are present in the pores, a fluid bulk density of $\rho_f = 1.1$ gm/cc may be required. A ρ_f of 1.0 gm/cc is appropriate where the

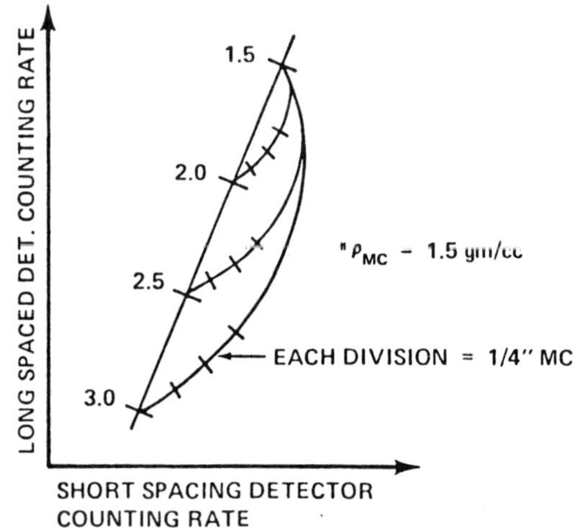

Figure 7-3. Variations of mudcake density. (Courtesy of Gearhart.)

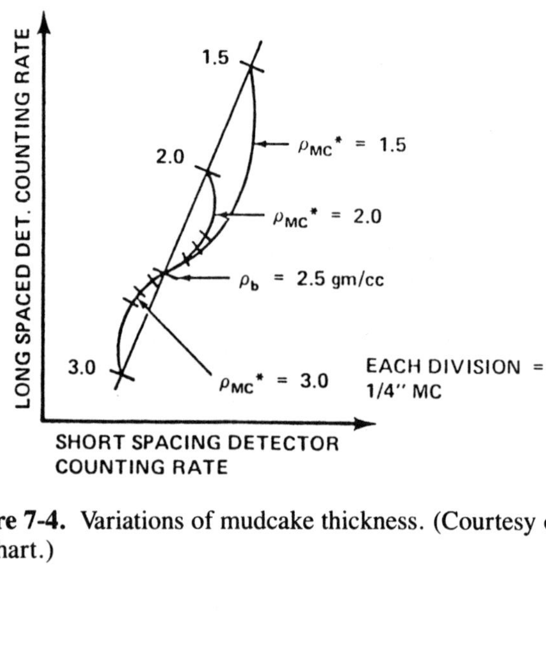

Figure 7-4. Variations of mudcake thickness. (Courtesy of Gearhart.)

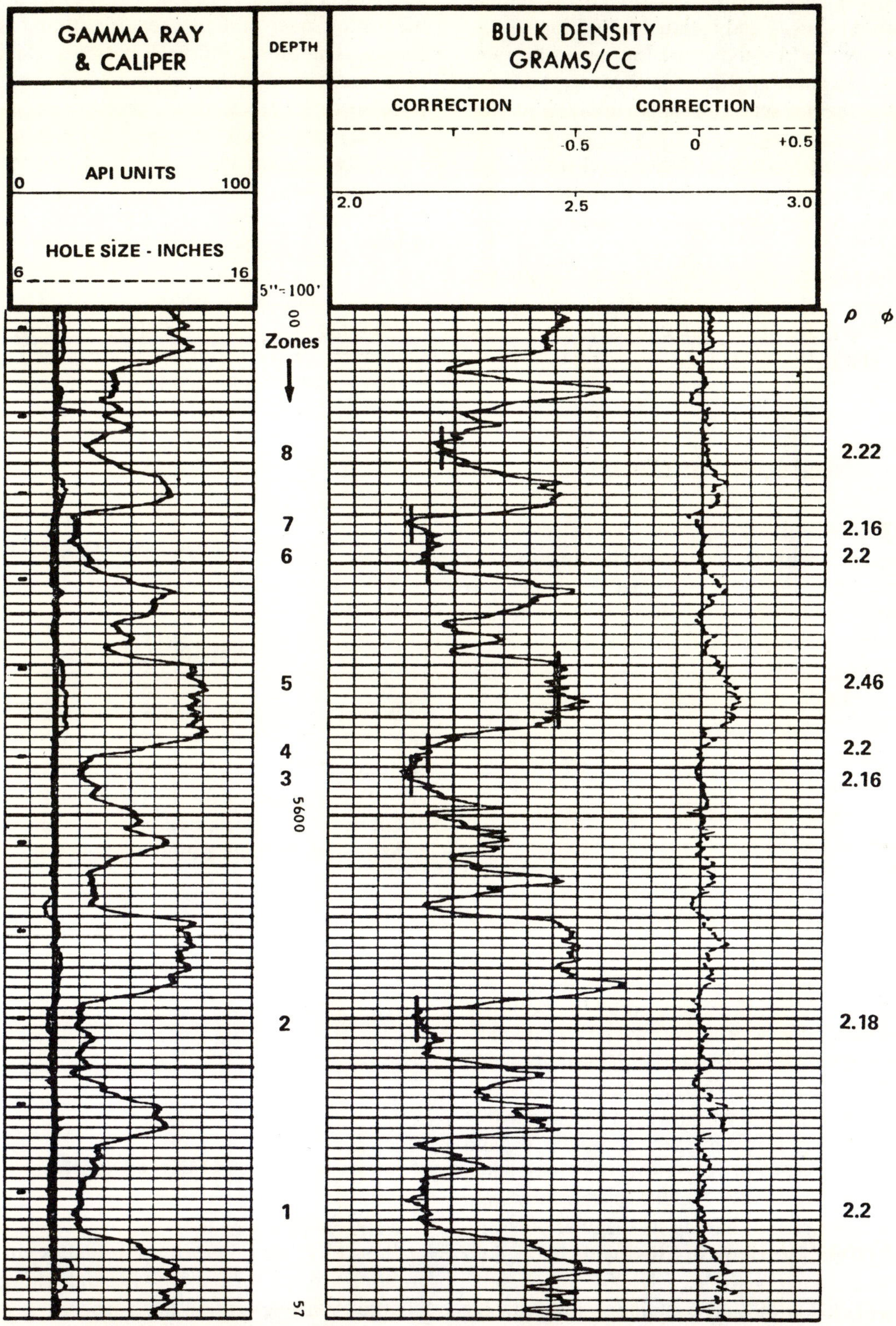

Figure 7-5. Reading density values from the curve. (Courtesy of Dresser Atlas.)

fluid is very fresh. For flushed gas or oil reservoirs even lower ρ_f values should be assumed, depending on hydrocarbon density and residual saturation. If a ρ_f value is usually used for the entire log, porosities will be optimistic. In some applications, hydrocarbons are indicated by the presence of abnormally high log porosities.

It should be noted that if oil-based mud ($\rho_f > 1.0$) is used in drilling a well, porosities will appear optimistic in some beds if there is a lower fluid density in the invaded zone.

Figure 7-6 presents the porosity equation in graphical form for the three common reservoir rock types, assuming a fluid density of 1.1 gm/cc. Figure 7-7 is similar to Figure 7-6, except $\rho_f = 1.0$ gm/cc is assumed.

Shaly Formations

The density-to-porosity equation generally suffers from the same problems as the sonic equation. In shaly sands, either the ρ matrix must be changed to accommodate the mixture, or porosity must be adjusted for shale content.

For laminated shales:

$$\Phi_e = \Phi_D - (\Phi_{sh} \times V_{sh})$$

$$\text{where } \Phi_{dsh} = \frac{\rho_{ma} - \rho_{sh}}{\rho_{ma} - \rho_f} \text{ in nearby shale}$$

For dispersed shales:

$$\Phi_e = \Phi_a - V_{sh}$$

As in all calculation procedures, the accuracy of the resulting values of Φ_e will be limited by the accuracy of the input variables (ρ_b, ρ_{ma}, ρ_f, ρ_{sh}, V_{sh}). In addition, Φ_e is affected in shaly zones by the way in which the shales are distributed. The effective porosity in a dispersed shaly formation will always be less than that in a laminated shaly formation with the same V_{sh} and Φ_a. This can be visualized very easily. In a laminated shaly sand with $V_{sh} = 30\%$, 70% of the formation is filled with clean sand, having its original porosity, while 30% of the formation is filled with shale. In a dispersed shaly sand with $V_{sh} = 30\%$, the entire formation is filled with sand containing intergranular volume Φ_a. However, most of this intergranular volume is filled with shale. If, for example, $\Phi_a = 30\%$, all intergranular volume would be filled with shale, in which case $\Phi_e = 0\%$.

Shale zones, unlike most formations, will expel water (and hence increase the density) as a function of increasing over-burden pressure, time, and tectonic forces. Therefore, unlike limestone, sandstone, and other competent rocks, no one density (ρ_{sh}) can be attributed to shale. Shallow, recently deposited shales (actually clays) may have $\rho_{sh} < 1.8$ gm/cc, whereas very old shales which have undergone complete diagenesis and lithification can have $\rho_{sh} > 2.6$ gm/cc. Probably the best way to estimate ρ_{sh} for use in density log calculations is to take the ρ_b read from the log in 100% shale formations near the zones of interest. The value of ρ_b becomes ρ_{shale}. This method of estimating ρ_{shale} will probably be more accurate in laminated shaly beds than in dispersed shaly beds, where ρ_{shale} is often hard to determine.

For structural shales:

$$\Phi_e = \Phi_a$$

This is simply due to the fact that structural shales are part of the matrix itself and occupy no pore space.

Fluid Effects

When residual hydrocarbon saturations are fairly high, calculated porosity values will be greater than the true porosity, and should be corrected for this effect. Specific values for the fluid density are often difficult to obtain. Corrected porosity values may be approximated with the following equations:

For oil zones:

$$\Phi = .9 \times \Phi_a$$

For gas zones:

$$\Phi = .7 \times \Phi_a$$

Neutron Logging

Neutron porosity tools use a radioactive source, such as plutonium-beryllium or americum-beryllium, to bombard the formation with high-energy neutrally charged particles called neutrons.

As high-energy neutrons collide with the various atoms of both formation material and fluids, they begin to lose their energy. The amount of loss can be stated as:

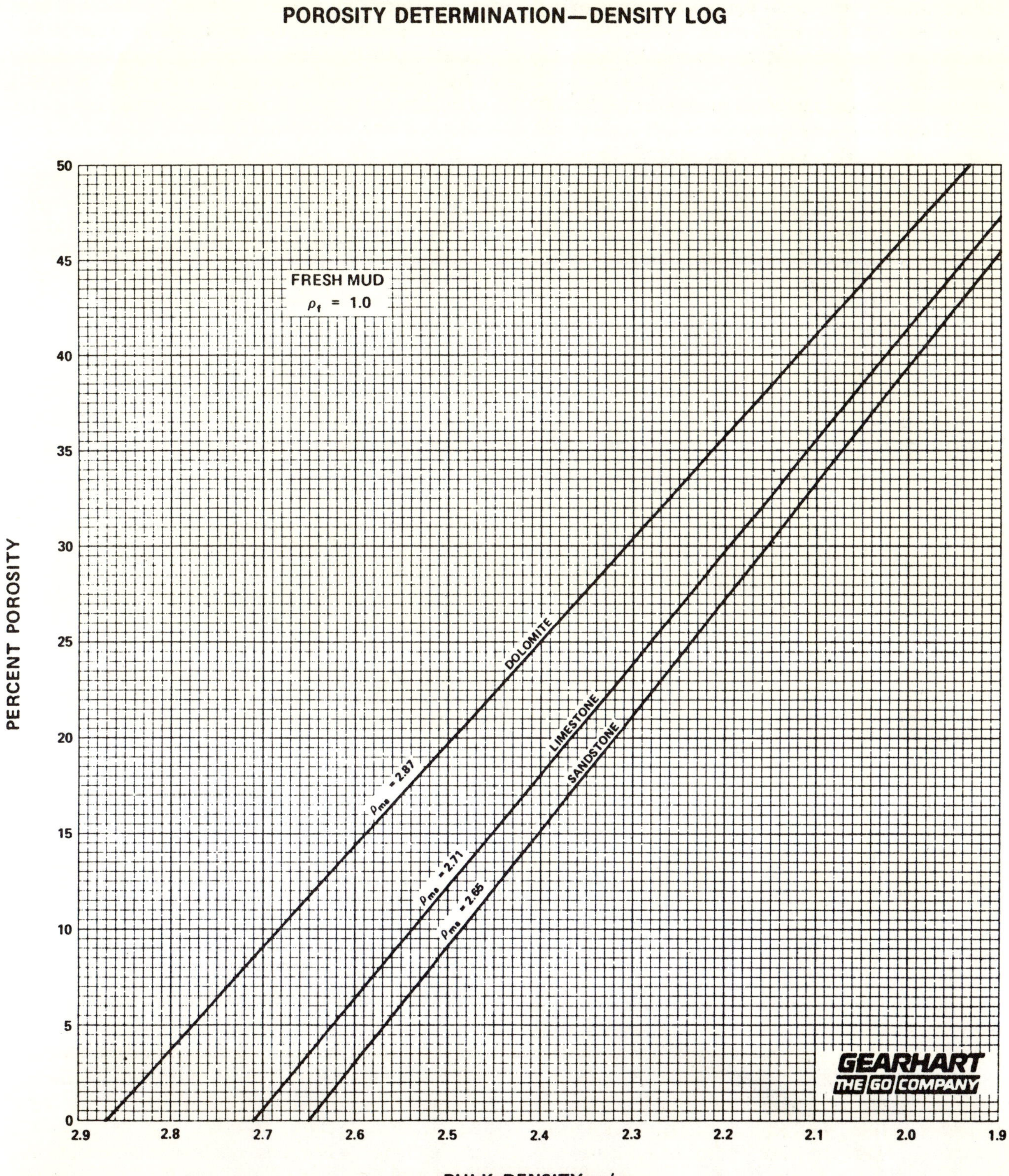

Figure 7-6. Porosity determination chart for ρ_f = 1.0. (Courtesy of Gearhart.)

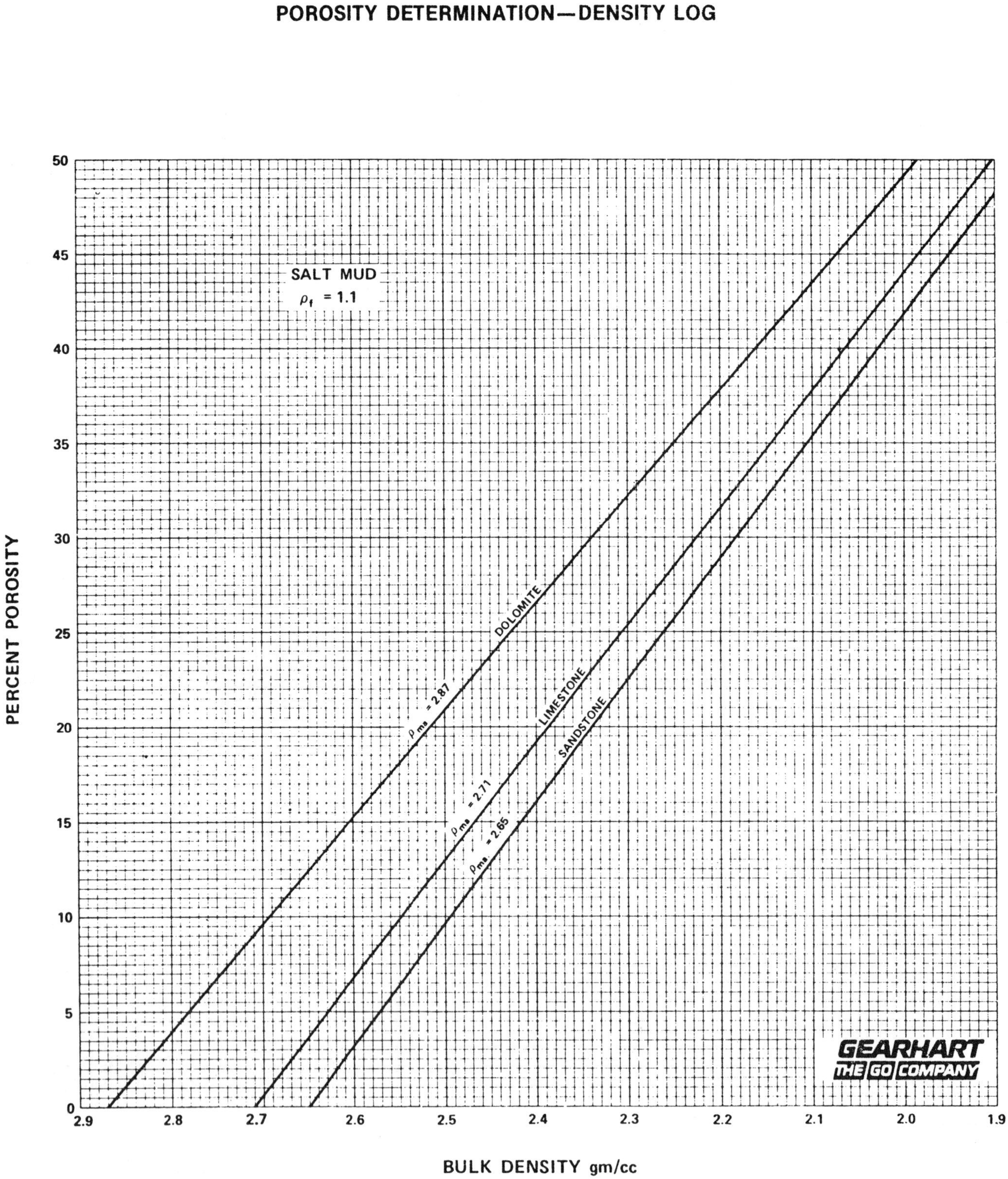

Figure 7-7. Porosity determination chart for $\rho_f = 1.1$. (Courtesy of Gearhart.)

$$FE_{loss} = \frac{4m}{(1 + m)^2}$$

where m is the mass of the struck nucleus in atomic mass units. FE represents the fractional energy loss. For carbon the loss is 28%, for oxygen the loss is 22%, for silicone it is 28%, and for lead, FE loss is 2%.

It should be apparent that the smaller the atom, the more energy the neutron will lose on collision. Simply viewed, consider a marble striking a billiard ball. The marble, while losing some of its energy, merely deflects from the larger ball and continues on a new path. When the marble collides with another marble, the first transfers a considerable amount of energy to the other.

Neutrons are about the same size as hydrogen atoms. As more hydrogen is encountered in the pore space, more neutrons collide, thereby losing their energy and becoming captured. The count rate is consequently reduced.

Pore space is usually filled with one or more types of fluids. Water, oil, and gas contain hydrocarbons. Water and oil have about the same amount of hydrogen, while gas has a lower hydrogen density. The neutron tool cannot differentiate between oil and water, but gas, having less hydrogen, can be detected.

The density tool measures the density of the formation, extracting porosity on the basis of the lesser the density per unit of volume the more pore space is available. The neutron tool measures the hydrogen in the pore space itself and is not affected by the matrix.

The curves presented on neutron logs will generally either be scaled in porosity units or count rate units, and in some cases, both.

Most current neutron porosity systems automatically convert the count rate output from the downhole instrument (or the ratio of count rates from dual-spaced tools) into porosity units prior to recording a curve on the log. The conversion is normally related to the response of the tool in test pit formations, including API porosity calibration formations at the University of Houston. The porosity conversion will be defined for tool response in a particular lithology type, usually either limestone or sand, and also can incorporate automatic caliper correction for borehole size variations. Other porosity curve corrections are normally made from correction charts by the log analyst, or by subsequent computer processing, based on the entire suite of logs from the well.

Neutron Log Corrections

Prior to 1976, neutron logs needed to be corrected for borehole size. With the advent of computers, the borehole correction was applied to the curve by the use of the caliper curve. It is therefore important to know that the caliper is functioning properly.

Salinity in both the borehole fluid and formation fluid must be accounted for. Chlorine is an excellent neutron absorber. As neutrons begin to lose their energy, they are easily absorbed by the chlorine atom, and consequently reduce the number of neutrons being counted by the tool. The lower the count rate, the higher the porosity, since the two are logically reciprocals.

Shale, having a tremendous amount of bound water, will also cause the count rate to decrease, thus increasing the porosity indication.

Mud and mudcake both have a high hydrogen content, therefore these elements must also be corrected for. Figure 7-8 depicts a correction chart for the dual-spaced compensated neutron log. The chart takes into account all of the factors just discussed.

The Neutron Tool as a Gas Locator

The sonic and density tools do not respond in the same way as the neutron tool in a gas zone. The sonic tool will have an apparent porosity greater than the effective porosity in a gas zone. The reason for this is the increased travel time caused by the gas. The density tool behaves in the same manner. The neutron tool is affected by the amount of hydrogen in the formation. Since gas is much less dense than either water or oil, it has fewer hydrogen atoms in a given volume. Because of this, the neutron tools tend to read too low of a porosity in a gas zone. When the neutron porosity and either sonic or density porosity are compared, any large differences in which the neutron porosity reads low and the sonic or density porosities read high, is an indication of gas (this separation is commonly known as the "football effect;" see Figure 7-9).

This relation may not always hold true for the following reasons. The neutron tool tends to read shale as high in porosity. This will offset the effects of gas, which tends to lower the apparent porosity. A shaly gas zone could read as a clean water zone. Also, invasion may displace gas away from the borehole beyond the depth of investigation of the neutron tool. Thus, even if the zone contains gas, the neutron tool could read mostly filtrate.

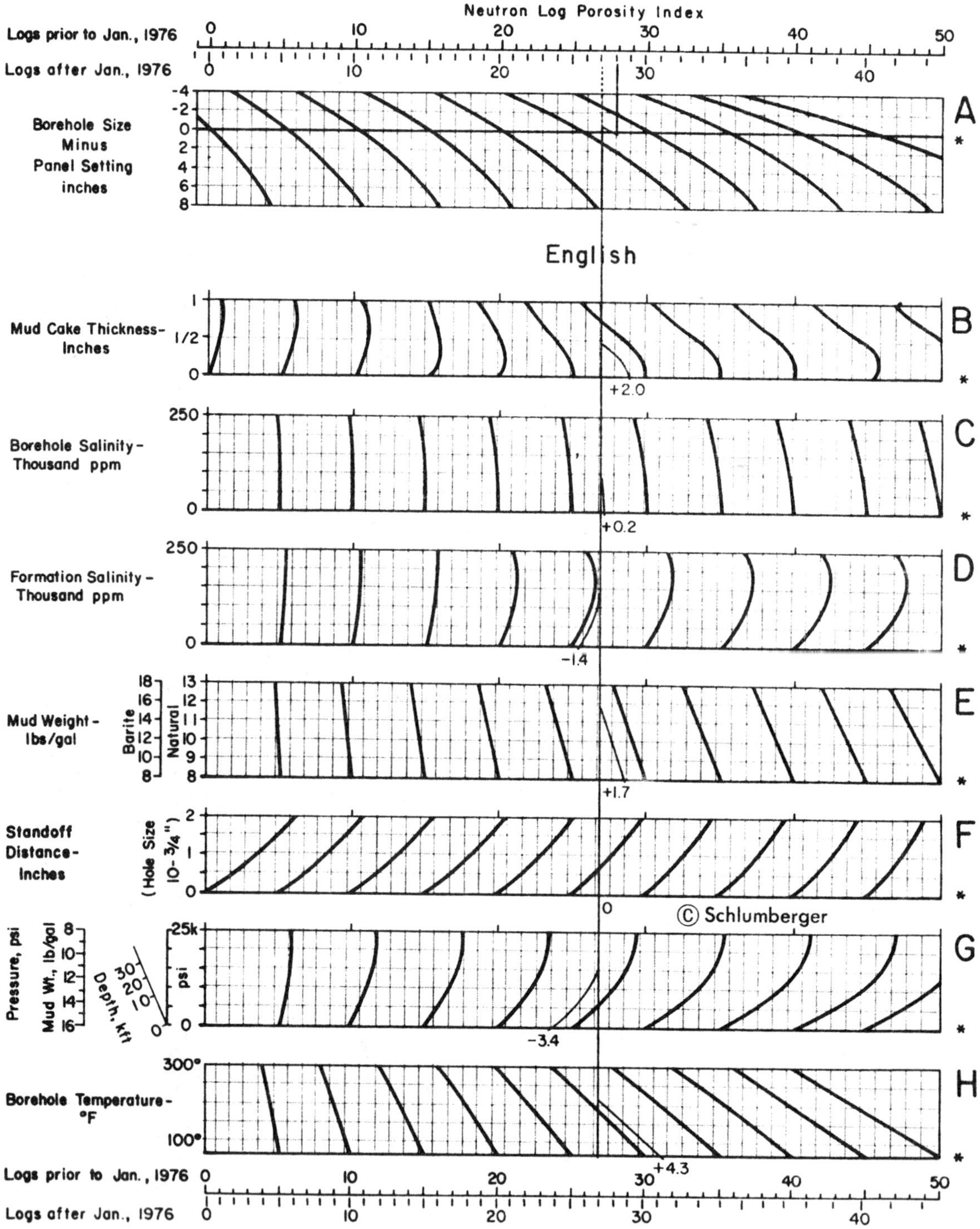

EXAMPLE 1. Auto. Caliper

GIVEN:

CNL Rdg. (corr. for bh)	.... 28 p.u.
Mud Wt.	.... 12 lb/gal
h_{mc}	.... ½ in.
T_{bh}	.... 225°F
Salinity (bh)	.... 100 kppm
Salinity (fm)	.... 150 kppm
Standoff	 0
Depth	.... 21,000 ft

SOLUTION:

Enter at 28 p.u.
Go to Index, Block A
Follow trend lines to −½ in.
Draw chart-base line at this point (27 p.u.)
Find corrections as shown:
$\Sigma\Delta\phi = +2.0 + 0.2 - 1.4 + 1.7 - 3.4 + 4.3 = 3.4$
$\phi_{cor} = 27 + 3.4 = 30.4$ p.u.

EXAMPLE 2. Separate Caliper

GIVEN:

CNL Rdg. (corr. for bh)	.... 30 p.u.
"BH − Cal."	.... +4 in.
Mud Wt.	.... 9.3 lb/gal
T_{bh}	.... 175°F
Salinity (bh)	.... 50 kppm
Salinity (fm)	.... 150 kppm
Standoff	 0
Depth	.... 12,500 ft

SOLUTION:

Enter at 30 p.u.
Draw chart-base porosity line through all blocks.
Find corrections as shown (skip Block B)
$\Sigma\Delta\phi = -4.0 + 0 - 1.5 + 0.7 + 0 - 2.5 + 3.3 = -4.0$
$\phi_{cor} = 30 - 4.0 = 26$ p.u.

Figure 7-8. Dual-spacing neutron log correction nomograph for open hole. (Courtesy of Schlumberger.)

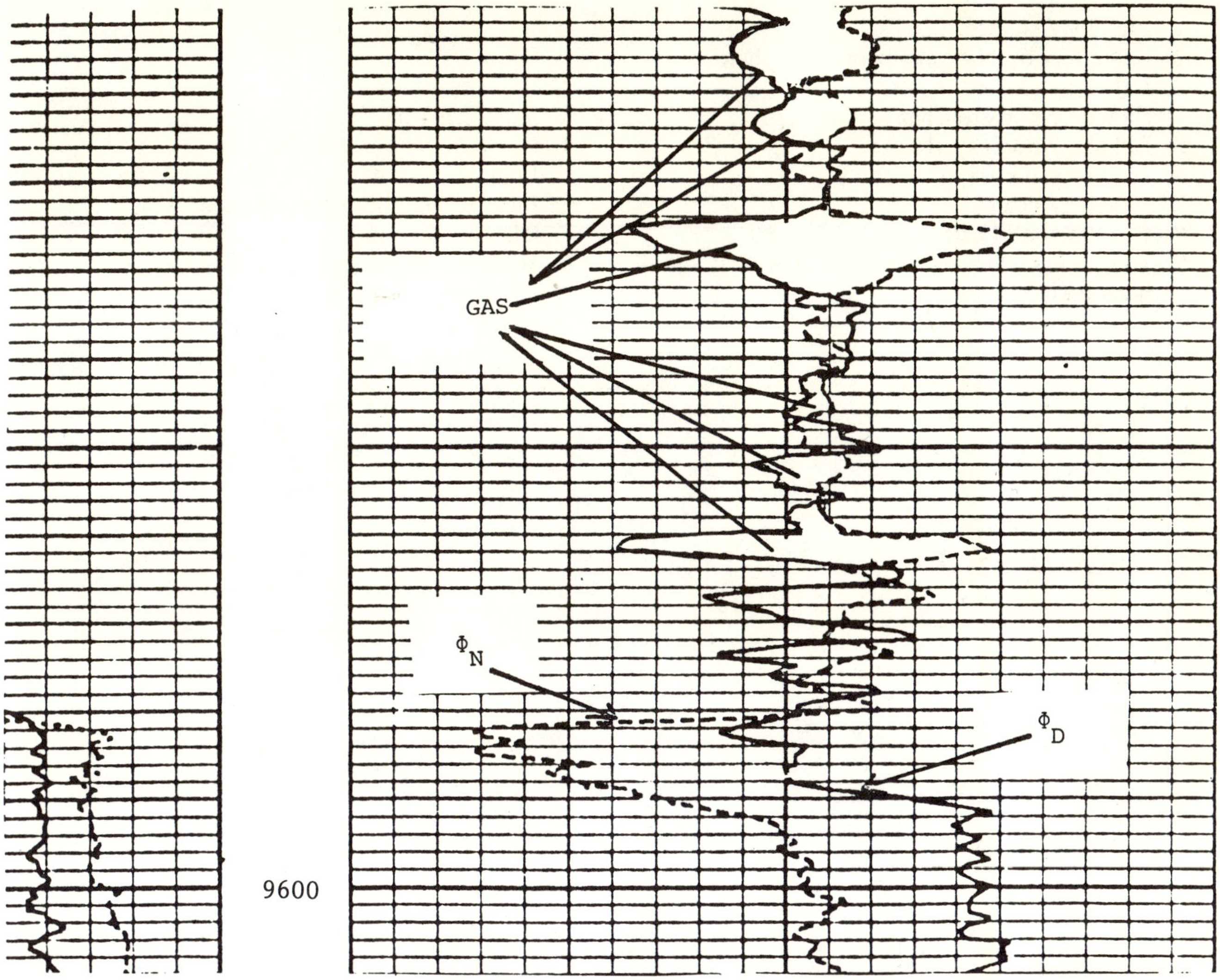

Figure 7-9. Illustration of the effect of gas on the density-neutron curves.

Neutron-Density Patterns

The neutron-density pattern of Figure 7-10 was made using a matrix of limestone.

Pattern

1. Lithology change
2. Gas effect in thin zone or possibly shale
3. Lithology change
4. Lithology/porosity change
5. Lithology/porosity change
6. Lithology/porosity change
7. Lithology change
8. Gas "football" effect
9. Water-saturated limestone with statistical variations on the curve

The patterns made on Figure 7-11 represent the density-neutron curves on a sandstone matrix.

Pattern

1. Shale
2. Shaly water sand
3. Shaly gas sand
4. Invaded low-porosity gas sand
5. Low-porosity gas sand with little or no invasion
6. Invaded high-porosity gas sand
7. Clean water sand
8. High-porosity oil sand
9. High-porosity gas sand

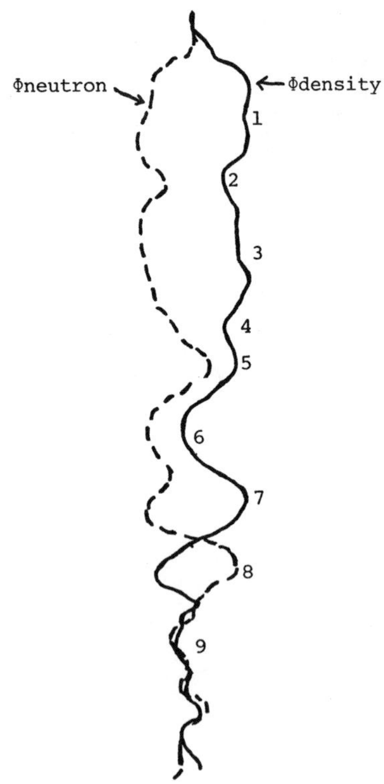

Figure 7-10. Neutron-density patterns for a limestone matrix.

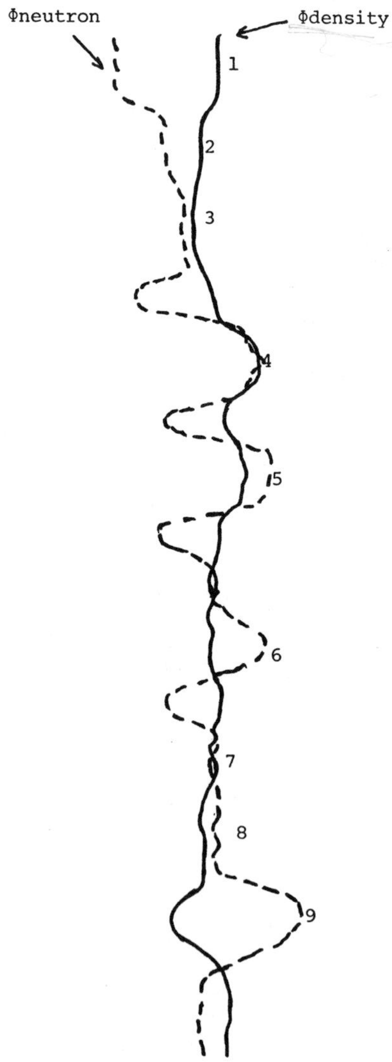

Figure 7-11. Neutron-density patterns for a sandstone matrix.

Crossplotting Neutron Density for Shale

One of the most effective means for obtaining shale volume and corrected porosity is by crossplotting the neutron and density logs. Construction of the plot is easy. Starting with a sheet of linear graph paper, draw two axes and scale them, according to your judgement, in porosity units. The horizontal scale is neutron porosity, and the vertical is density porosity. Next, draw a diagonal line at 45 degrees from the apex of the axes.

Figure 7-12 shows the completed step. Note that the porosity scales end at 50%. It is pointless to extend the scale, because porosities beyond 50% are not realistic.

The next step involves reading the neutron and density porosity in the shaliest portion of the log.

Assume:

$\Phi_n = 45\%$

$\Phi_d = 9\%$

in a thick clean (no sand) shale.

Now, plot these two porosities on the chart and draw a line from the point to the origin or apex of the chart. Figure 7-13 illustrates the step. This point is the 100% shale point, and the line represents 0% porosity.

Parallel to the 0% porosity line, draw corresponding porosity lines as in Figure 7-14.

Parallel to the 45° line bisecting the chart, draw the 100% shale line (at a 45° angle) as shown in Figure 7-15.

The last step in the construction of the chart entails dividing the area between the porosity-corrected line and the 100% shale line into equal parts. Each part corresponds to the amount of shale present in terms of percent. The area can be divided into any units that may be convenient to the user. On Figure 7-16, the area was divided into 4 equal parts, each being 25%.

Figure 7-16 completes the construction of the chart, and may be used for analyzing the zone of interest. A 22% density porosity, and a 30% neutron porosity would crossplot to 23% shale and a corrected porosity of 20%. If the formation contained gas, the plot would come out to the northwest quadrant of the chart. In such a case, the porosities must first be corrected for gas effect.

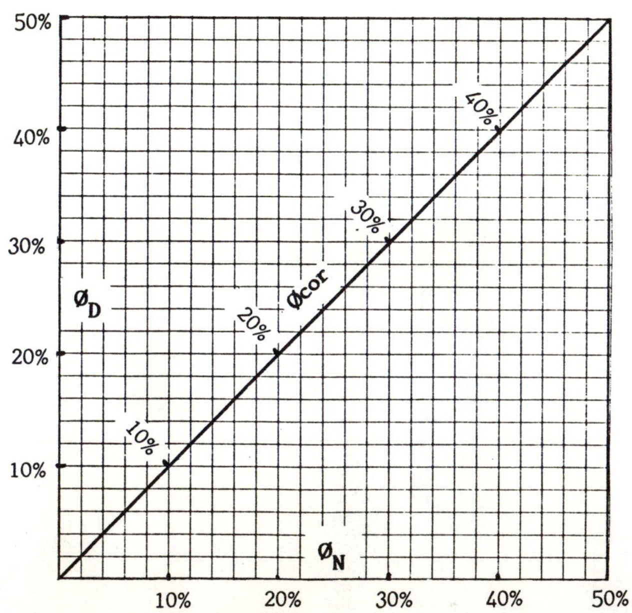

Figure 7-12. Construction of a neutron-density shale volume and porosity correction chart.

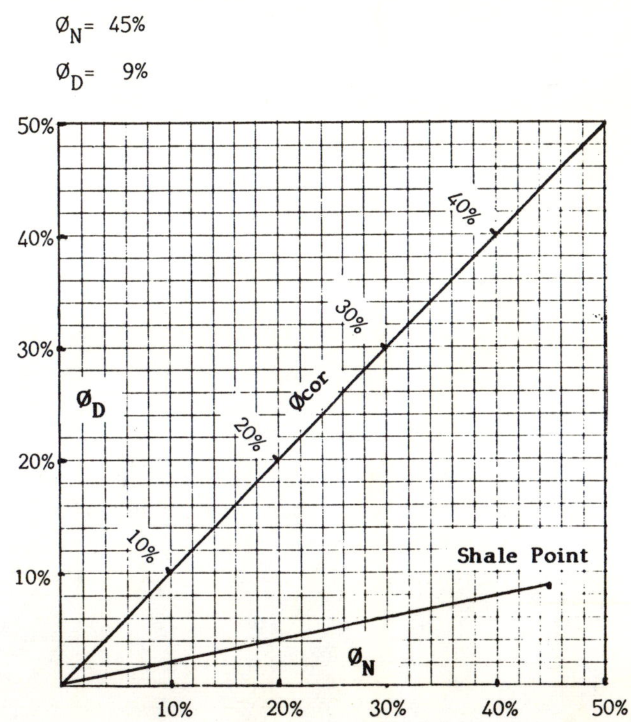

Figure 7-13. Plotting the shale point.

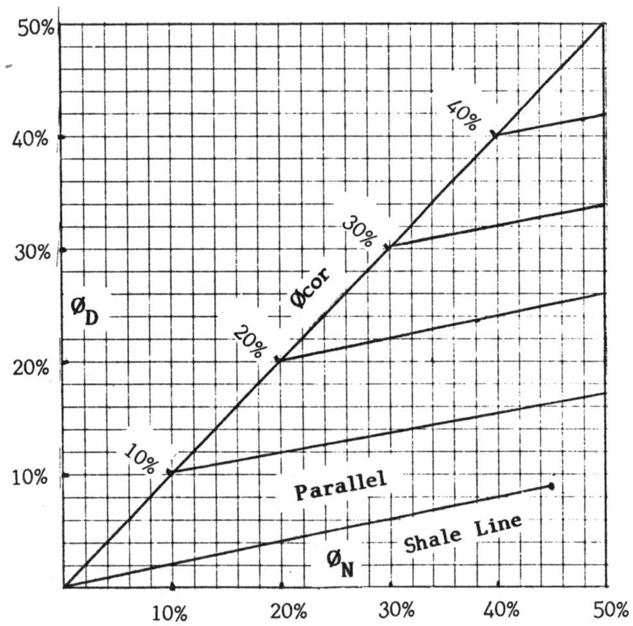

Figure 7-14. Porosity scales are drawn parallel to the shale line.

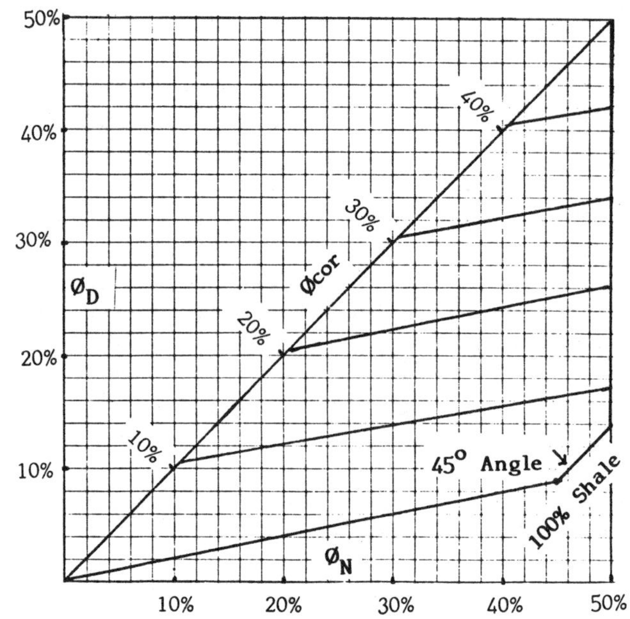

Figure 7-15. Plotting the 100% shale line at a 45° angle.

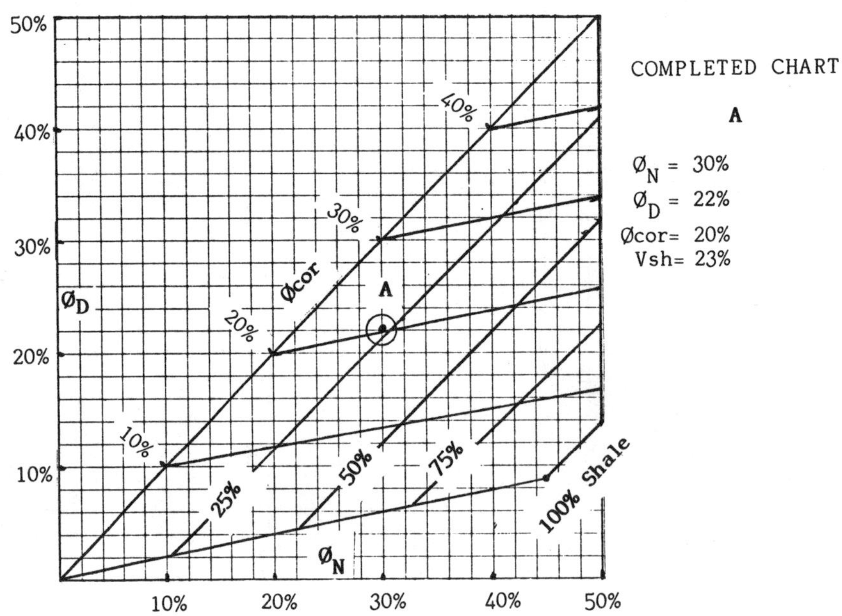

Figure 7-16. Completed chart illustrating a typical plot.

The Gamma Log

The gamma log is a measure of the natural radiation of the formation. Radioactive elements tend to concentrate in clays or shales. While shales and shaly sands are higher in radioactivity, clean sands and carbonates exhibit low levels of radiation also. The gamma ray curve differentiates between potential reservoir rocks and shales. Thus, the gamma log can be used to determine lithology. The gamma ray log can also be used to determine volume of shale in the formation.

Gamma Ray Properties

Gamma rays are high-frequency electromagnetic waves. Natural formation radiation is due to radioactive potassium 40, the uranium-radium series, and the thorium series. The strength of the received gamma rays is determined by the strength of the source, the density of the formation, and the distance the gamma radiation must travel to reach the tool.

Radioactive Element Deposition

A major portion of earth's radioactive potassium and uranium-thorium series elements were originally contained in igneous rocks. The radioactive elements present in the original igneous rocks will therefore be present in the clay (shale). Since shales are composed of small clay particles with large openings in the lattice structure, the radioactive elements are allowed to reside in the resulting shale. The permeability of this shale is generally so low that these radioactive elements are never washed out.

Quartz crystals have strong, well-ordered structure. This tends to eliminate any radioactive elements from the quartz structure. Since sandstones are created from the erosion of quartz, they generally exhibit low radioactive content. Limestones are also low in radioactive content.

Dolomite is formed from a chemical reaction between limestone and dissolved magnesium in migrating ground waters. Ground water contains a small amount of low-level radiation and in the process of dolomitization, some radioactivity is deposited. For this reason, dolomite has a small but higher amount of natural radiation than limestones.

Gamma Ray Curve Characteristics

Gamma ray tools have vertical resolutions 6 in. to about 2 ft, and depths of investigation between 6 in. and 1 ft. Increased hole diameter tends to decrease the number of gamma rays observed. KCL muds tend to increase the background radiation because of the radioactive potassium the mud contains.

The gamma ray curve, as with all radioactive curves, is statistical in nature. This means that even with the tool stationary, the gamma ray tool will record small fluctuations. To smooth these fluctuations, a filter is added which averages the counts over a small period of time. This is known as a "time constant." If the time constant is too long, the tool will be unable to resolve thin beds, since the data will be averaged out. Too short of a time constant will yield a gamma ray curve which fluctuates too much to be useful. Logging speed has an effect much like that of a time constant. Slow logging speeds allow the tool to see and average more counts for a given depth, giving a truer picture of the overall count rate. A fast logging speed does not allow much averaging and tends to show less detail. It is recommended that a time constant of 2 seconds and a logging speed of 1,800 ft/hr be used.

Use of Gamma Ray to Determine V_{sh}

The gamma ray log can be used to determine the volume of shale in a formation. Volume of shale can then be used to correct the apparent porosities found by the sonic, neutron, and density logs. This method of finding V_{sh} will generally yield good values in common formations. If, however, the formation is unusually radioactive, this method will not work.

In order to determine V_{sh}, the gamma ray index must be found from:

$$\text{Gamma ray index} = \frac{GR - GR_{\text{clean sand}}}{GR_{\text{shale}} - GR_{\text{clean sand}}}$$

V_{sh} is then found from a graph such as that shown in Figure 7-17.

$GR_{\text{clean sand}}$ is found from the cleanest sand in the well. That is, the lowest reading gamma ray in the well for a given lithology. GR_{shale} is found in thick shale zones near the zone of interest. This will normally be the highest reading gamma ray.

If a density log is available, V_{sh} can be found by:

$$V_{sh} = \frac{\rho_b \times GR - GR_{\text{clean}}}{\rho_{b,\text{shale}} \times GR_{\text{shale}} - GR_{\text{clean}}}$$

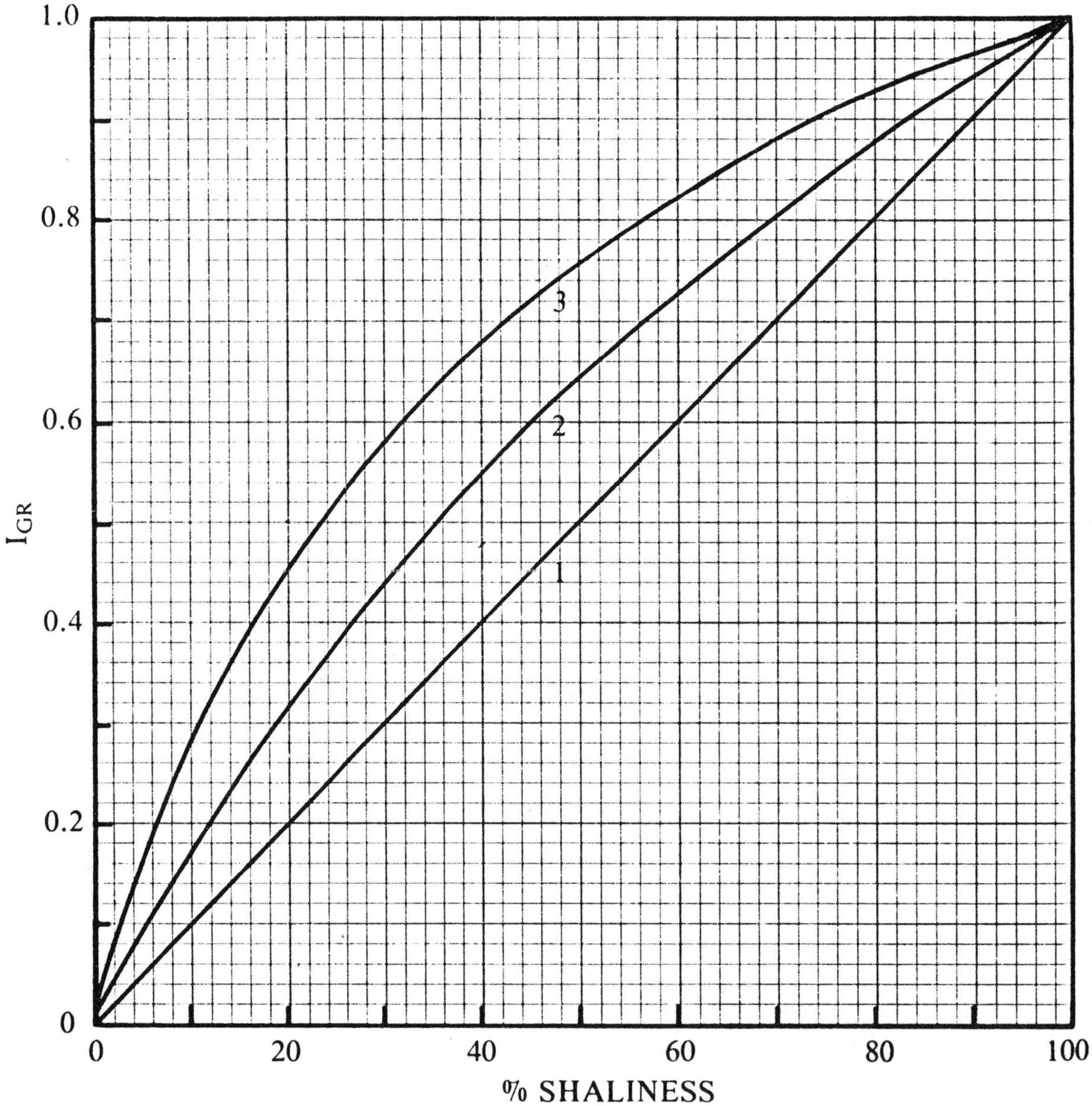

Figure 7-17. Shale volume versus gamma ray index. (Courtesy of Dresser Atlas.)

8

Resistivity-Porosity Crossplots

As stated previously, the value of R_w is the cornerstone of analysis. An incorrect R_w can introduce a significant error in water saturations. In situations where the water saturation is based on R_w's from the SP, it is wise to confirm the values used. In cases where an SP is not available, another tool for its determination must be employed.

Resistivity-porosity crossplots can be made in order to determine or confirm R_w. Besides R_w, this plot can also yield values for "a," "m," and matrix.

The Resistivity-Porosity Crossplot

The resistivity-porosity crossplot, sometimes called the Hingle plot, is a plot constructed on a linear-logarithmic scale which has been scaled proportional to the proper F-Φ relationship. The two most common are the Humble relationship and the Archie relationship.

$$F = \frac{1}{\Phi^2} \quad \text{Archie}$$

$$F = \frac{0.62}{\Phi^{2.15}} \quad \text{Humble}$$

Figures 8-1 and 8-2 show the graphs accordingly. Note that each resistivity scale has been proportioned accordingly to the corresponding F-Φ relationship. Also note that the porosity scale is linear. While a neutron, density, or sonic porosity can be used, it is also possible to use either Δt or bulk density. This feature allows us to crossplot without knowing matrix or fluid values. If a porosity is plotted, the matrix value must be constant (i.e., limestone matrix). The values used are apparent and need not be corrected. The value of R_w is assumed to be constant.

To fully understand the crossplot, let us plot one. From a thick zone, the following resistivity values were chosen from the deep induction curve. Along with these readings, a corresponding Δt is read. Δt has been chosen because we are unsure about the matrix travel time. The lithology is carbonate.

Data Point	Ra	Δt
1	66	61.0
2	48	62.7
3	32	62.5
4	23	62.5
5	24	59.7
6	60	53.5
7	46	54.8
8	60	52.0
9	17	57.5
10	9	58.0
11	60	52.0
12	100	51.5
13	260	49.3

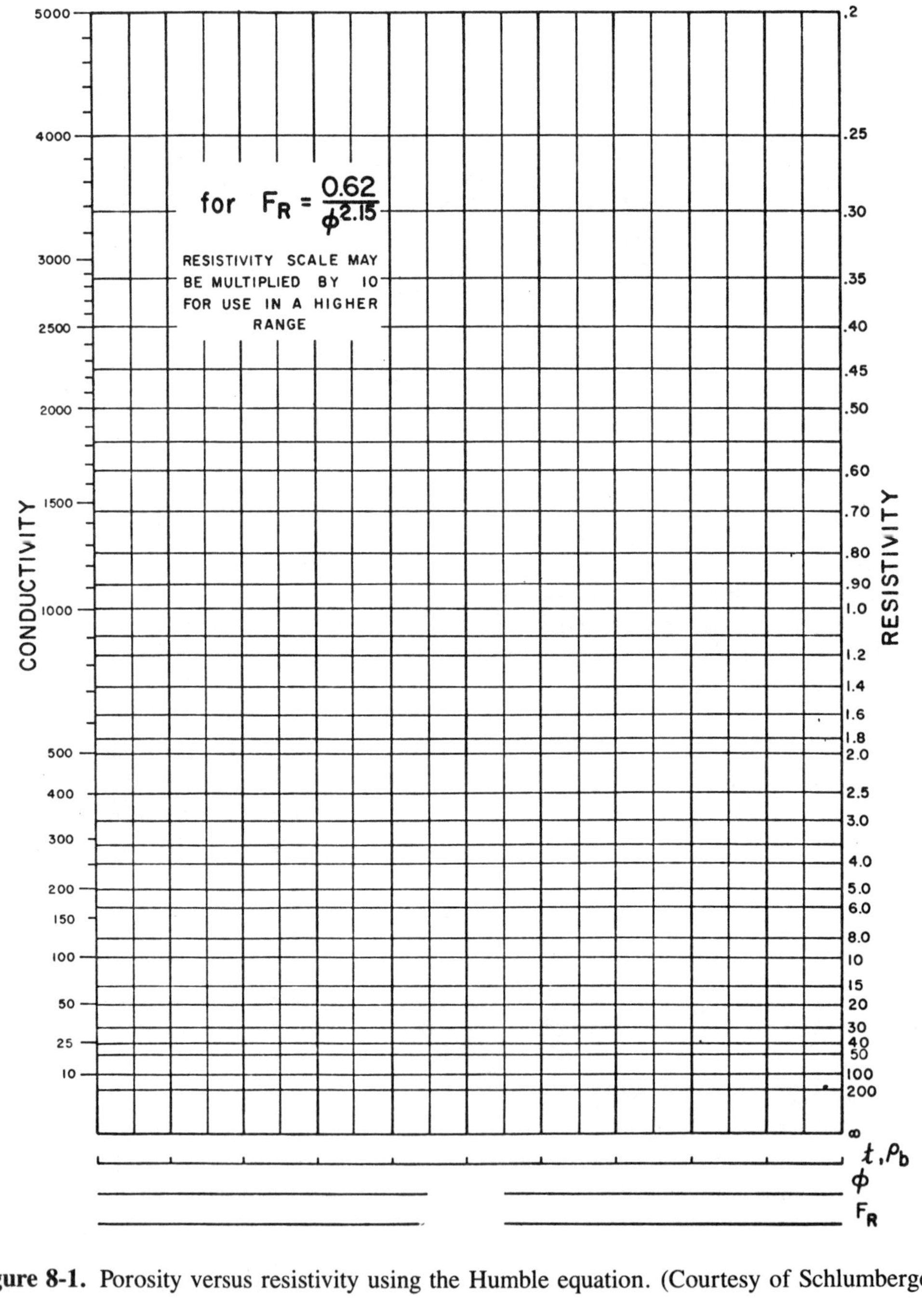

Figure 8-1. Porosity versus resistivity using the Humble equation. (Courtesy of Schlumberger.)

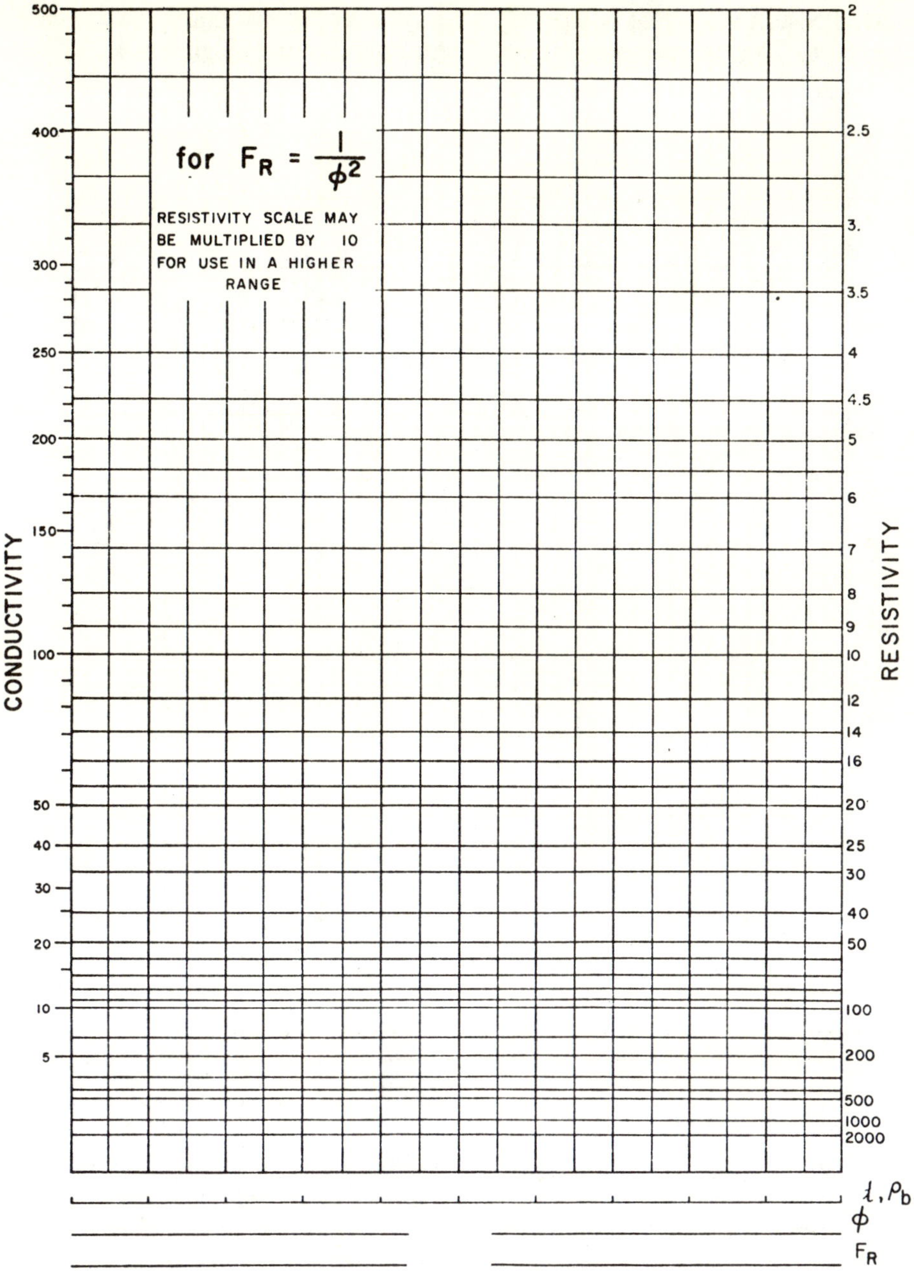

Figure 8-2. Porosity versus resistivity using the carbonate equation. (Courtesy of Schlumberger.)

The first step is to choose the appropriate chart. Since we do know that we are in carbonate lithology, the $F = \frac{1}{\Phi^2}$ chart is chosen.

Secondly, we must choose a scale that adequately covers the Δt range. 40 to 70 microseconds is most suitable. For Δt each linear division is 3 microseconds apart. If one of our data points had a 72 microsecond reading, then we could have scaled it from 43 to 73 microseconds, in order to afford it the range.

Each point is plotted on the graph, as in Figure 8-3. A straight line is drawn through the most northwesterly, or upper left, points. In this case, three points determine the line (Figure 8-4).

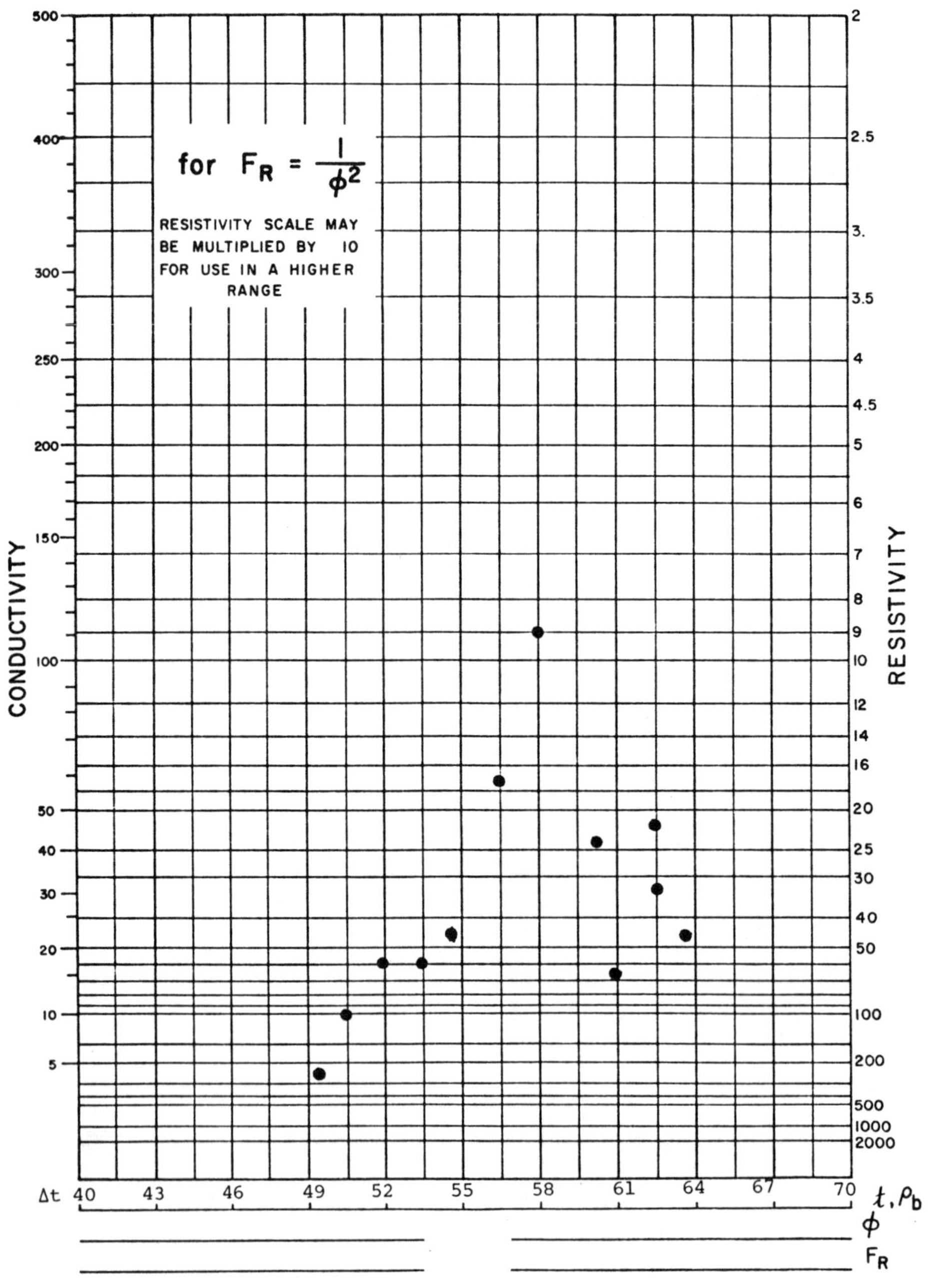

Figure 8-3. Porosity versus resistivity chart showing data plotted. (Courtesy of Schlumberger.)

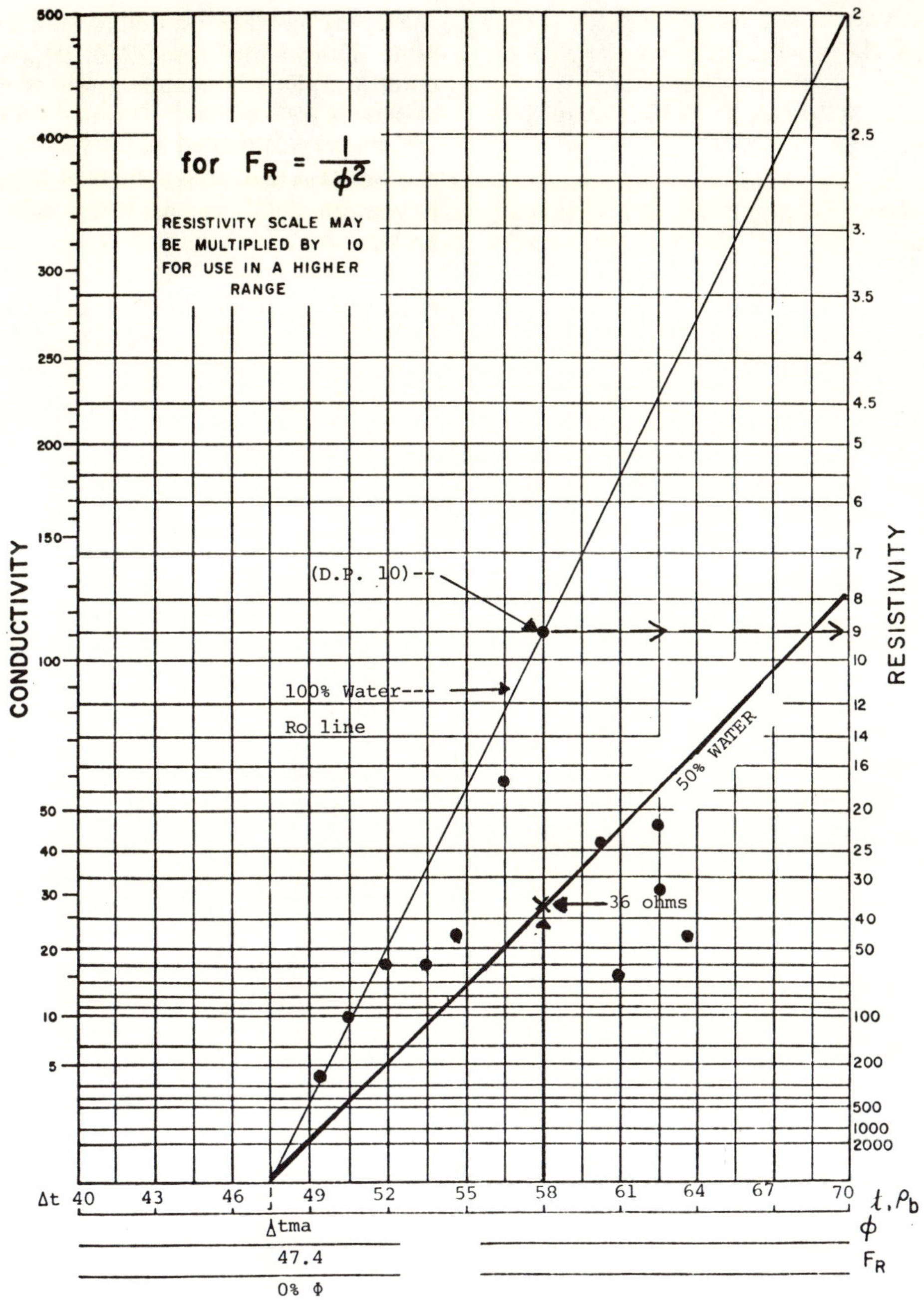

Figure 8-4. Porosity versus resistivity chart showing 100% and 50% water lines. (Courtesy of Schlumberger.)

This is the 100% water-saturated portion of the zone. Any point below this line represents a water saturation of less than 100%. The "y" intercept indicates a value of R_o, which is 2 ohm meters. The "x" intercept (Δt line) shows the value of Δt matrix! This point can be considered 0% porosity. The value 47.4 microseconds indicates a limestone lithology. Accurate porosities can now be calculated using the Wyllie transform.

$$\Phi = \frac{\Delta t - \Delta t_{ma}}{\Delta t_f - \Delta t_{ma}}$$

where $\Delta t_f = 189$
$\Delta t_{ma} = 47.4$

Choosing data point 10 (see D.P. 10, Figure 8-4), we calculate porosity:

$$\Phi = \frac{58 - 47.4}{189 - 47.4} = 0.075 \text{ or } 7.5\%$$

We now calculate F:

$$F = \frac{1}{\Phi^2} = \frac{1}{0.075^2} = 177.8$$

Extending a line from the R_o line to the resistivity scale, as shown by the dashed line on Figure 8-4, R_o for this point is 9 ohms.

Since $R_o = FR_w$, transposing

$$R_w = \frac{R_o}{F}$$

Thus

$$R_w = \frac{9}{177.8} = 0.05 \text{ at formation temperature}$$

The SP indicated an R_w of 0.05 ohms at formation temperature. Note that any point on the 100% R_o line could have been used with the same results.

Carbonate reservoirs are usually not commercially productive above the 50% water saturation level. For this reason, it is of value to draw in a 50% water saturation line.

Water saturation is defined as:

$$S_w = \sqrt{\frac{R_o}{R_t}}$$

If we transpose:

$$R_t = \frac{1}{S_w^2} R_o$$

$$\text{If } S_w = 50\%, \text{ then } R_t = \frac{1}{.25} R_o = 4\ R_o$$

Data point 10 has an R_o of 9 ohms at 58 μs (or 7.5% porosity). R_t at 50% water saturation is then 4 R_o, or $4 \times 9 = 36$ ohms.

Plot the point at the intersection of 58 μs and 36 ohms. Draw a line from 0% Φ (Δt_{ma} 47.4) through the point just plotted, extending it to the resistivity scale as shown in Figure 8-4. This is the 50% water saturation line. Any other Δt could have been chosen. If we choose 70 μs then R_o is 2 ohms. 4 R_o is $4 \times 2 = 8$, and as you can see, that is just where the line intersects the scale.

This method will work well as long as the lithology remains constant, R_w remains constant, and there is a 100% water-saturated zone. The effects of shale, lack of compaction, and residual hydrocarbon saturation frequently preclude the use of this crossplot.

Field Example

To better understand the crossplot, we will work out the following example.

Figure 8-5 shows a dual-induction sonic log. This is an offset well of the well used in Chapter 6. We are somewhat familiar with its parameters. The first step is to construct a table of R_{deep} and Δt.

Depth	R_{deep}	Δt
4620	6	76
4622	5	78
4624	4	80
4626	3.5	83
4630	3.2	84
4632	3.2	82
4634	3.5	80
4638	3.7	80
4640	3.5	80
4642	3.4	79
4644	3.3	80
4650	3.5	82
4656	2.5	81
4658	2.5	83
4660	3.0	81
4662	3.5	78
4664	4.0	75
4666	5.0	75
4668	10.0	72

Some points were omitted because they repeat and would plot on the same point.

The lithology is carbonate so the $1/\Phi^2$ chart is selected for the plot. The resistivity range of our data is from 2.5 to 10 ohms. This range fits well on the resistivity scale. The sonic range is 72 to 84 microseconds. A scale of 60 to 90 microseconds would fit the data well.

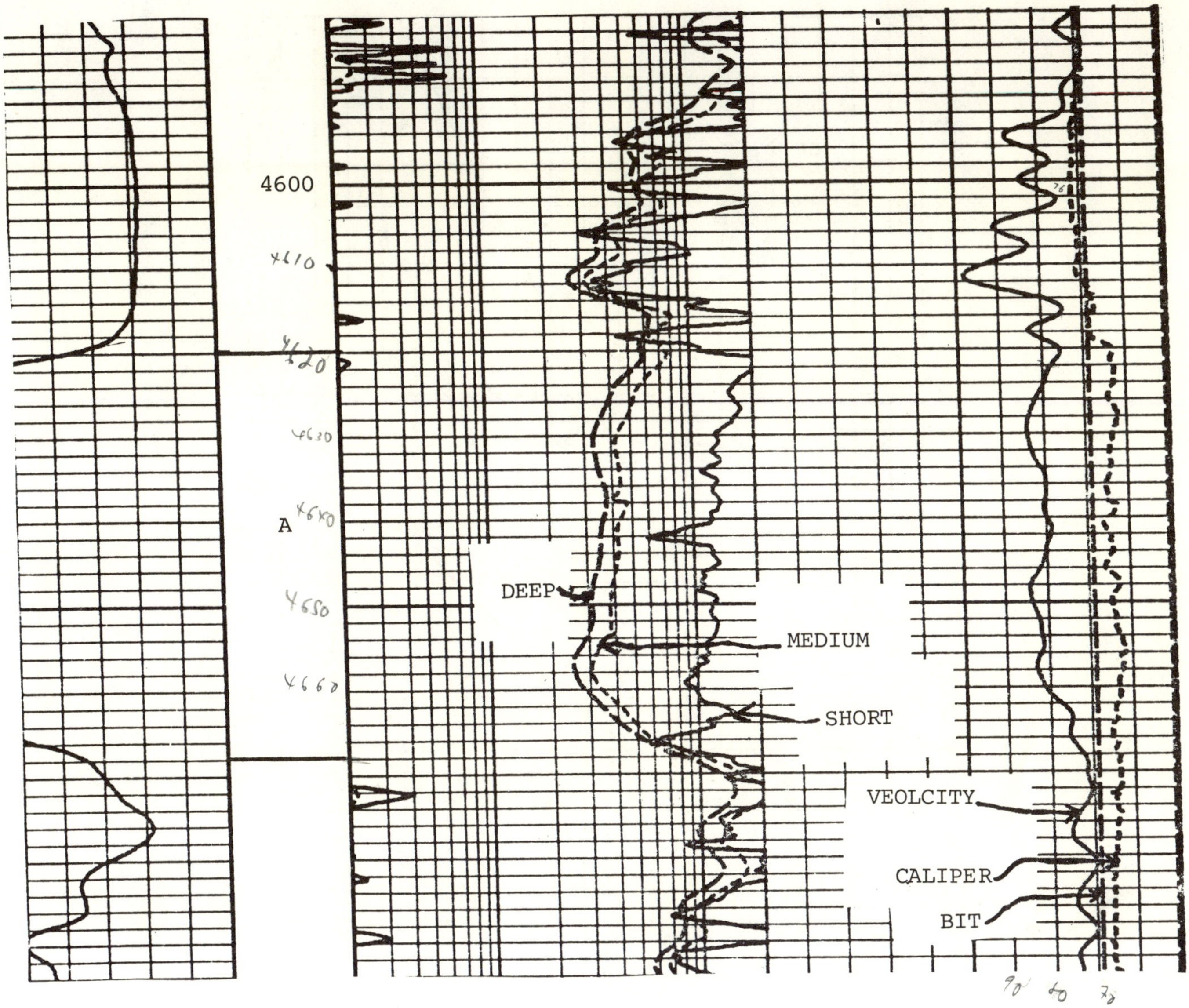

Figure 8-5. Zone of interest chosen for a porosity versus resistivity plot.

Figure 8-6 shows the data plotted and the chart scaled.

The 100% water saturation line is drawn in to fit the best northwesterly points (in this case, the three circled points). This line intersects the sonic scale at 61 microseconds. This becomes the matrix travel time or the zero porosity point.

Arbitrarily choosing the bottommost circled point, we now proceed to extract the R_w. First we find out what this point's porosity is.

$$\Phi = \frac{\Delta t - \Delta t_{ma}}{\Delta t_f - \Delta t_{ma}}$$

$$\Phi = \frac{75 - 61}{189 - 61} = 0.109 \text{ or } 10.9\%$$

Next, we calculate F:

$$F = \frac{1}{0.109^2} = 84$$

Reading the resistivity scale for this point gives us an R_o of 5 ohms.

$$R_o = F\,R_w$$

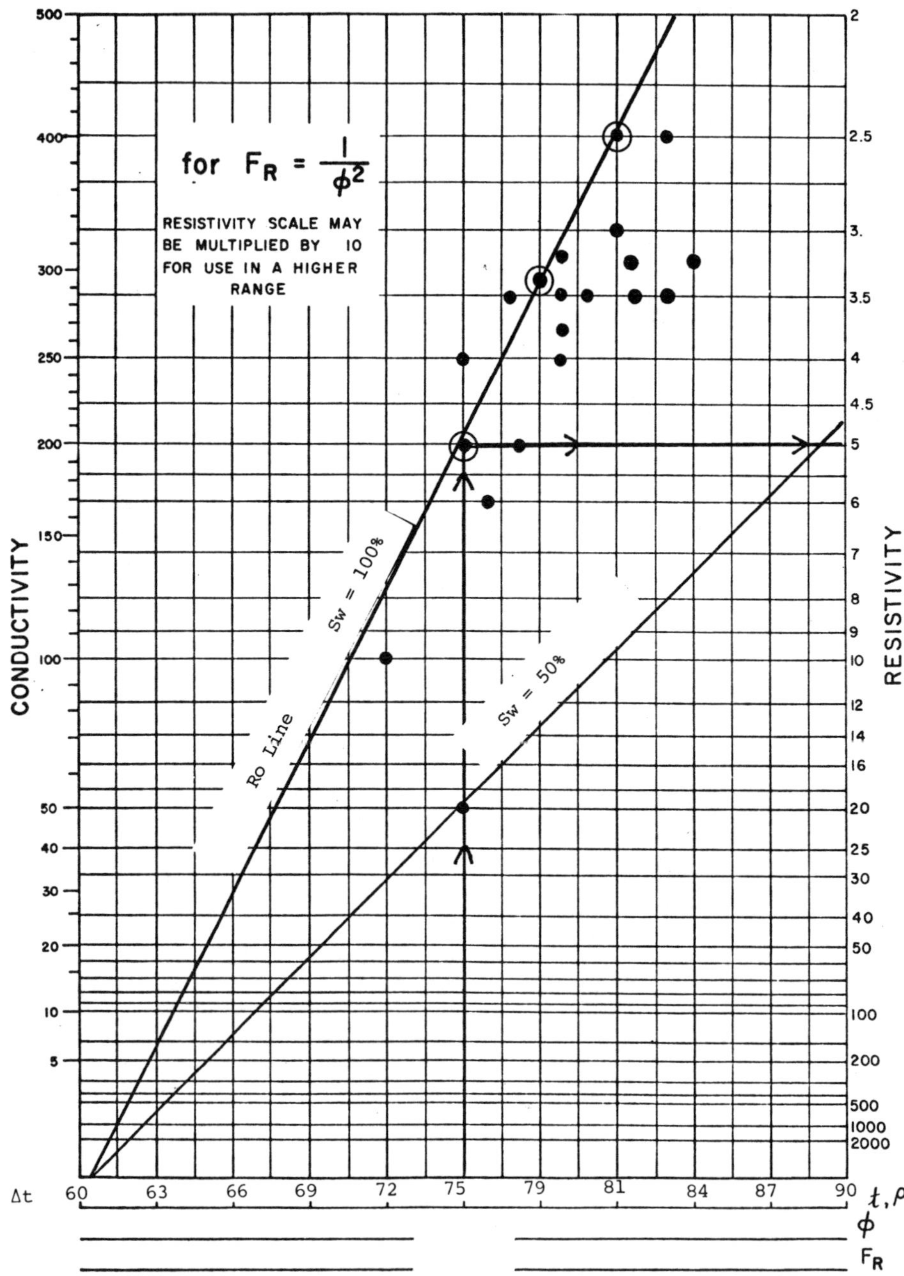

Figure 8-6. Porosity versus resistivity with the zone of interest plotted. (Courtesy of Schlumberger.)

Transposing:

$$R_w = \frac{R_o}{F} = \frac{5}{84} = 0.06 \text{ ohms}$$

at formation temperature, which is 150°F.

The final step is to draw in the 50% water saturation line. Recall that:

$$S_w = \sqrt{\frac{R_o}{R_t}}$$

Transposing:

$$R_t = \frac{1}{S_w^2} R_o$$

If S_w = 50%, then:

$$R_t = \frac{1}{0.5^2} \times R_o = \frac{1}{0.25} \times R_o = 4\ R_o$$

Using the same circled points:

$$R_o = 5$$

$$R_t = 4 \times 5 = 20$$

Plot 20 ohms at 75 microseconds and draw the line from 61 microseconds through the point to the resistivity scale.

It now becomes very apparent that this zone is not going to be producible. The entire zone is above 50% water saturation.

For the sake of accuracy, it is best to cross-check our results. Since the log has a good SP, we can check R_w against it.

R_{mf} at formation temperature (105°) is 1.6 ohms. Using the chart on Figure 8-7, we obtain R_{mfeq} and R_{weq}.

$$R_{mfeq} = 0.6$$

$$R_{weq} = 0.047$$

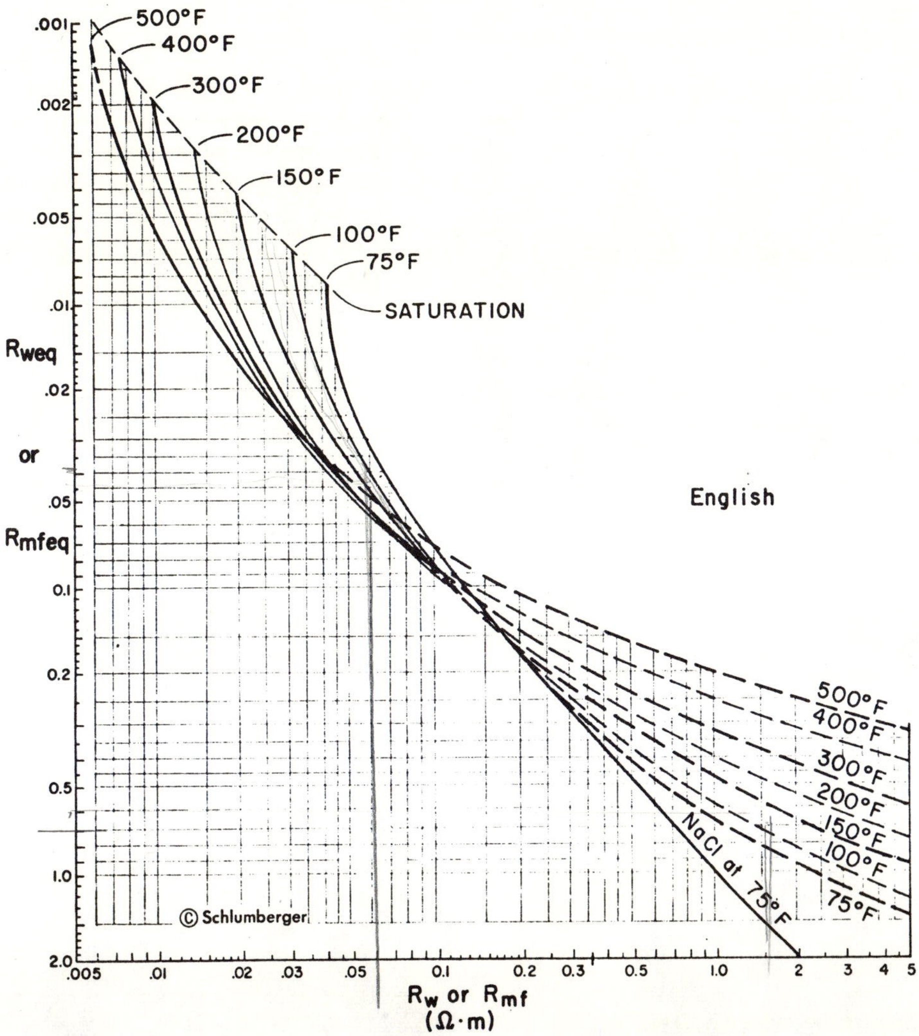

Figure 8-7. R_w versus R_{weq} and formation temperature. (Courtesy of Schlumberger.)

We solve for SP:

$$SP = -(60 + 133T) \log \frac{R_{mfeq}}{R_{weq}}$$

where T = formation temperature
log = the common logarithm

$$SP = -(60 + 0.133 \times 150) \log \frac{0.6}{0.047}$$

$$SP = -79.95 \log 12.766 = -88.4 \text{ mv}$$

Figure 8-8 shows the SP to be −104 mv. This is a difference of 16 mv, indicating that the crossplot is, for the most part, correct. The difference can be accounted for by several sources. The 16 mv, of added SP may be due to a slight human error in the crossplot. The plot assumes that the matrix is constant (it may not be). The plot also assumes that the value of "m" is 2 (it may not be).

R_w from the SP yields a value of 0.05. Which, then, is correct: 0.05 or 0.06?

The Log-Log R vs. Φ Plot

There is really no need to go any further with the preceding example. An R_w of 0.05 could not make this zone productive. For the sake of instruction, or perhaps curiosity, we may attempt to find out why there was a difference. For this, we will construct a resistivity vs. porosity crossplot on 3-cycle log-log paper.

The foundation behind the construction of this chart is the resistivity index. Recall from Chapter 6 that the ratio of R_t/R_o gave us a number which, if it were greater than 3, the zone would be at less than 50% water saturation. The ratio is known as the resistivity index (RI).

$$RI = \frac{R_t}{R_o}$$

and

$$S_w = \sqrt{\frac{R_o}{R_t}}$$

So, we can say that:

$$S_w = RI^{1/n}$$

where n = 2 (saturation exponent)

We can define R_o as:

$$R_o = F\, R_w$$

and

$$F = \frac{a}{\Phi^m}$$

If we combine these equations and take their logarithms:

$$\log R_t = -m \log \Phi + \log R_w + \log RI$$

The equation shows that a log-log plot of R_t vs. porosity yields a straight line of slope −m.

The slope represents the cementation exponent m, and furthermore, where the line intersects the 100% porosity scale, also renders the value of R_w!

Now we have the means of not only verifying the value of R_w, but the cementation exponent is thrown into the bargain.

To construct the plot, we add density-porosity to the previous table. From the neutron density log of Figure 8-9, we obtain porosity.

Depth	R_{deep}	Φ^d
4620	6	6.5
4622	5	6.2
4624	4	12.0
4626	3.5	17.8
4630	3.2	15.0
4632	3.2	13.5
4634	3.5	15.0
4638	3.7	13.0
4640	3.5	12.0
4642	3.4	13.0
4644	3.3	14.0
4650	3.5	14.0
4656	2.5	15.0
4658	2.5	17.0
4660	3.0	17.0
4662	3.5	13.0
4664	4.0	9.0
4666	5.0	8.0
4668	10.0	7.0

The chart in Figure 8-10 shows the data plotted. The resistivity scale was chosen so that all of the points could fall on the chart, a scale of 0.01 to 10 in three decades. If the resistivity was greater than 10, then an additional decade (cycle) would have been necessary. The porosity scale is from 1% to 100% porosity.

After plotting all of the data, a line is drawn through the 100% water wet points. Choosing the points, as you can see, can be a formidable task. I have chosen the same three points used in the previous chart (Figure 8-4). The 100% water saturation line is drawn through the circled points and extended to the 100% porosity scale. As you can see, at this 100% porosity

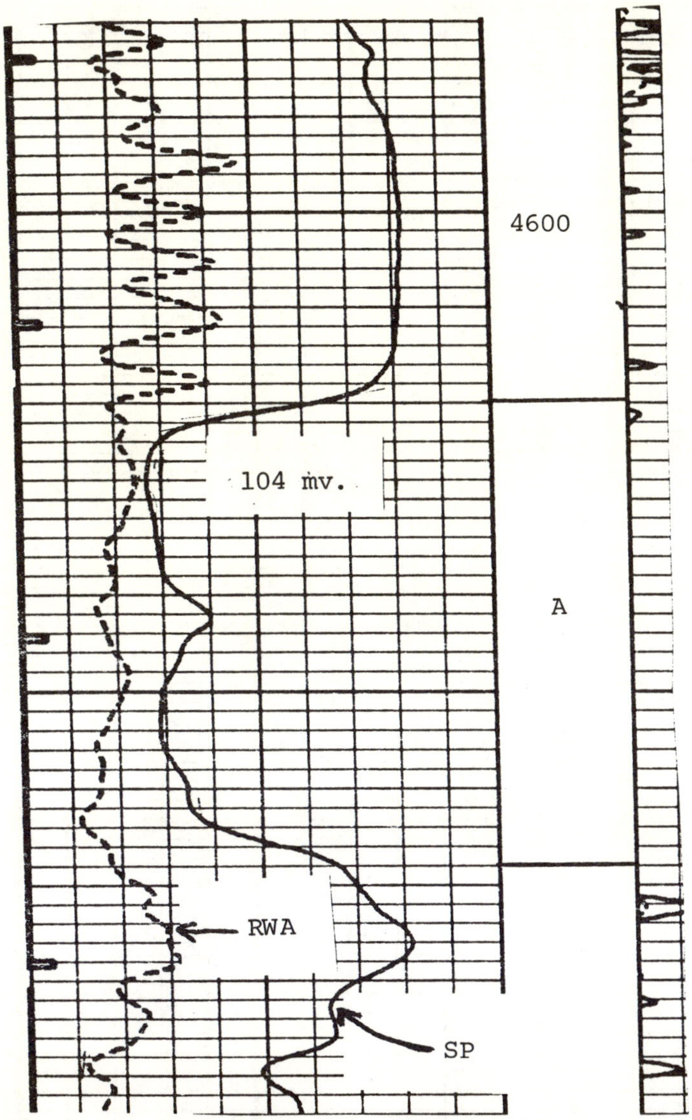

Figure 8-8. SP chosen for R_w computation. (SP scale 20 mv per division.)

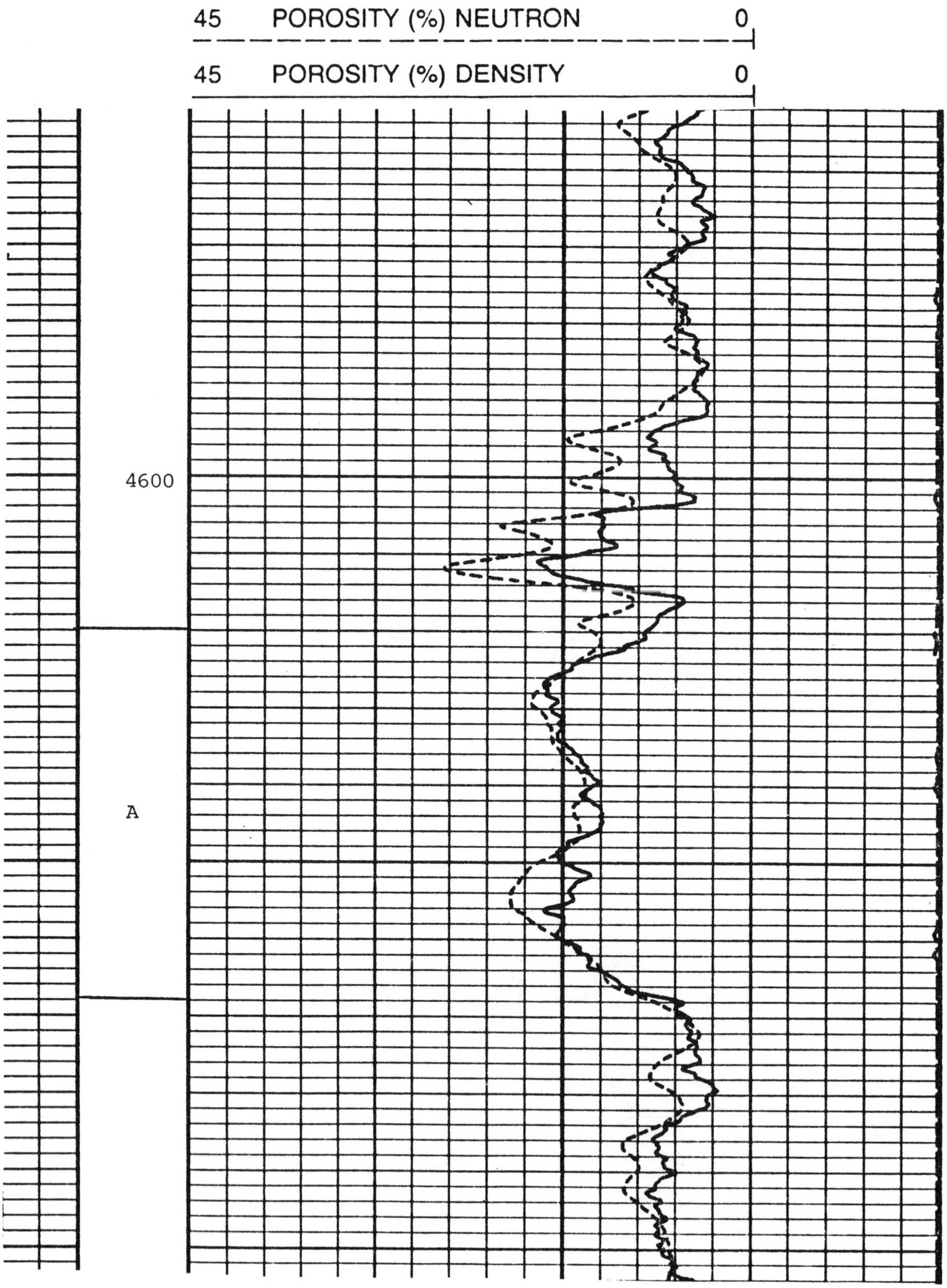

Figure 8-9. Neutron-density curves in the zone of interest.

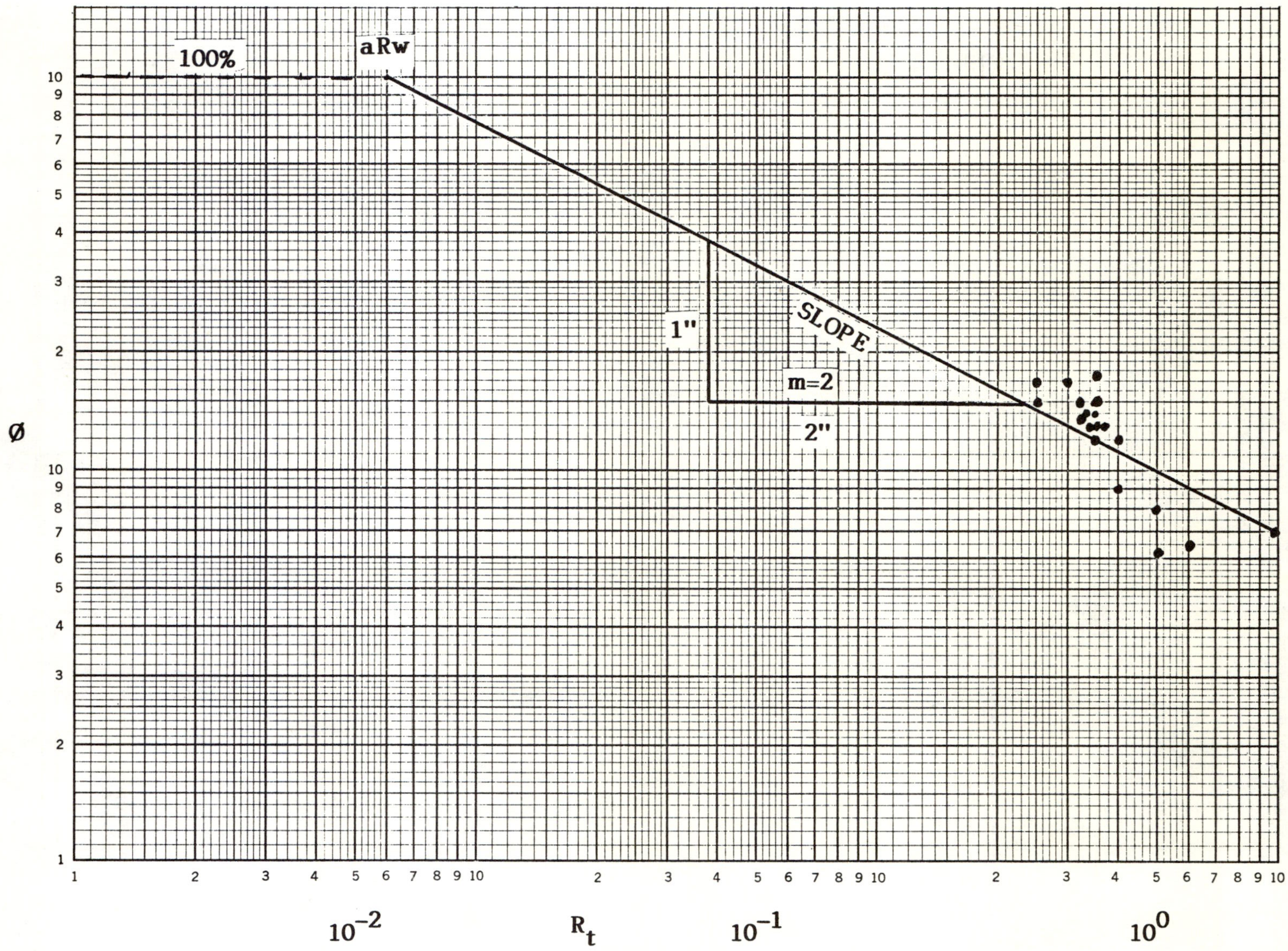

Figure 8-10. Log-log plot of porosity versus resistivity. The density matrix is too high. The correct matrix value will cause the points below the line to plot on or above the line. See Figure 8-11.

point, a corresponding R_w is read. This is 0.06 ohms at formation temperature. It agrees with the previous chart.

The slope of the line is the cementation exponent.

$$m = \frac{y}{x}$$

The easiest method of calculating the slope is to use an engineering ruler scaled in tenths (or use a metric ruler). Pick any point on the line and draw a 1-inch vertical line. This will give you a value of 1 for the variable x or a 1 in the denominator. Next draw a horizontal line at a right angle until it intersects the 100% S_w line. Read the distance of this line. The reading is the value of m. On Figure 8-10, x is drawn to 1-inch and y measures 2 inches.

$$m = \frac{2}{1} = 2$$

Our initial assumption that $F = \Phi^{1/2}$ was indeed correct.

The value of R_w is 0.06.

Why, then, does the SP indicate 0.05? Upon further investigation, the offset well shows an R_w from the SP to be 0.062 at 132°F. This value agrees with our findings. 150°F was used to compute R_w from SP. It is now obvious that the bottom hole temperature was estimated by the field engineer on this job, rather than measured. The correct temperature should have been 133°.

A value of 61 μs for Δt seems a bit out of place. The average value for a limestone is 48 μs. An unconsolidated sandstone is 59 μs. The high reading could be due either to miscalibration of the tool, or to shale in the zone of interest.

One method of cross-verification is to re-plot the data points on log-log paper, substituting porosity for:

$$\Delta t - \Delta t_{ma}$$

If the matrix value of the data is not valid, then it will not plot a straight line. Figure 8-11 illustrates the point.

The same technique can be applied to the density by plotting:

$$\rho_b - \rho_{ma}$$

Miscalibration

Miscalibration of the tools can be detected by the use of the previously mentioned plots. There are three errors that can occur.

1. Tools are not properly zeroed.
2. Tool sensitivity is in error.
3. A combination of 1 and 2.

The first error is easiest to detect. If the induction tool is improperly zeroed (sonde error), the plotted curve exhibits a marked curvature in the lower resistivity scale. If the porosity tool has a zero error, the curvature will occur in the higher resistivity scale.

Correction for this miscalibration is a simple matter of adding a correction constant to the resistivity values in order to make the plot linear. Graphically, it means to straighten the line out.

The second case presents the sensitivity error. This type of error is more difficult to detect. The entire curve is shifted on the resistivity scale by the logarithm of the error (logE). Since linearity is unaffected the shift may go undetected. The best possible way to detect such errors is by comparison of this technique with that of another.

The third case presents both errors. Since errors 1 and 2 do not conflict with each other, the error can be adjusted by Cases 1 and 2.

Conclusion

It has been demonstrated that the resistivity-porosity crossplot is a powerful tool for the analyst. Care must be taken in its construction, as error can cause a significant change. I propose to you that no one single plot be used, but rather a combination of two or more. Points describing a single line of 100% water saturation are often hard to pick. Perhaps the first or Hingle plot is the easiest to use. A combination of methods can often uncover mistakes, errors, and miscalibrations. In this chapter, it was discovered that the bottom hole temperature was incorrect.

When a crossplot fails to define an answer, keep in mind the assumptions and investigate the failure.

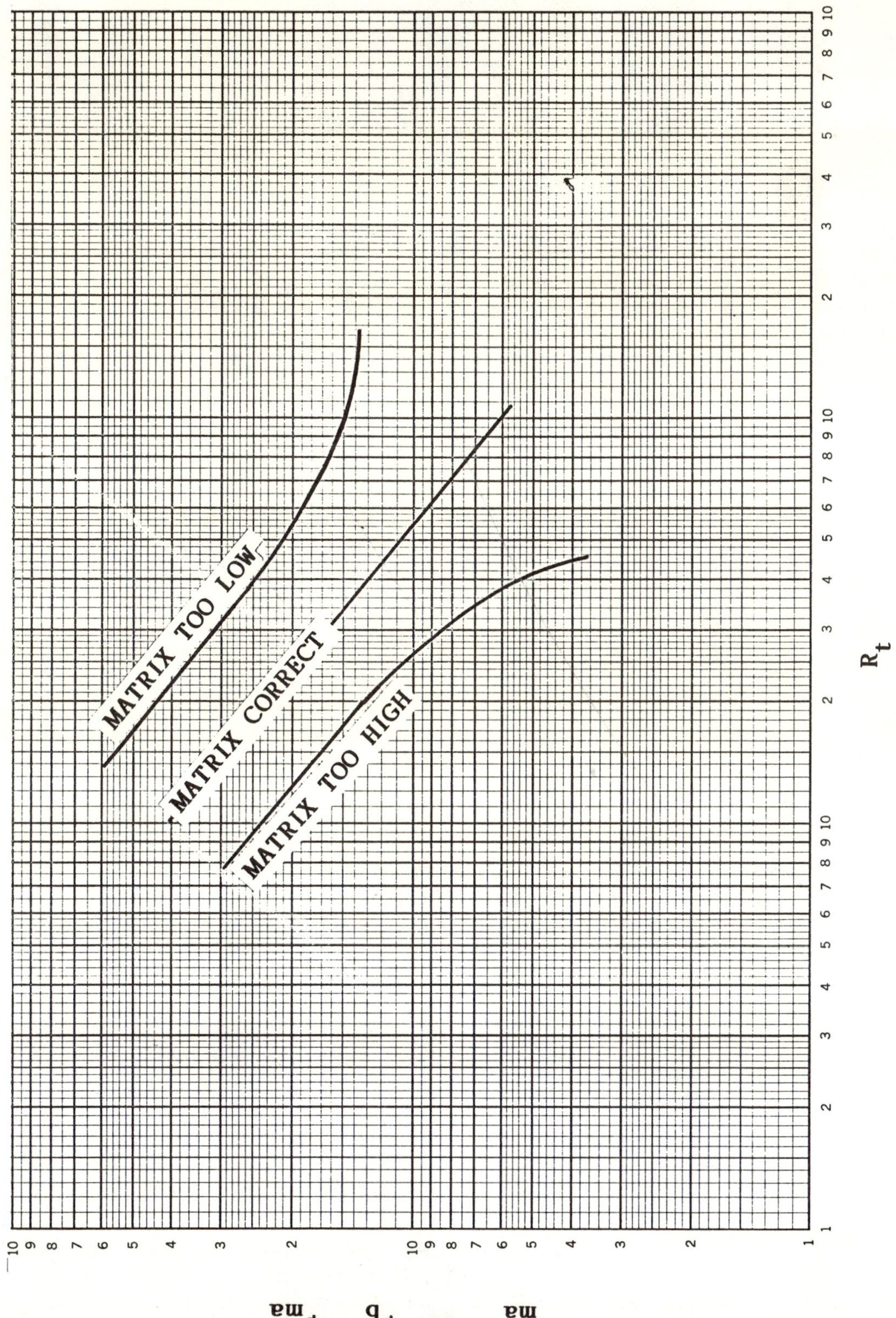

Figure 8-11. Illustration of incorrect matrix values.

9

Porosity Crossplots

Chapter 8 attempted to extract an R_w value as well as a matrix value from the resistivity porosity crossplot. It became evident that at times this technique is rather limited in that it is difficult to predict the 100% S_w line.

This chapter presents the two-porosity crossplot which will enable you to select the appropriate value of matrix to be used. With this added information, you can return to the resistivity vs. porosity crossplot and extract the R_w value with more confidence.

A unique and advanced method of finding these values for shaly sands will be discussed later in the text.

Porosity crossplots employ two porosity devices. Since there are three basic porosity tools, the following plots are available:

- Neutron-density
- Density-sonic
- Neutron-sonic

Each combination has its particular advantage and disadvantage. Of the three, perhaps the neutron-density crossplot is the most popular one in use.

These crossplots are relatively easy to use. One assumption is made when using these plots, that is, that the lithology of the formation is either pure or a binary mixture. The formation must be a pure dolomite, limestone, or sandstone. It can also be a mixture of two components.

Porosity, for the most part, can be used as apparent from the log. If gas or shale is present, then it must be corrected before the plot is made.

Construction of the plot is made by each wireline company based on the response of the particular tool used. The analyst must restrict the analysis of the chart which was made for that particular tool. One cannot crossplot a *compensated neutron log* with a *sidewall neutron chart.* Crossplotting a Schlumberger tool on a Dresser chart is also invalid.

Neutron-Density Crossplots

The charts in Figures 9-1 and 9-2 show the typical form of the two-porosity crossplot. The bottom horizontal scale represents the neutron porosity calculated in a limestone. The left-hand vertical scale is the bulk density from the compensated density curve. The right hand vertical scale is the density porosity scale calculated in a limestone lithology. The only differ-

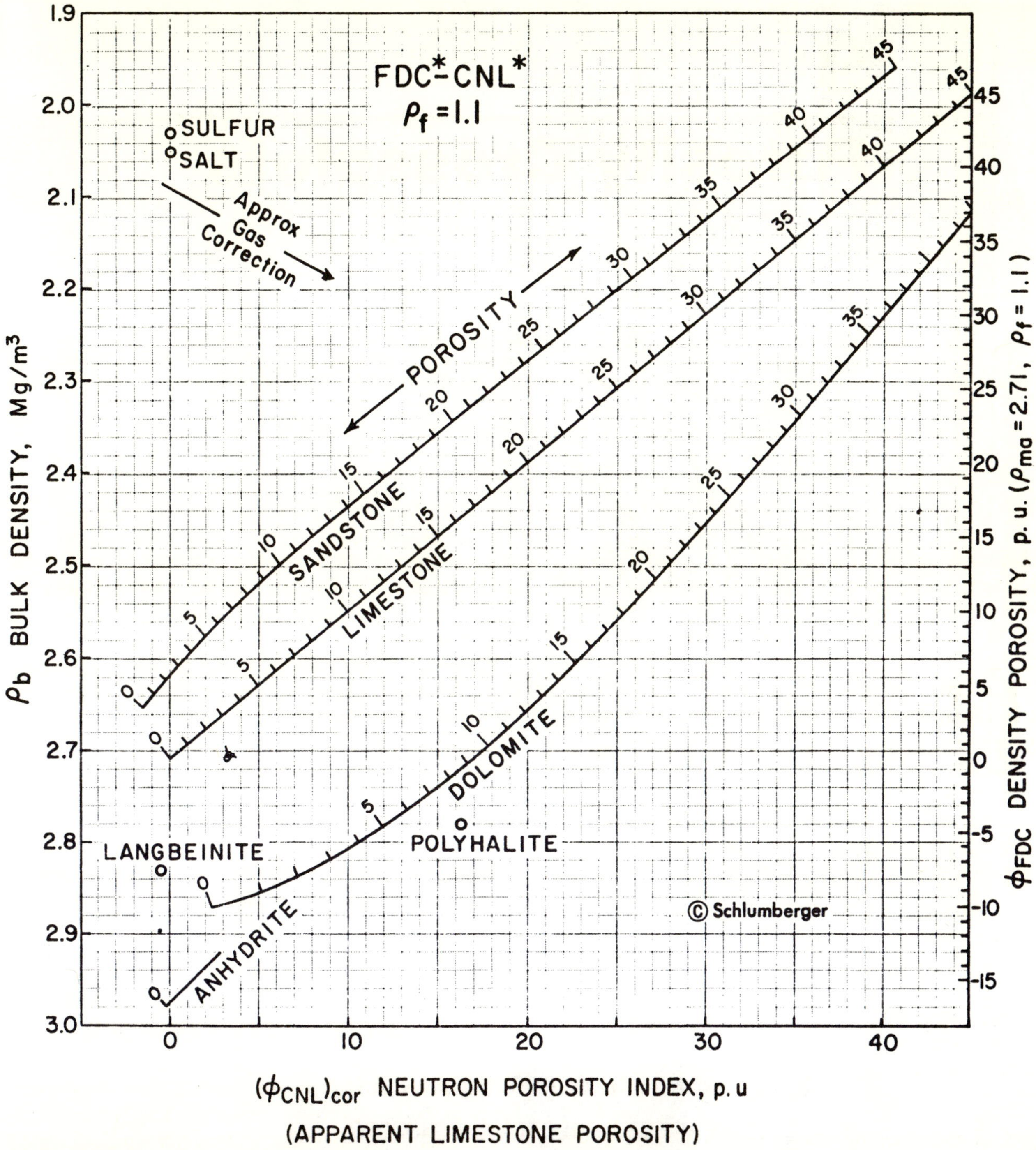

Figure 9-1. Porosity and lithology determination from formation density log and compensated neutron log. (Courtesy of Schlumberger.)

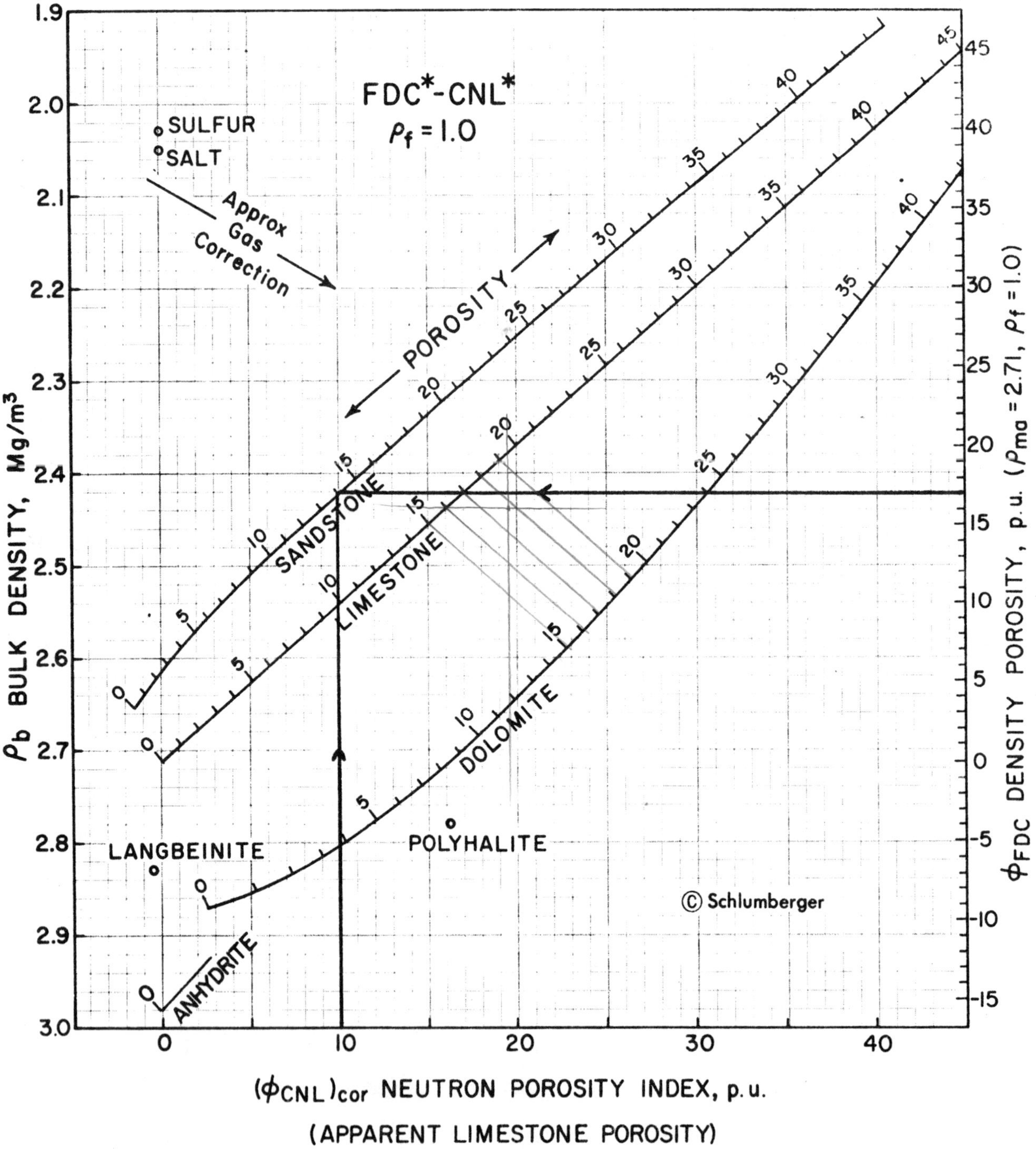

Figure 9-2. Crossplot of a pure sandstone. (Courtesy of Schlumberger.)

ence in these two charts is ρ_{fluid}. Figure 9-1 represents a fluid density of 1.1, while Figure 9-2 is 1.0.

Since both the neutron and density tools have a shallow depth of investigation, the value chosen will depend mostly on the invading fluid or mud filtrate.

If a salty mud is used, the chart of Figure 9-1 is chosen, with a fluid density of 1.1, or else use the chart for a fresh mud with a density of 1.0.

A data point is chosen on the log in the zone of interest. The neutron-density is read at the point and plotted on the chart. The following points are plotted on Figures 9-2, 9-3, and 9-4.

Figure	Φ_n	Φ_d
9-2	10%	17%
9-3	10%	10%
9-4	10.2%	−5%

Figure 9-2 shows the lithology to be a pure sandstone. The crossplot porosity for this lithology is 14%.

Figure 9-3 shows a pure limestone lithology with a crossplot porosity of 10%. This is so because the chart is constructed using limestone porosity.

This brings up a good point. When a chart scale calls for a limestone porosity, then one must use lime. If the log is recorded on a sandstone or dolomite lithology, it must first be converted to lime before entering the charts. This is usually the case because most tools are calibrated in a standard limestone at the API test pit at the University of Houston.

Back to our example, a 100% water wet lime will cause both the neutron-density readings to be the same (see page 96). In this event, the crossplot porosity is also the same. Note that the lithology line on the chart is linear for lime.

Figure 9-4 shows a dolomite lithology with a 4% crossplot porosity.

Once the crossplot porosity has been established, the density matrix value can be checked for validity. In the case of the sandstone example, the density porosity read 17%. Calculating bulk density:

$$\rho_b = \Phi\rho_f + (1 - \Phi)\rho_{ma}$$
$$\rho_b = .17 \times 1 + (1 - .17)2.71 = 2.42 \text{ (lm)}$$

Now, calculating porosity using a sandstone matrix:

$$\Phi = \frac{\rho_{ma} - \rho_b}{\rho_{ma} - \rho_f}$$

$$\Phi = \frac{2.65 - 2.42}{2.65 - 1.0} = 13.9\%$$

which certainly agrees with the crossplot porosity of 14%. As long as the lithology remains constant, this matrix value can be used and density porosity can be used without crossplots.

The effect of oil will be slight, but that of gas requires more complex analysis and will be deferred for now.

Figure 9-5 has the data point 20% neutron and 14.9% density plotted. This case is that of a binary mixture of dolomite and lime.

The crossplot porosity is 17.3%. Lines are drawn (dashed) from one porosity scale to the other, as shown in the figure, which establish the crossplot porosity for the pair.

The lithology is either a limey-dolomite or a dolomitic-lime. Crossplots do not tell us which there is most of. In order to find out, we measure the distance between the lime line and the data point, and the distance between the lime and dolomite. A good engineering ruler scaled in tenths of an inch is most helpful in this task.

The lime-to-data point distance of Figure 9-5 is .4 in. The lime-to-dolomite distance is 1.15 in. These measurements are conducted at the data point parallel to the porosity lines.

Since we have no knowledge of which lithology there is more of, we calculate volumes of each of the pair as vol. 1 and vol. 2.

Case 1

$$\text{vol. } 1 = \frac{0.4}{1.15} = 0.349 \text{ or } 34.9\%$$

and

$$\text{vol. } 2 = 1 - .349 = 0.651 \text{ or } 65.1\%$$

The matrix value is then expressed as:

$$\rho_{ma} = \text{vol.}1 \times \rho_{ma}1 + \text{vol. } 2 \times \rho_{ma}2$$

Letting vol. 1 be dolomite, and vol. 2 be lime, we solve for ρ_{ma}:

Matrix density for dolomite = 2.87

Matrix density for lime = 2.71

(text continued on page 123)

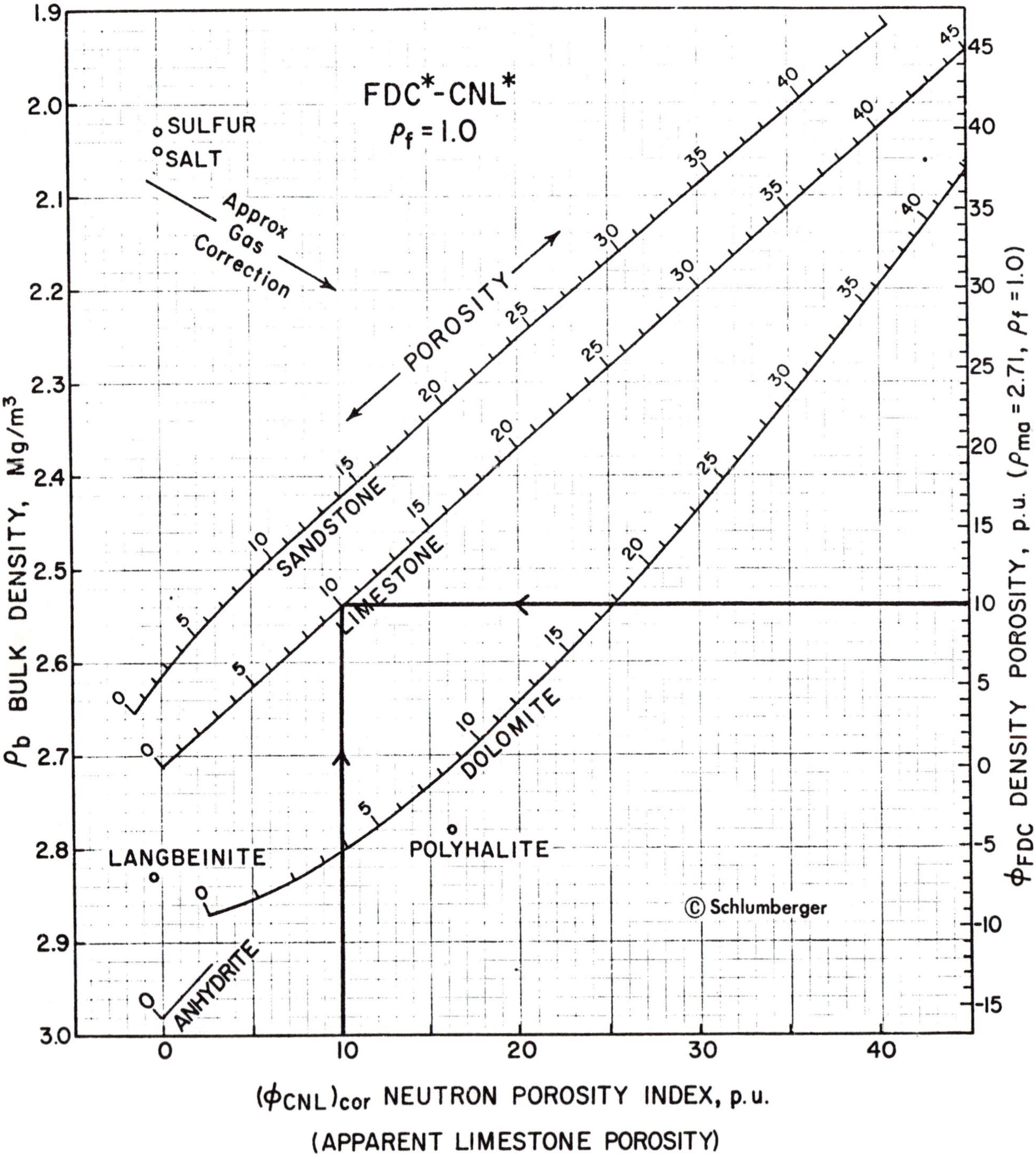

Figure 9-3. Crossplot of a pure limestone. (Courtesy of Schlumberger.)

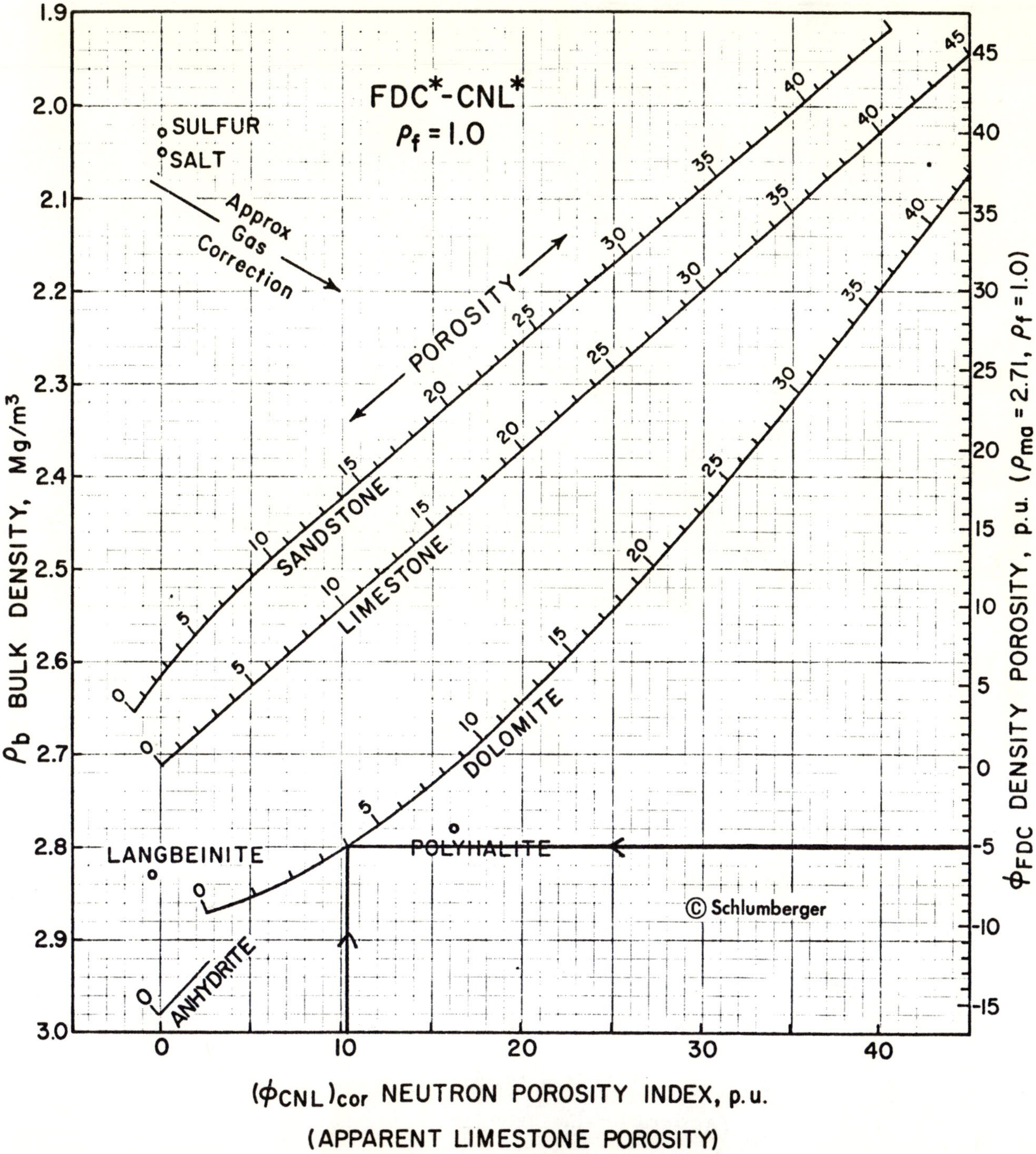

Figure 9-4. Crossplot of a pure dolomite. (Courtesy of Schlumberger.)

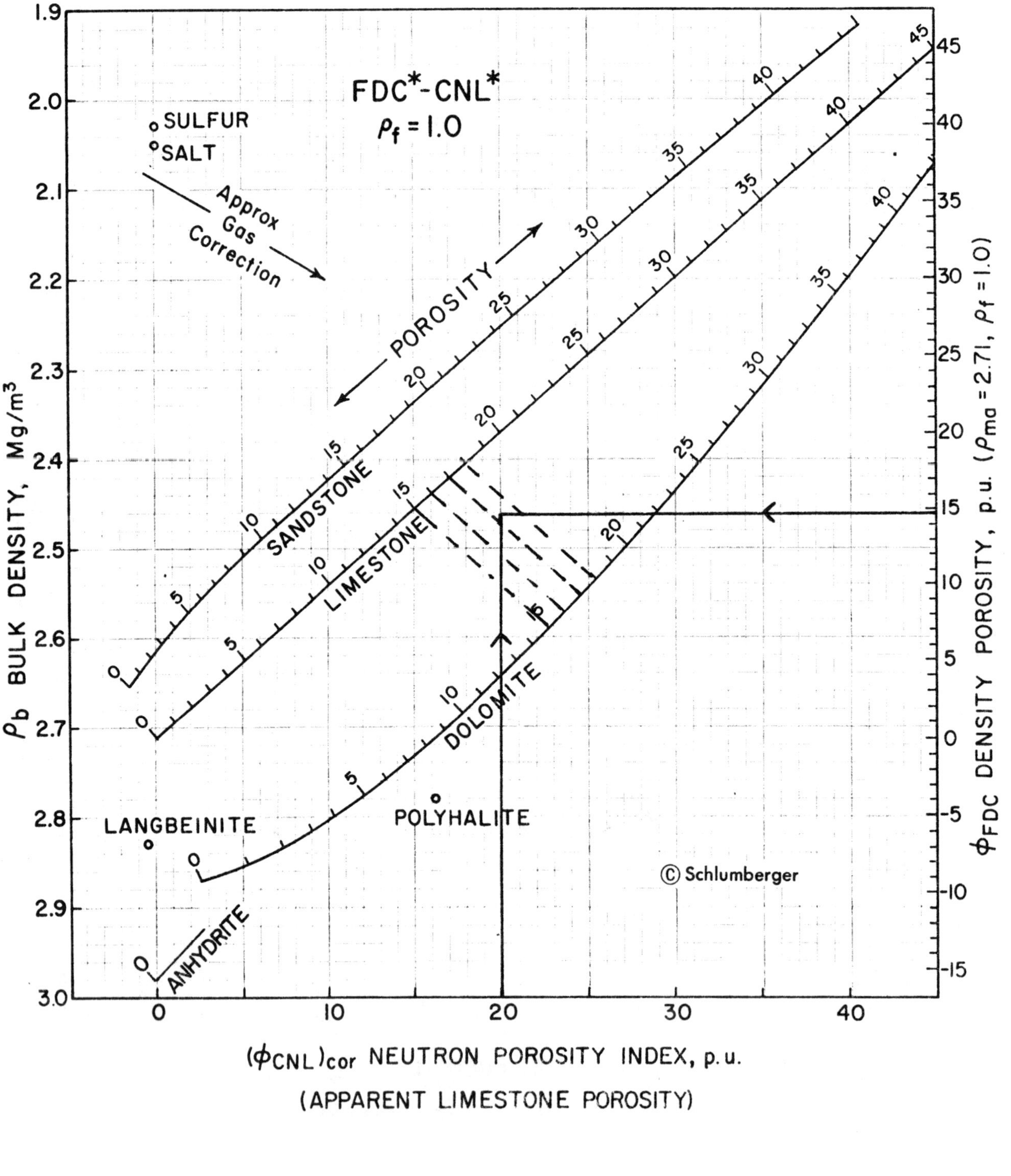

Figure 9-5. Crossplot of a mixture of limestone and dolomite. (Courtesy of Schlumberger.)

$$\rho_{ma} = 0.349 \times 2.87 + 0.651 \times 2.71$$

$$\rho_{ma} = 2.766$$

To confirm the validity, we compute the porosity using the new matrix value. The bulk density for 14.9% density porosity in a lime is 2.455 g/cc.

$$\Phi = \frac{2.766 - 2.455}{2.766 - 1.0} = 0.176 \text{ or } 17.6\%$$

Case 2

Now we investigate the case where vol. 1 is lime and vol. 2 is dolomite:

$$\rho_{ma} = 0.349 \times 2.71 + 0.651 \times 2.87$$

$$\rho_{ma} = 2.814$$

Porosity is:

$$\Phi = \frac{2.814 - 2.455}{2.814 - 1.0} = 0.198 \text{ or } 19.8\%$$

The crossplot porosity is 17.3; therefore, Case 1 is the correct case, where the matrix value is 2.766, because it closely agrees with the crossplot. The lithology is mostly dolomite with 34.9% lime.

Plotting another point:

$$\rho_b = 2.28 \ (25.2\% \ \Phi)$$

$$\Phi_n = 12.0\% \text{ lime}$$

We see that the point (Figure 9-6) plots above the sandstone line. The indication here is gas. No real crossplot is possible until both porosities are corrected for gas. If desired, an approximate porosity can be used as illustrated in Figure 9-6.

Porosity lines are extended from the limestone line to the sandstone line, through the same intersections of porosity and beyond to the data point. An approximate value for this data point is 20% (dashed line).

In gas-saturated zones, a gas-corrected porosity may be calculated using a root mean square equation.

$$\Phi_e = \sqrt{\frac{\Phi_n^2 + \Phi_d^2}{2}}$$

For this example:

$$\Phi_e = \sqrt{\frac{0.12^2 + 0.252^2}{2}} = 0.197 \text{ or } 19.7\%$$

indeed, very close to the approximate porosity of Figure 8-6. 9-6.

Thus far, all of the apparent porosities that have been used were in limestone porosity. If the log had been recorded in a sandstone matrix, it would first have to be converted to limestone.

For the density we can avoid such conversions by using bulk density instead of porosity. Otherwise, convert porosity to bulk density using the equation previously found in this chapter, using a sandstone matrix. Then, if you wish, convert it to limestone porosity.

The use of equations for the neutron porosity is more involved and it is simpler to use a chart. Figure 9-7 shows a limestone porosity converted to sandstone. If you reverse the process, then a sandstone matrix porosity can be converted to a limestone matrix porosity. The same procedure holds true for logs recorded in a dolomite matrix.

If one intends to crossplot extensively, ask the field engineer to hand you a log calculated in a lime matrix. With today's computers, the task is a simple one, and may save you a great deal of time.

At the beginning of this chapter, it was stated that all porosity values should be corrected for shale before they are entered into the crossplot. While this remains true for a binary mixture, such as sand-lime, or a lime-dolomite, it does not remain so if the binary mixture is sand-shale or lime-shale.

A neutron-density crossplot for shale-corrected porosity from these charts is not highly recommended. However, there is valuable information contained in such a plot.

Suppose that in a nearby bedded shale, we read a bulk density of 2.6 and a limestone neutron porosity of 40%. If these porosities were plotted, the data point would plot on the southeast quadrant of the chart, as seen in Figure 9-8. This point is then termed the shale point.

Plotting a binary mixture of shale and lime, or sand, at several points in the zone will yield a tendency for the points to follow a direction. The direction is not necessarily a straight line, but some direction can be seen. The following interpretation can be made.

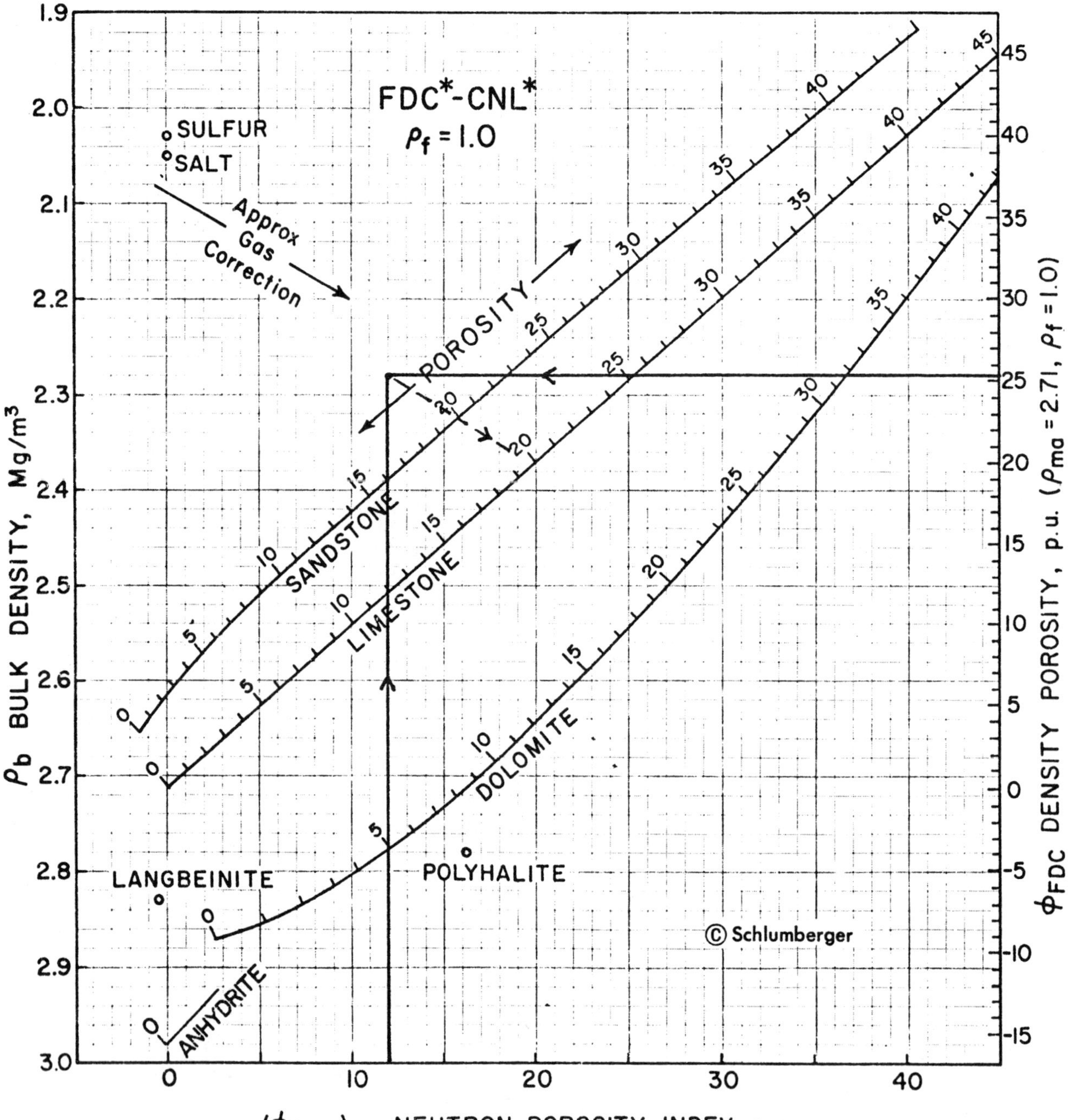

Figure 9-6. Crossplot illustrating the effects of gas. (Courtesy of Schlumberger.)

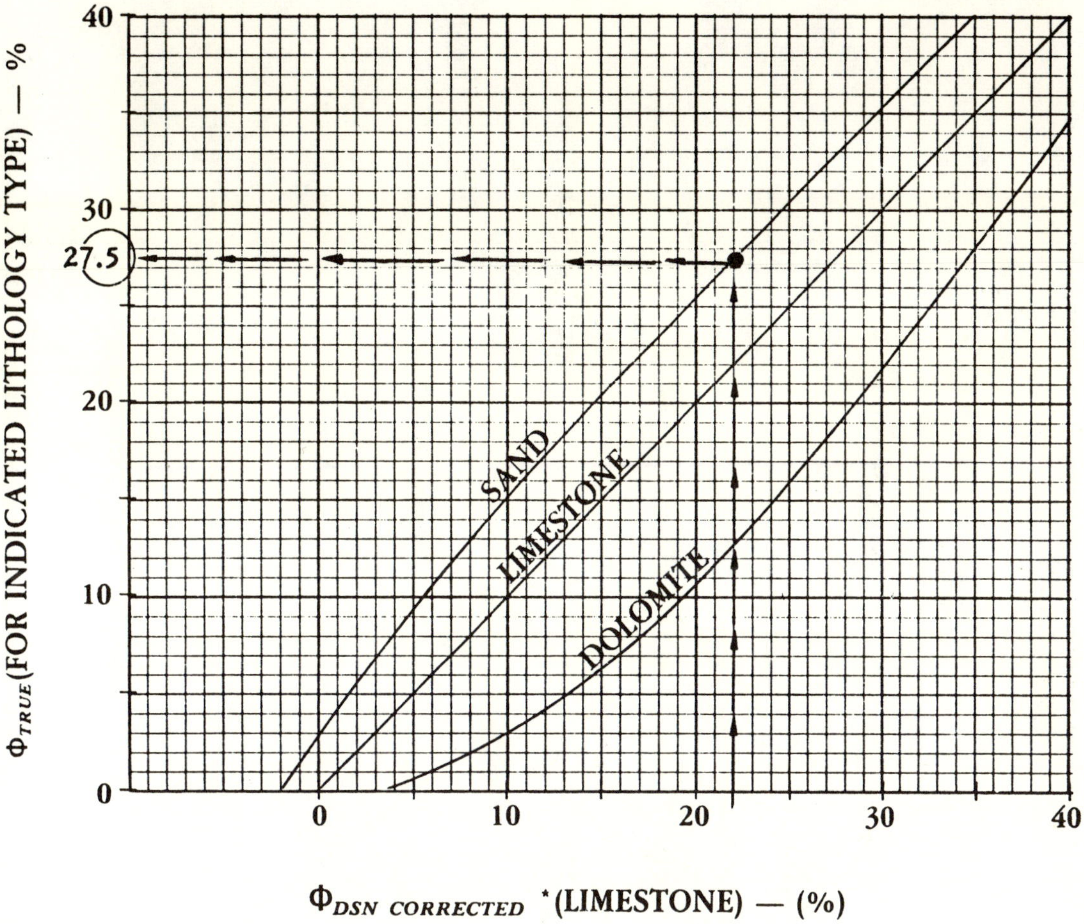

Figure 9-7. Lithology conversion for neutron porosity. (Courtesy of Welex.)

1. If the points follow the direction of line B (Figure 9-8) towards the shale point, then the shale is laminated.
2. If the points follow the direction of line A, then the shale is structural.
3. If the points follow the direction of line C, then the shale is dispersed.

The information provided by this plot is of great importance because structural shale has no effect on effective porosity, while dispersed shale cuts off the effective porosity drastically.

Sonic-Neutron Porosity Crossplots

Travel time and neutron porosity can be crossplotted in the same fashion as the density-neutron. Figure 9-9 shows a point plotted.

$$\Delta t = 67\ \mu$$

$$\Phi_n = 19\%$$

It is midway between the limestone and dolomite lithology lines. The crossplot porosity is 15%. The sonic matrix value is calculated in similar manner as the density matrix.

$$\Delta t_{ma} = \text{vol. lime } \Delta t_{ma} \text{ lime} + \text{vol. dol } \Delta t_{ma} \text{ dol}$$

The average value of lime matrix is 47.6 μ, and the average for dolomite is 43.5 μ. The fractional component is 50%.

The new matrix value is calculated:

$$\Delta t_{ma} = 47.6 \times 0.5 + 43.5 \times 0.5$$

(text continued on page 128)

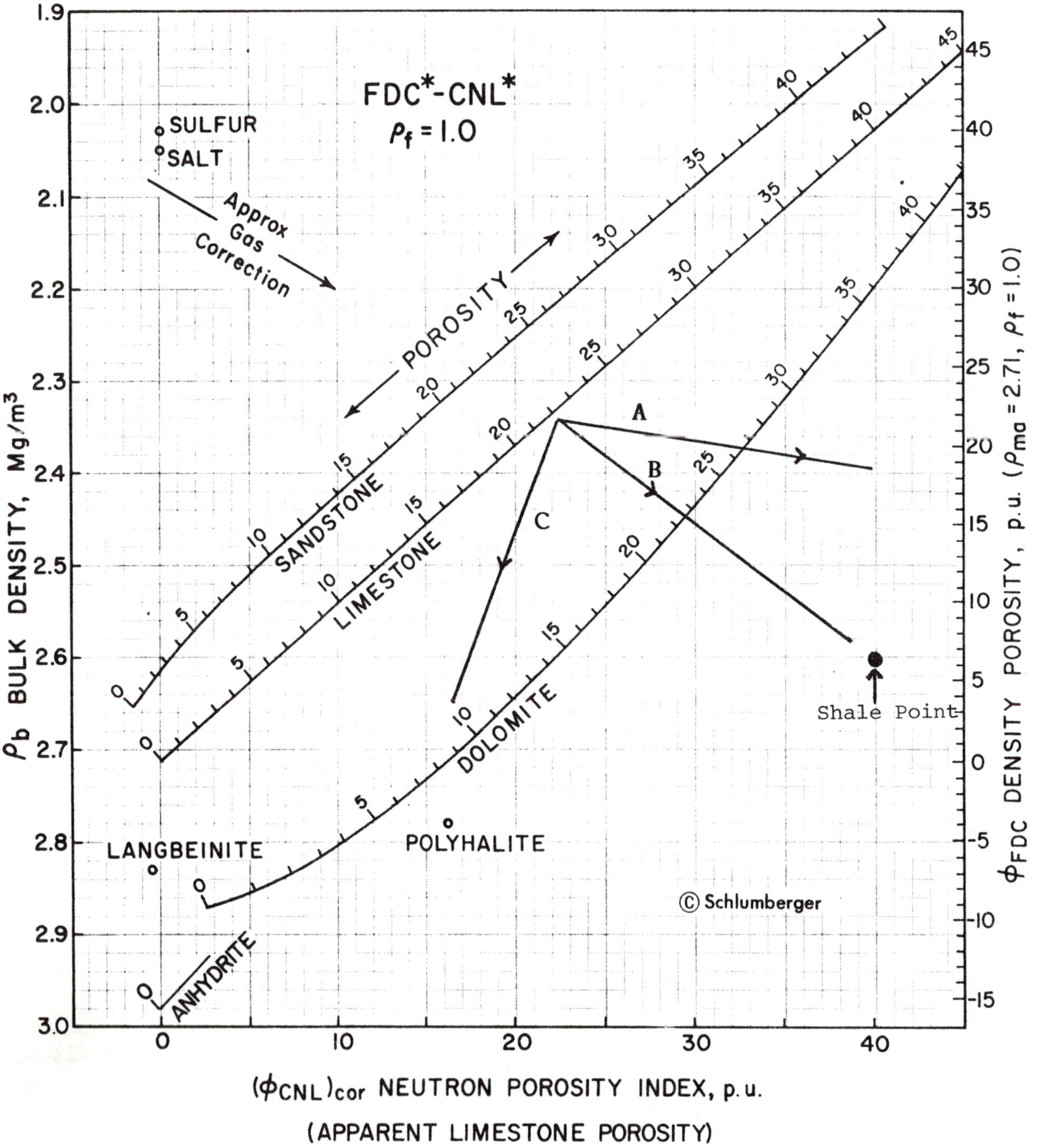

Figure 9-8. Crossplot illustrating the plots of three types of shale. (Courtesy of Schlumberger.)

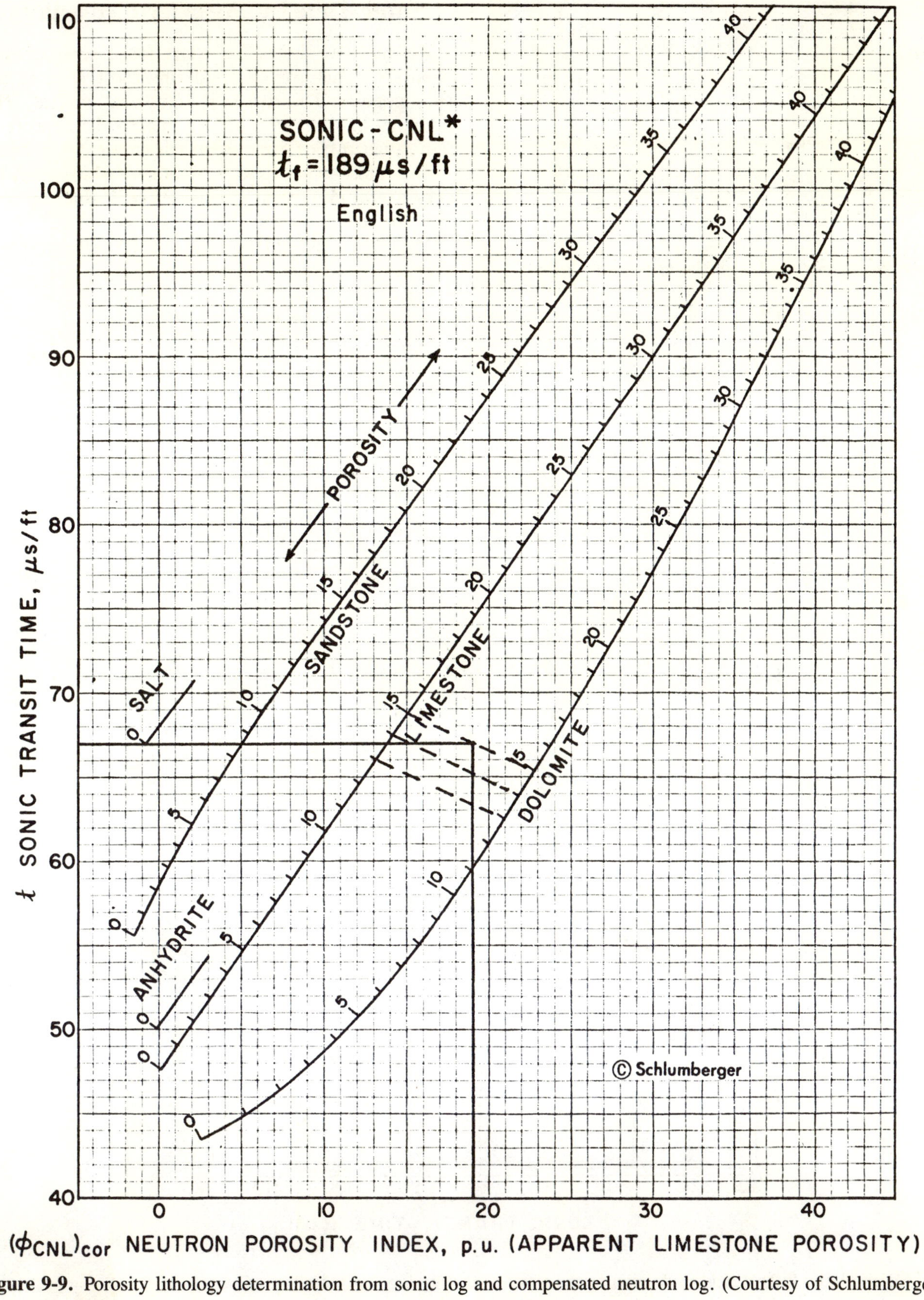

Figure 9-9. Porosity lithology determination from sonic log and compensated neutron log. (Courtesy of Schlumberger.)

$\Delta t_{ma} = 45.5$

Porosity is:

$$\Phi = \frac{\Delta t - \Delta t_{ma}}{\Delta t_f - \Delta t_{ma}}$$

$$\Phi = \frac{67 - 45.5}{189 - 45.5} = 14.9\%$$

The value agrees well with the crossplot porosity.

Sonic-Density Porosity Crossplots

Figure 9-10 shows the chart for a sonic-density crossplot. This crossplot, for the most part, is not very popular. The reason is somewhat evident. The lithology lines are bunched up, making its interpretation rather difficult and often inaccurate.

This plot, when used in another fashion, can be very useful in mineral identification. It shall be covered in a later chapter.

The chart is used in the same fashion as the previous charts.

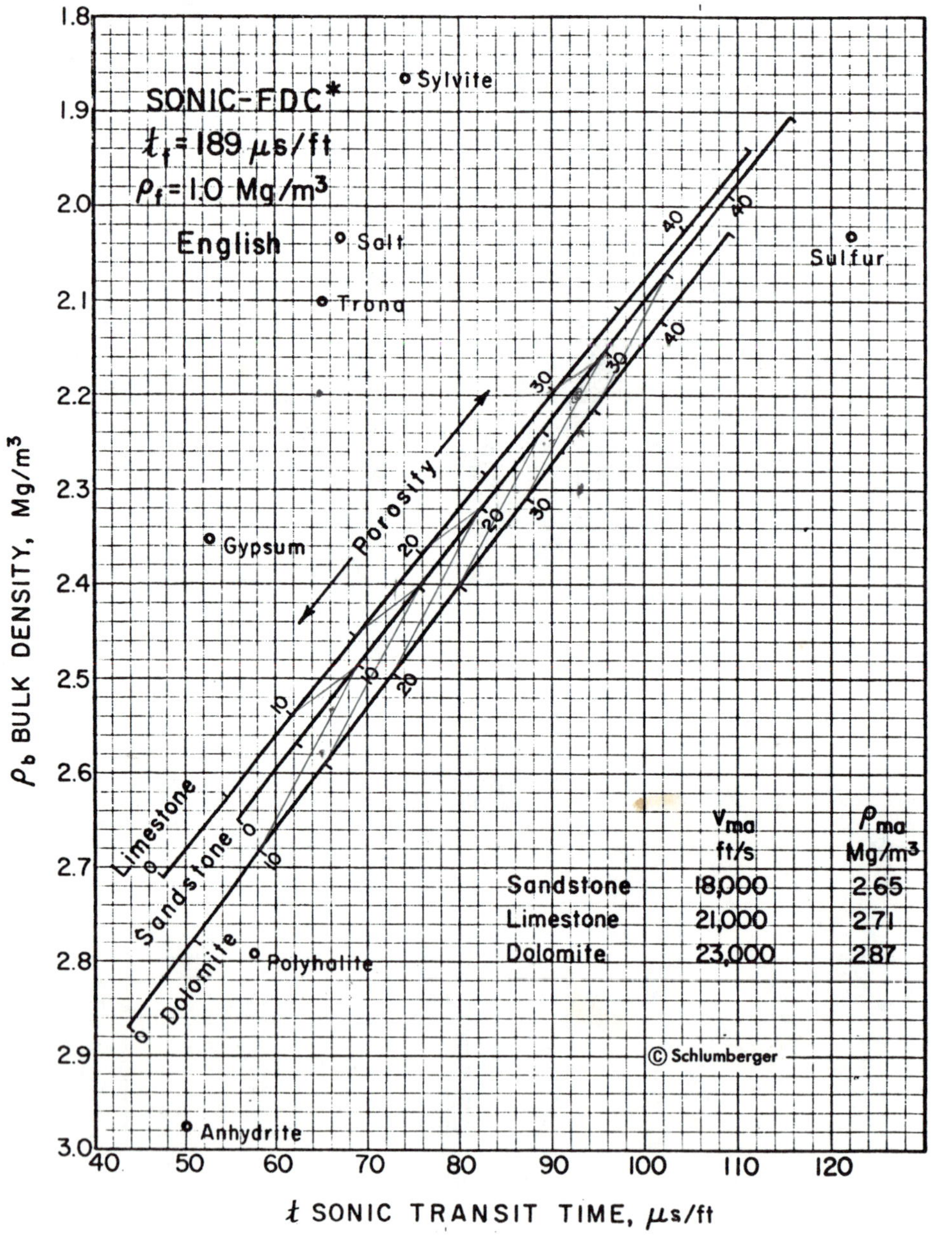

Figure 9-10. Lithology identification from formation density log and sonic log. (Courtesy of Schlumberger.)

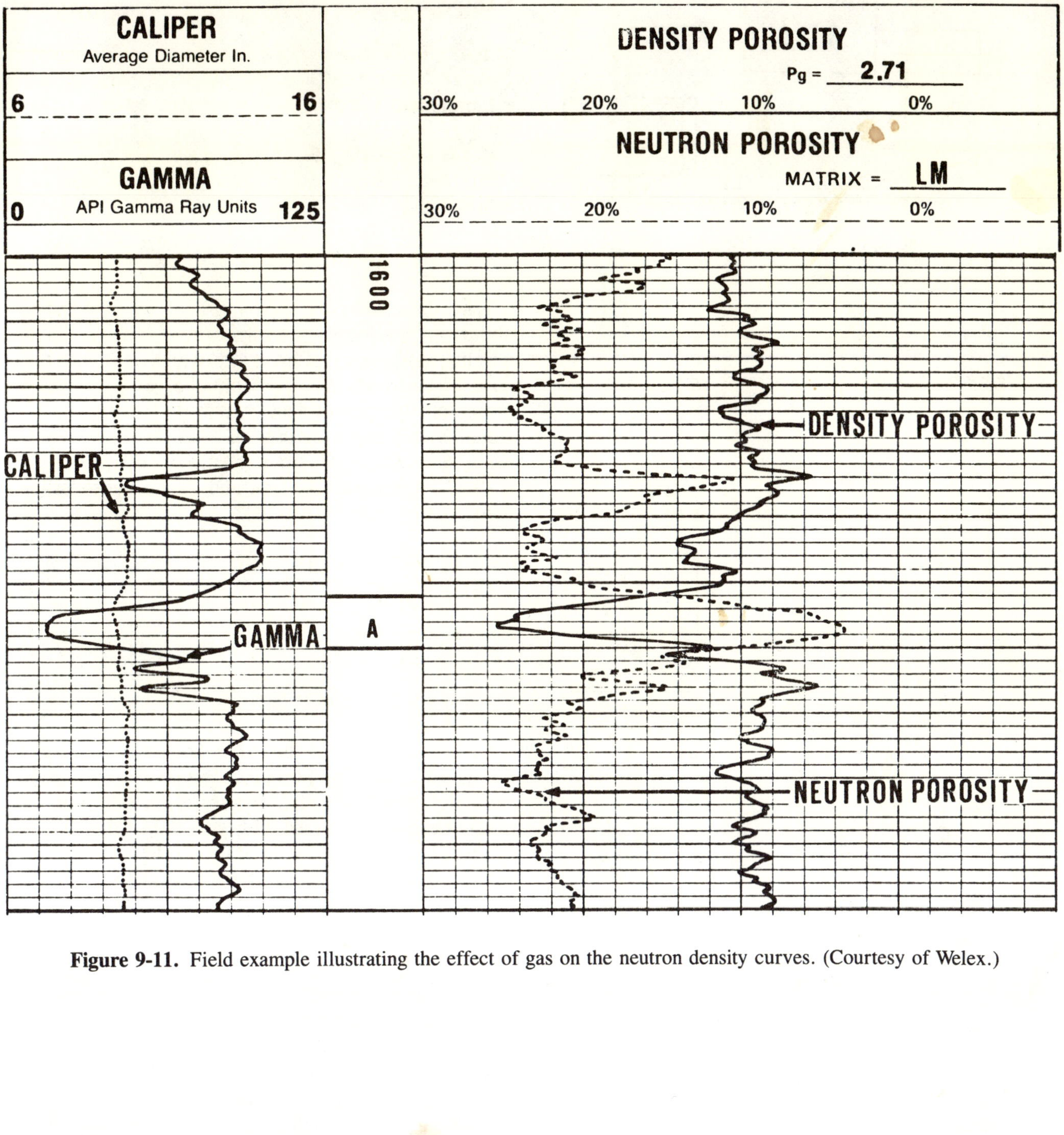

Figure 9-11. Field example illustrating the effect of gas on the neutron density curves. (Courtesy of Welex.)

Gas Plots

Figure 9-11 shows the density-neutron curves crossing (football effect), indicating the presence of gas in zone A.

A crossplot of this zone would yield a similar plot as that of Figure 9-6. A chart correction for gas is at best a guesstimate. Correcting by the use of a mean square can yield a reasonable value of porosity, but it does not account for matrix in the case of a binary mixture. These two points are presented in flow chart form in Figure 9-12.

Figure 9-13 presents the flow diagram of a more complex method for total correction.

The first step is to determine if there is any shale using the gamma ray and SP as indicators. If there is, we correct accordingly.

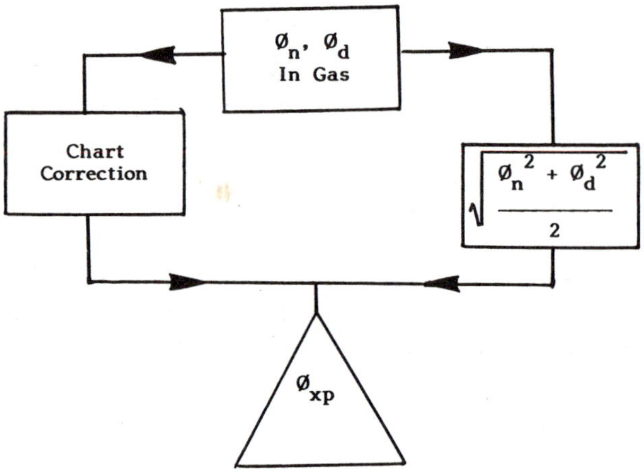

Figure 9-12. Flow chart illustrating gas correction.

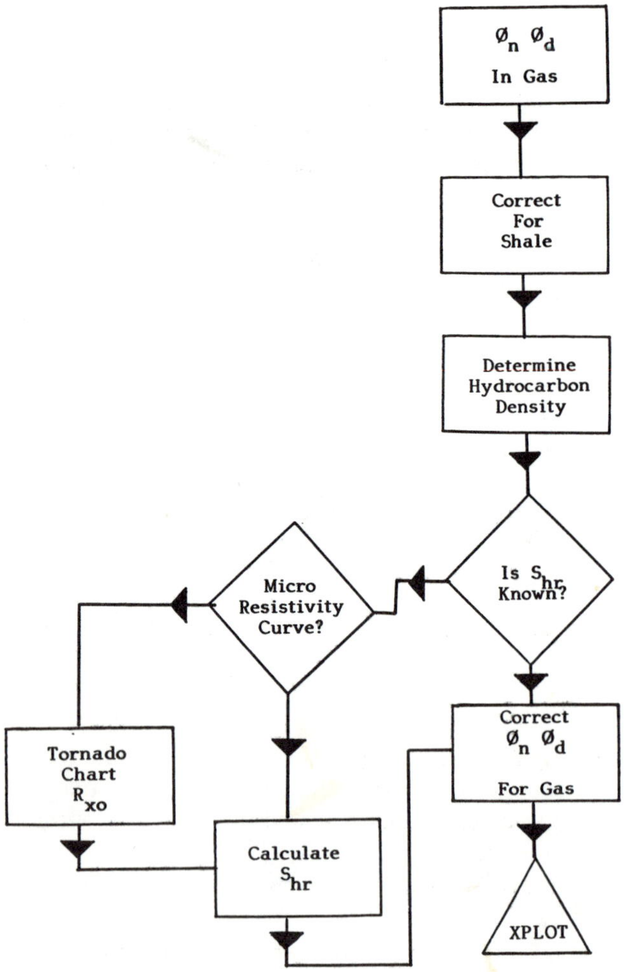

Figure 9-13. Flow chart illustrating corrections before crossplotting.

- Dispersed

$$\Phi_e = \Phi_a - V_{sh}$$

- Laminated

$$\Phi_e = \Phi_a - (V_{sh}\ \Phi_{sh})$$

- Structural

$$\Phi_e = \Phi_a$$

With the correction accomplished, we proceed to obtain an estimate of hydrocarbon density. Some wireline companies offer charts which facilitate this task. Figure 9-14 shows a Schlumberger chart which crossplots the ratio of neutron-density porosity vs. residual hydrocarbon saturation (S_{hr}). If this chart is used, then the flow diagram must be modified slightly so that we compute S_{hr} first and obtain hydrocarbon density second.

$$\rho_g = \frac{\log (D/1{,}000)}{5.473 - 0.0278\ (D/1{,}000)}$$

(Figure 9-15)

where D is the depth in feet

Once the density of the gas is known, we can plot the neutron-density porosities on a gas saturation crossplot, like the one in Figure 9-16. One note of caution is due. The chart on Figure 9-16 calls the saturation value the saturation of the flushed zone, when in fact the figure obtained is residual gas saturation.

$$S_{hr} = 1 - S_{xo}$$

Residual hydrocarbon saturation is equal to one minus the flushed zone saturation (S_{xo}).

To prove the point, we plot 30% density and 16% neutron porosities on Figure 9-16. The gas saturation is 40% (solid line). The same porosities plotted on Figure 9-17 show S_{xo} to be just a bit higher than 70%.

The Welex chart of Figure 9-16 gives us the value of S_{hr}, whereas the Schlumberger chart of Figure 9-17 is S_{xo}. To obtain S_{hr} from the Schlumberger chart, we solve:

$$S_{hr} = 1 - S_{xo}$$

The point is not at all clear on the Welex chart and shows poor judgment on the use of nonstandard la-

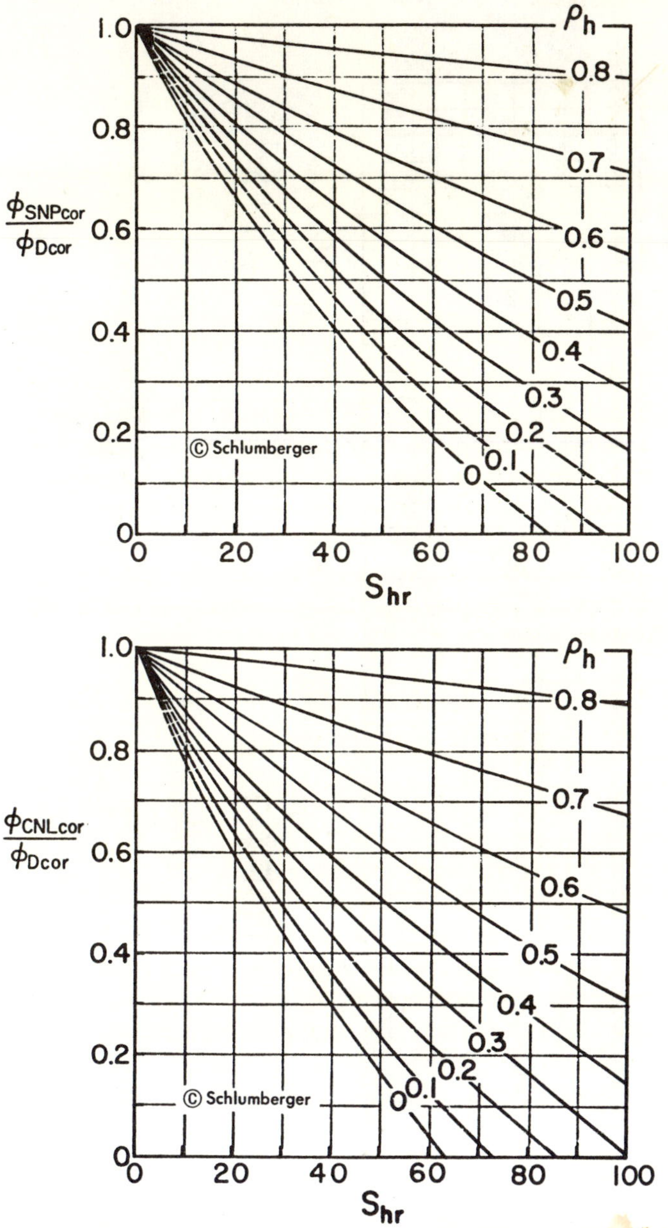

Figure 9-14. Hydrocarbon density charts. (Courtesy of Schlumberger.)

bels, which is a good source for confusion and miscalculation. Other than syntax, the chart is valid.

The disparity between the two charts also points out that one company's tools cannot be mixed with another's charts. If this were not the case, one should read 40%, while the other 60%.

On the Welex chart, the density of the gas is 0.1 for the solid lines and 0.2 for the dashed. On the Schlumberger chart, the circled numbers represent a gas density of 0.1.

Both charts crossplot the gas-corrected porosity, which again, can be computed using a mean average.

Thus far the discussion has presented the easiest route because it provides a value of residual hydrocarbon saturation. If the lithology is monolithic or pure, then we can take the crossplot porosity indicated and continue with a conventional analysis. If a binary mixture exists, then we need to correct for lithology as well as for gas.

If S_{hr} is needed (i.e., to use the charts of Figure 9-14 or simply to compute its value and bypass the gas saturation crossplots), the following equation may be used.

$$S_{hr} = 1 - \sqrt{\frac{F \times R_{mf}}{R_{xo}}}$$

where R_{mf} is at formation temperature and R_{xo} is the resistivity of the flushed zone obtained from one of the following tools:

- Microlaterolog
- Proximity
- F_oR_{xo}

or other R_{xo} resistivity tools. If an R_{xo} tool is not available, as is frequently the case, then it must be obtained from a tornado chart, using the dual-induction log.

For a review on the use of the tornado chart, see page 77.

Having obtained the values for S_{hr} and ρ_g, we correct the two porosities for gas.

Density Correction

$$\Phi_{dcorr} = \Phi_d - \Delta\Phi_d$$

$$\text{where } \Delta\Phi_d = \frac{1.07\ \Phi\ S_{hr}[1.11(1 - \rho_g) - 0.03]}{\rho_{ma} - 1}$$

and matrix is 2.71, since the plots are constructed for lime.

Neutron Correction

$$\Phi_{ncorr} = \Phi_n - \Delta\Phi_n$$

$$\text{where } \Delta\Phi_n = \Phi_n S_{hr}(2.2\ \rho_g - 1)$$

And the last step is to crossplot on a conventional density-neutron crossplot chart such as the one in Figure 9-2, for lithology.

(text continued on page 135)

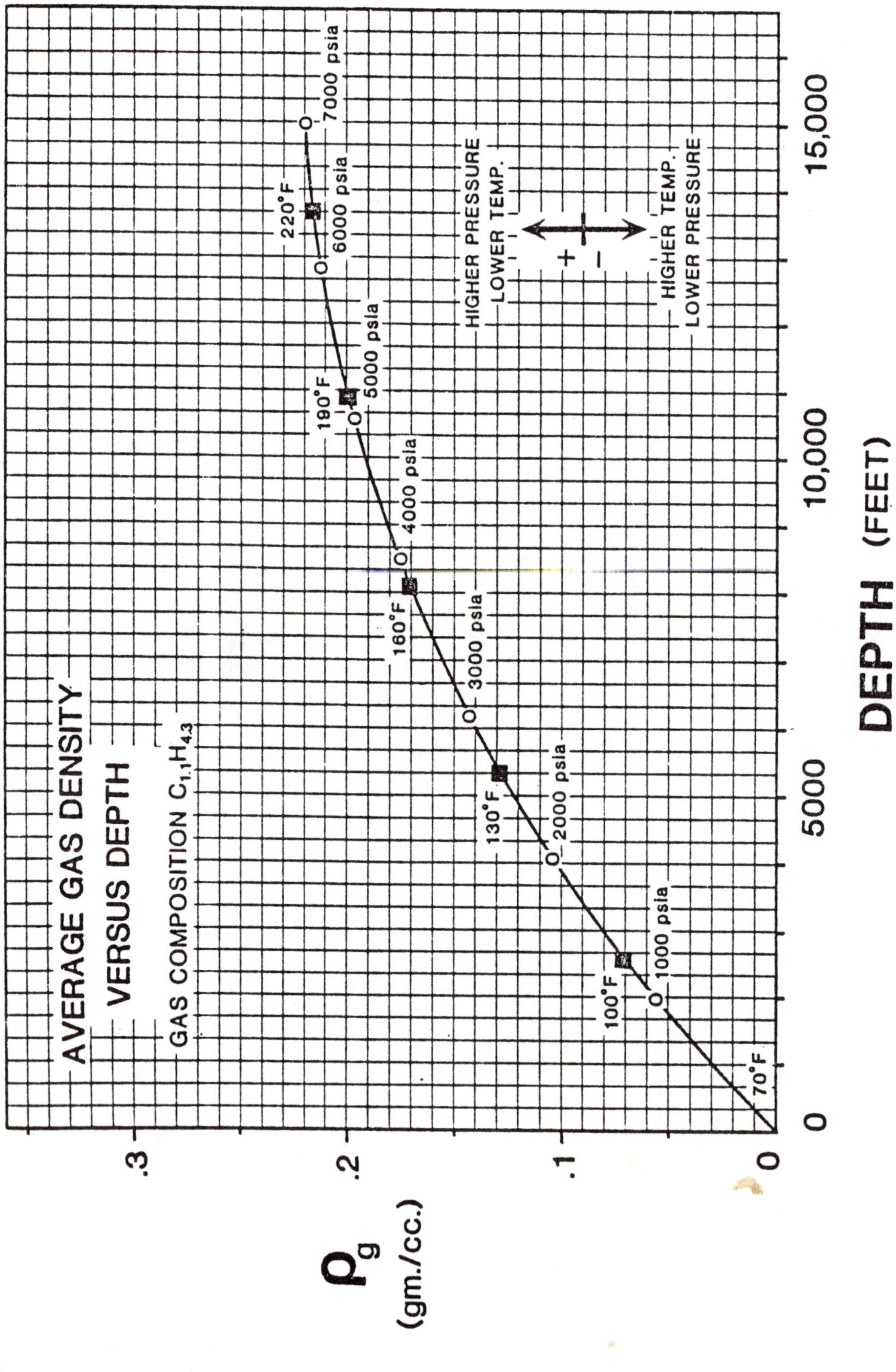

Figure 9-15. Estimation of gas density as a function of depth. (Courtesy of Welex.)

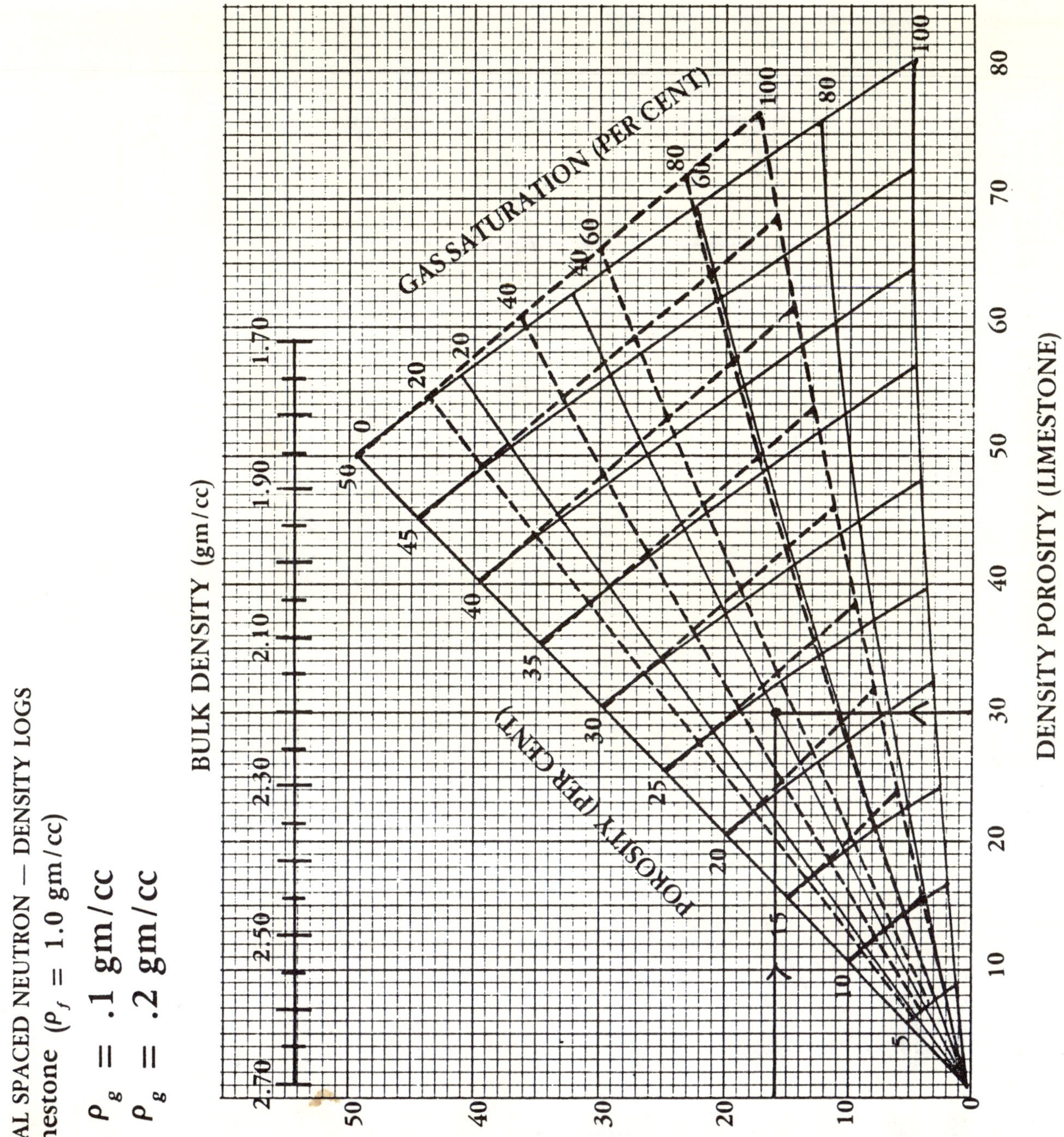

Figure 9-16. Porosity and gas saturation crossplot. (Courtesy of Welex.)

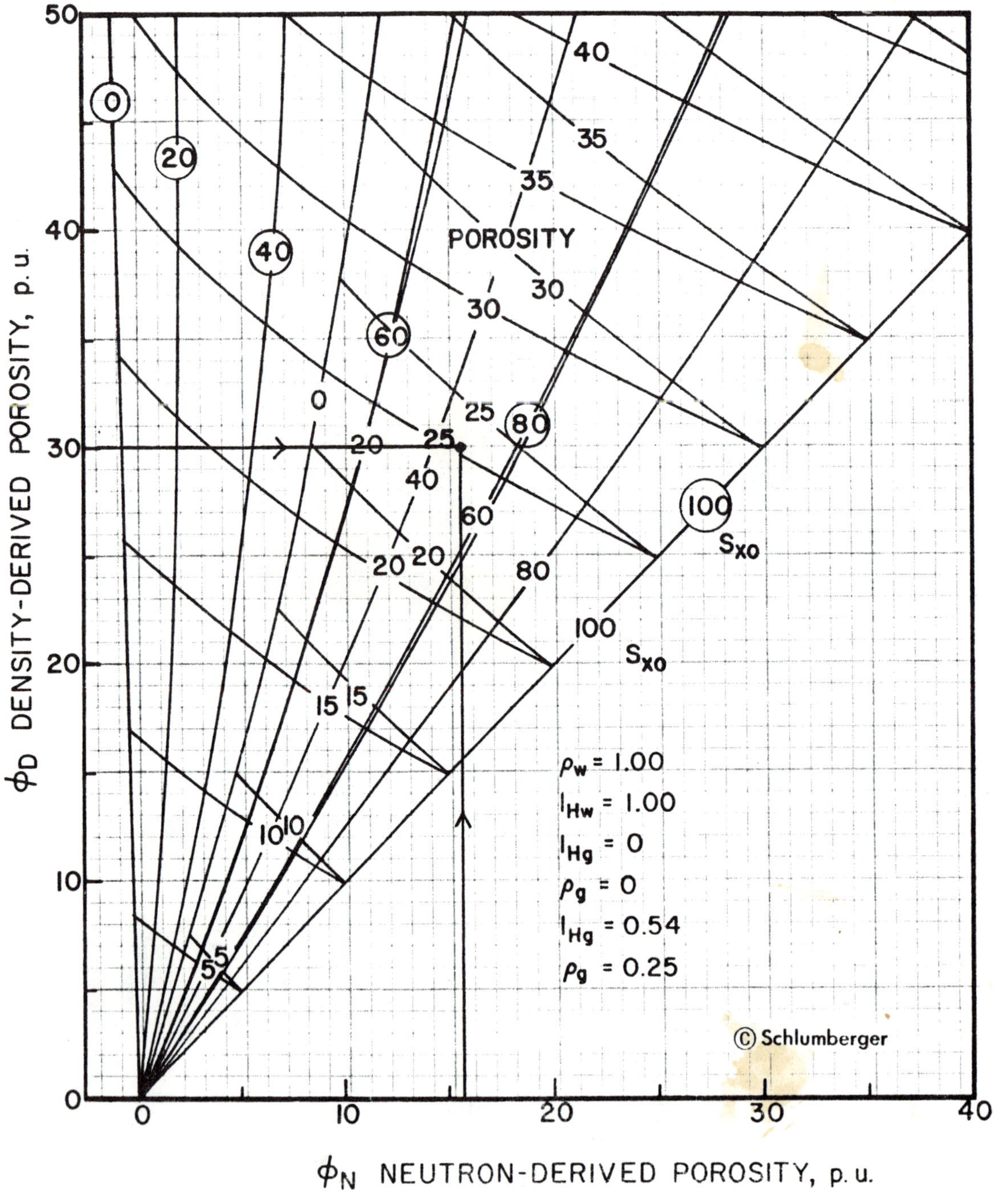

Figure 9-17. Porosity and gas saturation crossplot. (Courtesy of Schlumberger.)

Shale-Free Example

1. Given: Formation depth = 4,000 ft

$$\Phi_n = 10\%$$
$$\Phi_d = 24\%$$

2. Plot 4,000 on Figure 9-18 and read a gas density of 0.1, or compute:

$$\rho_g = \frac{\log(4{,}000/1{,}000)}{5.473 - (4{,}000/1{,}000) \times 0.0278}$$

$$\rho_g = 0.1$$

3. Using the solid lines on the gas chart found in Figure 9-19 (solid lines for $\rho_g = 0.1$), we plot the neutron-density porosities and obtain a residual hydrocarbon saturation (S_{hr}) value of 52%, with a crossplot porosity of 17.5%.
4. Correct the density for gas:

$$\Phi_{dcorr} = \Phi_d - \Delta\Phi_d$$

where $\Delta\Phi_d =$

$$\frac{1.07 \times .24 \times .52(1.11 \times 0.9 - 0.3)}{2.71 - 1}$$

$$\Delta\Phi_d = 0.076$$

$$\Phi_{dcorr} = .24 - 0.076 = 0.164 \text{ or } 16.4\%$$

5. Correct the neutron for gas:

$$\Phi_{ncorr} = \Phi_n - \Delta\Phi_n$$

where $\Delta\Phi_n = .1 \times .52(2.2 \times .1 - 1)$

$$\Delta\Phi_n = -0.04$$
$$\Phi_{ncorr} = 0.1 - (-0.04) = 0.14 \text{ or } 14\%$$

Using the conventional crossplot we see that the lithology is a sand-lime mixture. Figure 9-20 shows the point plotted. The encircled point represents the original uncorrected porosity plot.

Notice that the gas saturation chart shows porosity to be 17.5% (Figure 9-19), while the corrected crossplot (Figure 9-20) shows it to be 15.1%. The gas plot assumes the matrix to be a simple lithology which, in this case, is wrong.

If a gas saturation chart is not available, or you wish to compute S_{hr} independently, we continue with this example.

Since most people do not run R_{xo} tools but do run dual-inductions, we will use the tornado chart.

Given

$$R_{deep} = 17.1 \text{ ohms}$$
$$R_{medium} = 22.6 \text{ ohms}$$
$$R_{shallow} = 58.0 \text{ ohms}$$

in the same zone, we compute the ratios:

$$R_m/R_d = 22.6/17.1 = 1.32$$

$$R_s/R_d = 58/17.1 = 3.4$$

We plot these points on the tornado chart found in Figure 9-21. From the chart:

$$R_t/R_d = 0.97$$

$$R_{xo}/R_t = 5$$

$$R_t = R_d \times 0.97$$

$$R_t = 17.1 \times 0.97 = 16.6$$

$$R_{xo} = R_t \times 5$$

$$R_{xo} = 16.6 \times 5 = 83$$

With R_{xo} in hand, we calculate the other parameter needed for S_{hr}.

$$F = \frac{1}{\Phi^2}$$

where $\Phi = \sqrt{\dfrac{\Phi_n^2 + \Phi_d^2}{2}}$

$$\Phi = \sqrt{\frac{.10^2 + .24^2}{2}} = .18$$

$$F = \frac{1}{.18^2} = 30.8$$

$$R_{mf} @ T_f = 0.6 \text{ ohms}$$

(text continued on page 140)

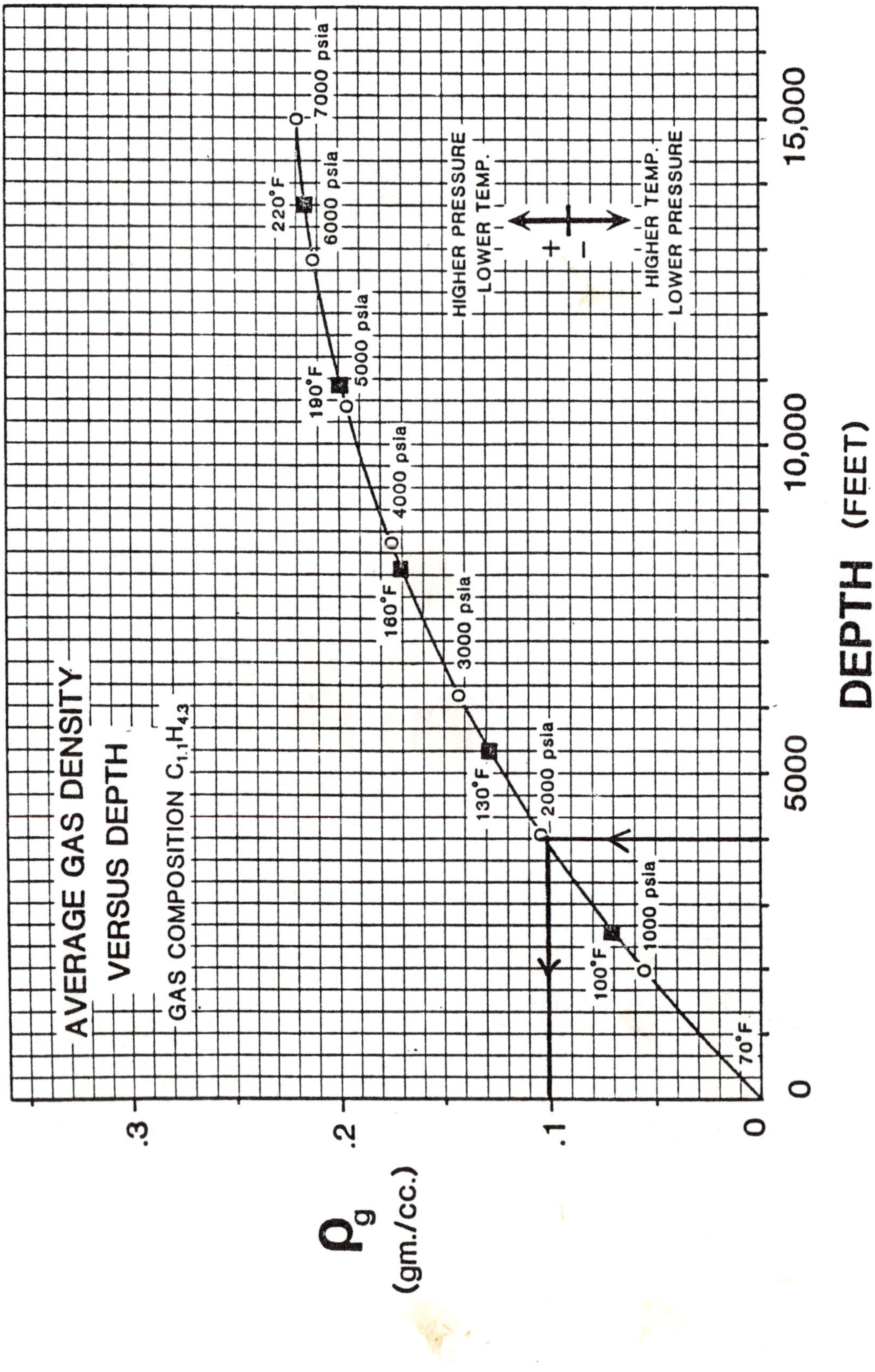

Figure 9-18. Plot showing hydrocarbon density as a function of depth. (Courtesy of Welex.)

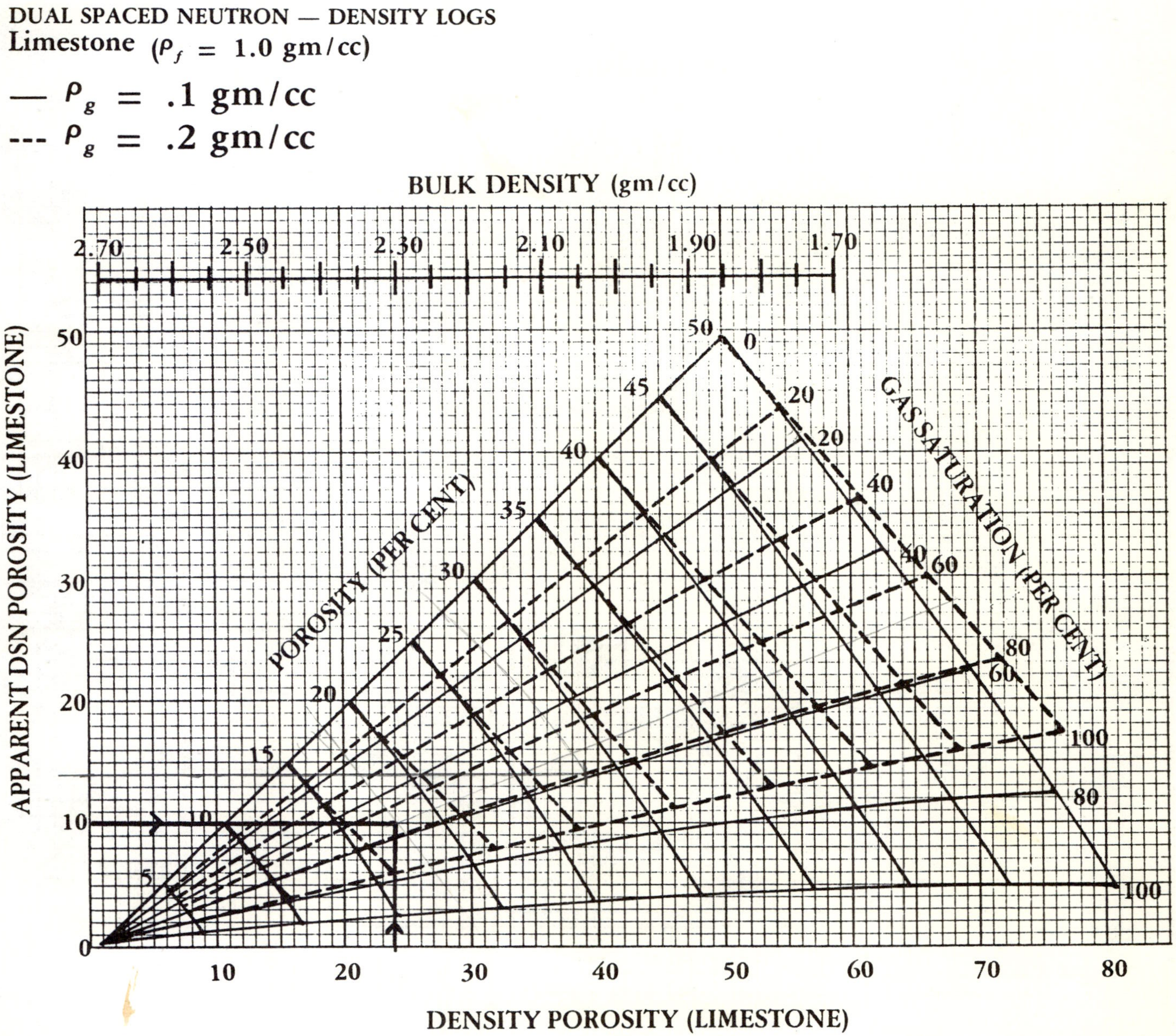

Figure 9-19. Plot showing gas saturation and corrected porosity. (Courtesy of Welex.)

DUAL SPACED NEUTRON–COMPENSATED DENSITY CROSSPLOT

LIQUID FILLED BOREHOLES
FRESH WATER SATURATED FORMATIONS

(ρ_f = 1.0 gm/cc)

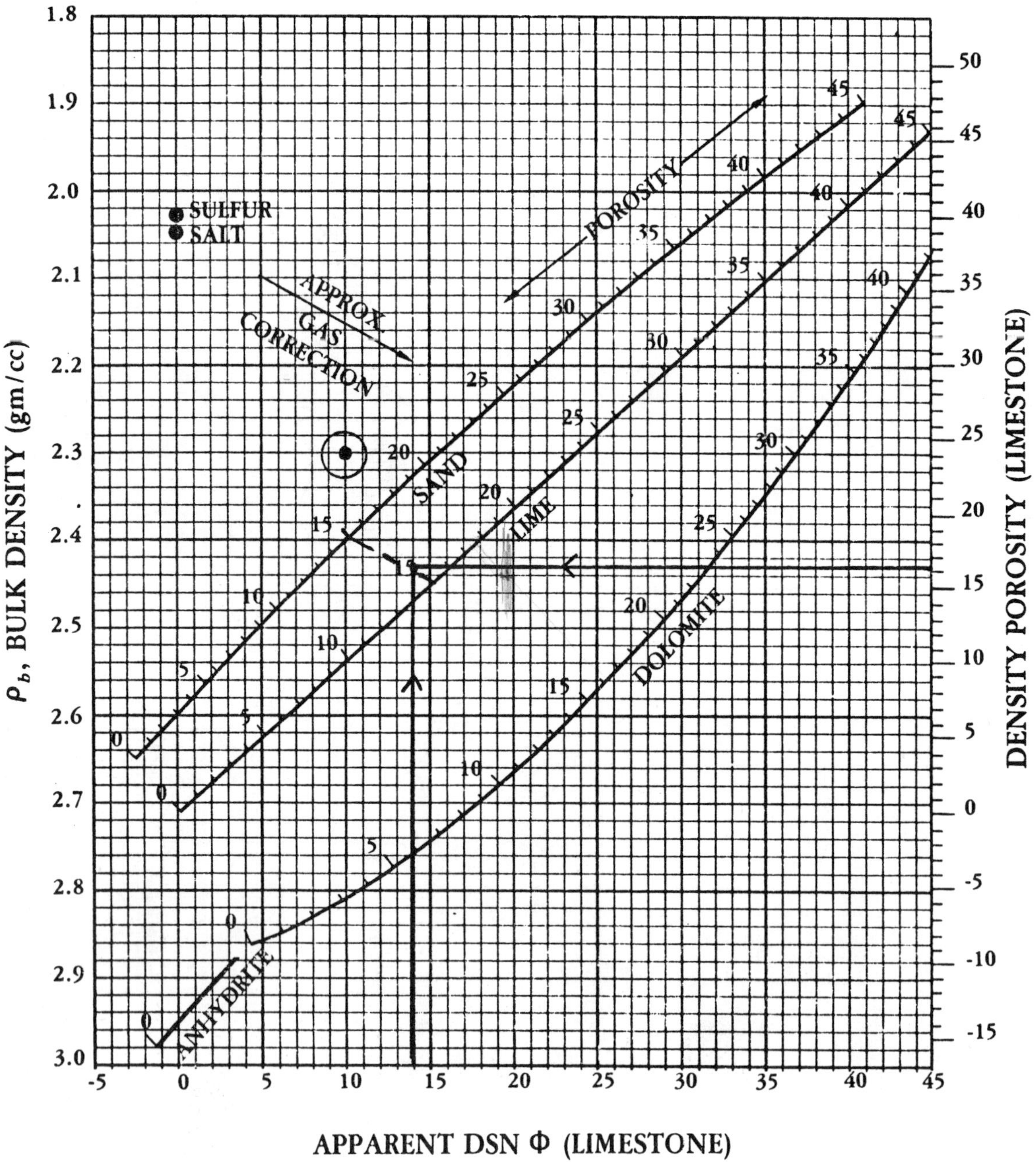

Figure 9-20. Crossplot showing a point plotted in a formation containing gas.

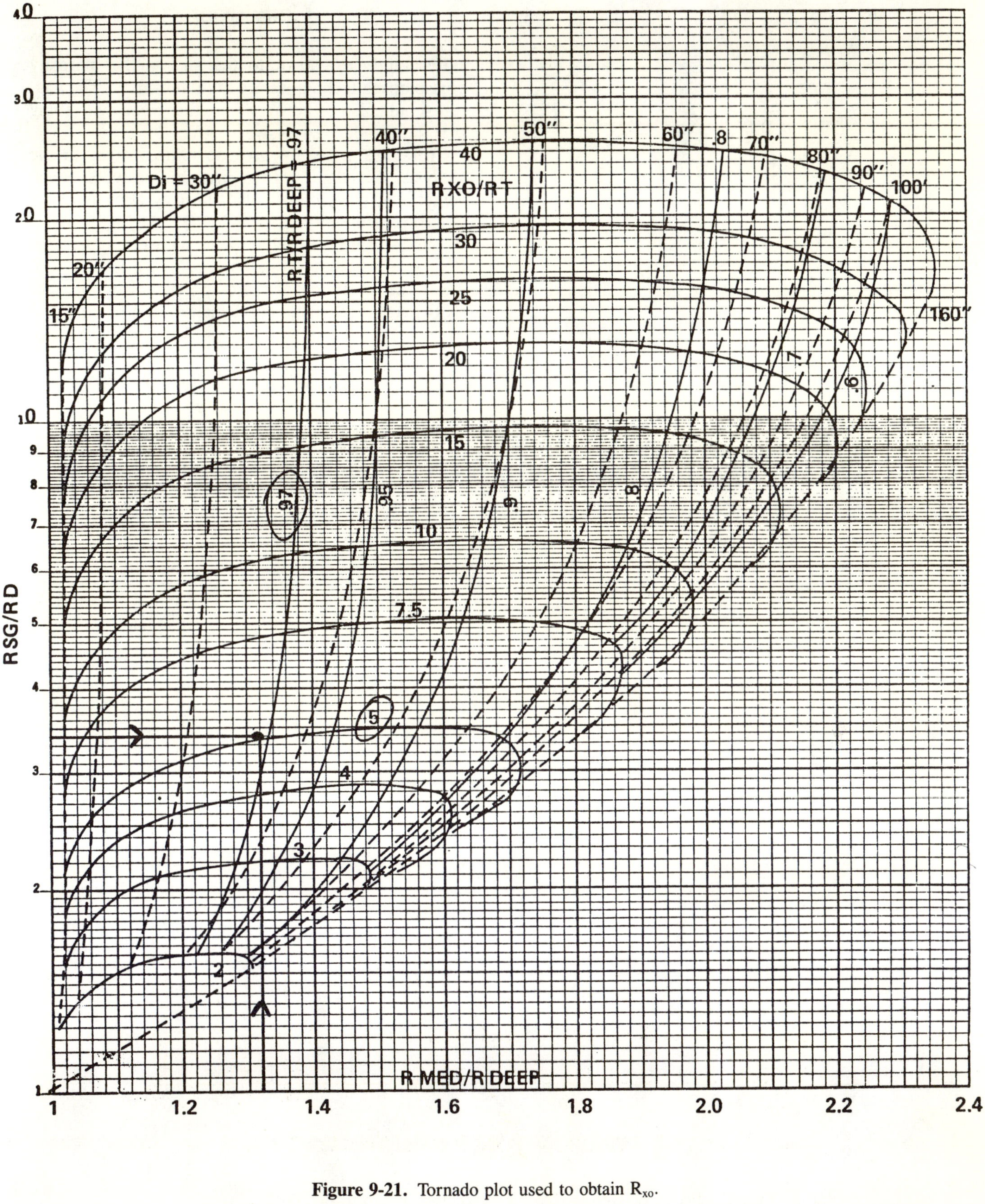

Figure 9-21. Tornado plot used to obtain R_{xo}.

$$S_{xo} = \sqrt{\frac{F\ R_{mf}}{R_{xo}}}$$

$$S_{xo} = \sqrt{\frac{30.8 \times 0.6}{83}} = 0.47$$

$$S_{hr} = 1 - S_{xo}$$

$$S_{hr} = 1 - 0.47 = 0.53 \text{ or } 53\%$$

which closely agrees with the gas saturation plot of Figure 9-19.

Conclusion

It should be evident to the analyst that porosity crossplotting techniques are quite powerful. The analyst should also be keenly aware that when everything is right, the derived answers agree.

Crossplots not only provide more insight to the analysis, but also confirm previous answers or assumptions. The plots are also alarm systems that alert you to possible pitfalls.

The crossplots of Chapter 8 are a generic type. That is, any wireline tool may be used to construct them. The plots in this chapter are specific. Schlumberger tools mandate Schlumberger charts, and so on.

10

Mineral Crossplots

The true title of this chapter should be "miscellaneous crossplots." There is a collection of odd plots which are of value to the analyst, but each one in itself does not represent a system of analysis. For the most part, these are tools that we use in special cases.

As the density and neutron tools gain more popularity, the sonic tool has been pushed out of the limelight. To a certain extent, it is understandable, since the sonic tool ignores secondary porosity and crossplots for lithology are not very definitive.

The sonic tool, however, shines in the area of mineral identification. Mineral identification is perhaps overlooked by the analyst, primarily because the target for the well is petroleum. Mineral identification should be paid some attention for two reasons. One is that it is sometimes the cause for a zone that looks highly producible, and the second is the potential for profit in the mineral.

The cornerstone of mineral identification is the sonic-density crossplot. If we extend the Δt and bulk density scales of the plot, we obtain the crossplot of Figure 10-1A and 10-1B. Over eighty minerals can be found in sedimentary evaporites. The exception is trona. A listing of some minerals exhibiting potential for profit (a subject beyond the scope of this book) follows:

- Sylvite—ore of potassium
- Langbeinite—potassium sulfate used as fertilizer
- Carnalite—potassium ore, noncommercial
- Kainite—potassium and magnesium ore, also used for fertilizer
- Polyhalite—potassium salt
- Anahydrite—anhydrous form of gypsum
- Gypsum—source of cement, plaster of paris
- Halite—rock salt
- Keiserite—Epsom salts
- Trona—source of carbonate and bicarbonate

Sulfur

Sulfur is another mineral which is of commercial importance. It is easily distinguished on the sonic density plot of Figure 10-1, characterized by a low density and a high travel time.

Sulfur does not exist, as is a common belief, in thick beds. Sulfur occurs as an infilling of the pore space of limestone. It may also be found in the caprock of some salt domes.

This being the case, a method of obtaining the volume of sulfur is presented:

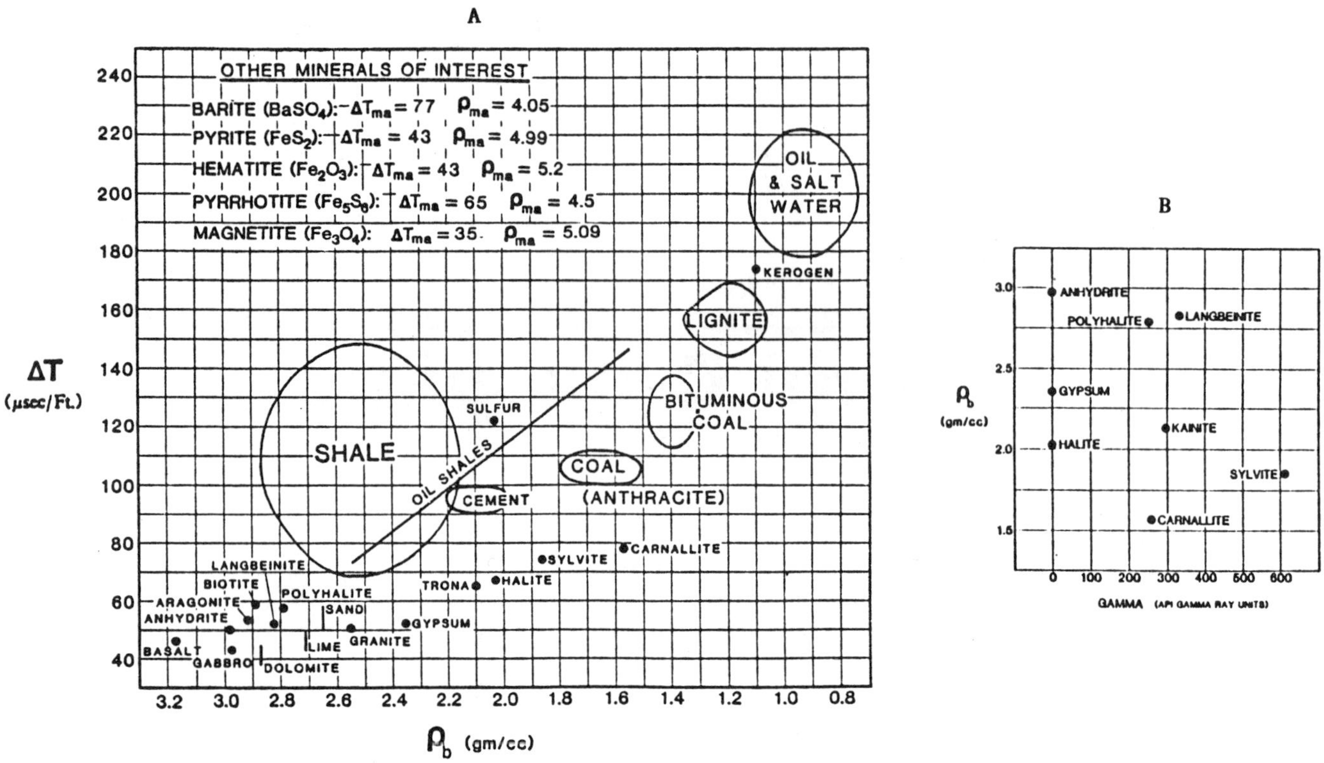

Figure 10-1. Mineral plot. (Courtesy of Welex.)

$$\Phi_1 = \frac{0.53\ \Phi_d - 0.4\ \Phi_s}{0.13}$$

$$V_{sulfur} = \frac{\Phi_s - \Phi_1}{0.53}$$

or

$$V_{sulfur} = \frac{\Phi_d - \Phi_1}{0.4}$$

where V = volume
Φ_d = density porosity in limestone
Φ_s = sonic porosity in limestone

The amount of sulfur expressed as a percentage of matrix is:

$$S_{\%ma} = \frac{V_{sulfur}}{1 - \Phi_1}$$

This latter equation is of more use to the mining engineer.

Coal

Figure 10-1 clearly shows the plot of coal. The crossplot not only defines coal, but also the grade of coal.

Lignite coal is a low grade coal of brown color. Sometimes lignite still retains particles of wood from which it was formed. It is easily ignited and has a smokey flame.

Bituminous coal, having a higher percentage of coal, is considered to be a higher grade coal. This coal (also known as soft coal) burns easily and is black in color.

Anthracite, having the highest percentage of coal, is of course the highest grade. Anthracite is harder to ignite, but burns steadily with a small flame that produces high heat and little smoke.

The ash content of coal can be high. The potential of coal with a high content of ash to be analyzed as a petroleum reservoir of good quality is great. Given a coal with an ash content of 70% would indicate a bulk density of 2.24, which translates to a porosity of 27.5%. One may easily assume gas, and correct the porosity to 19.3%, which is reasonable. Saturations will be optimistic, and the hole will be cased and perforated, but then nothing will produce. Coal resistivities can range from less than one ohm to well over 1,000 ohms. Coal has fooled many analysts, so care must be taken.

Metallic Ore

As can be expected, metal ores have a low resistivity (very conductive) and a high density (heavy). Ores of commercial importance are listed in Table 10-1.

M & N Crossplots

The M & N crossplot is a three-porosity crossplot that is used to determine a complex mineral. The usual two-porosity crossplot assumes a binary mixture, whereas the M & N crossplot can find up to three minerals. Figure 10-2 shows the chart for the plot.

The values of M & N are the ratios of the three porosity devices, and calculated from the equations listed below:

$$M = \frac{\Delta t_f - \Delta t}{\rho_b - \rho_f} \times 0.01$$

$$N = \frac{\Phi_{nf} - \Phi_n}{\rho_b - \rho_f}$$

where 0.01 = a scaling factor
Φ_{nf} = 1.0

If the lithology is pure, a point will plot on top of a point, indicating the lithology. For example, if N is 0.64 and M is 0.81, the lithology is pure silica, or sandstone. If the lithology had two components, such as a dolomitic lime, the point would plot on a straight line, between the dolomite point and the calcite point. Figure 10-3 shows the points plotted.

If there are three minerals in the matrix, the point will plot on the inside of a triangle formed by the three points of the minerals involved. Figure 10-4 shows the three possible triangles.

Most minerals on the chart have a circle and dot connected by a dashed line. This merely indicates the range that is possible for each mineral. In the case of silica two lines are included. The 18,000 ft/sec line indicates a matrix value of:

$$\frac{10^6}{18{,}000} = 55.5\ \mu s/ft$$

and the 19,500 ft/sec line:

$$\frac{10^6}{19{,}500} = 51.3\ \mu s/ft$$

In the cases of dolomite, three lines represent three porosity ranges.

1. 5.5% to 30%
2. 1.5% to 5.5% and > 30%
3. 0% to 1.5%

Line 1 is apt to be used most frequently.

Neutron porosity must be used in limestone units. Working an example, the log shows:

Δt = 70 μs
ρ_b = 2.48 g/cc
Φ_n = 15% in limestone matrix

Table 10-1
Resistivities and Densities of Commercially Important Ores

Material	Density ρ_{ma}	Sonic Δt_{ma}	Ohms R_t	Principle Ore
Hematite	5.2	43	10^{-3} to 10^6	Iron
Magnetite	5.09	35	10^{-4} to 10^{-1}	Magnetic iron
Pyrite	4.99	43	10^{-6} to 10^{-1}	Fool's gold
Siderite	3.72	—	10 to 10^3	—
Barite	4.05	77	—	Barium
Galena	6.14	—	10^{-5} to 10^{-2}	Lead
Chalcocite	5.21	—	0.5 to 10^{-6}	Copper
Chalcopyrite	—	—	10^{-3} to 10^{-1}	Copper
Bornite	4.78	—	0.1 to 10^{-5}	Copper
Cinnabar	6.8	—	10^6 to 10^{10}	Mercury
Bauxite	2.57	—	10^2 to 10^6	Aluminum

(text continued on page 147)

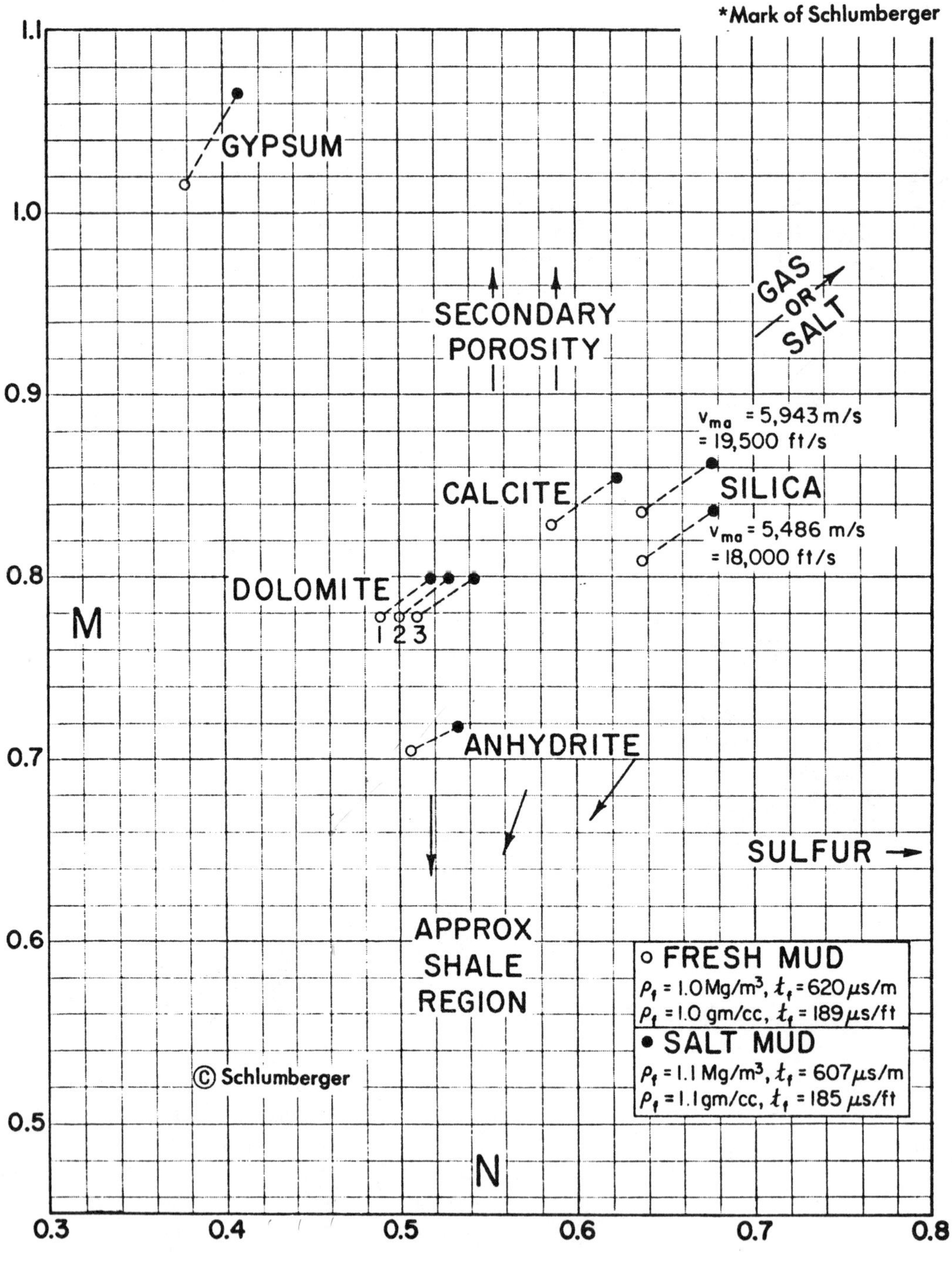

Figure 10-2. M-N plot for mineral identification. (Courtesy of Schlumberger.)

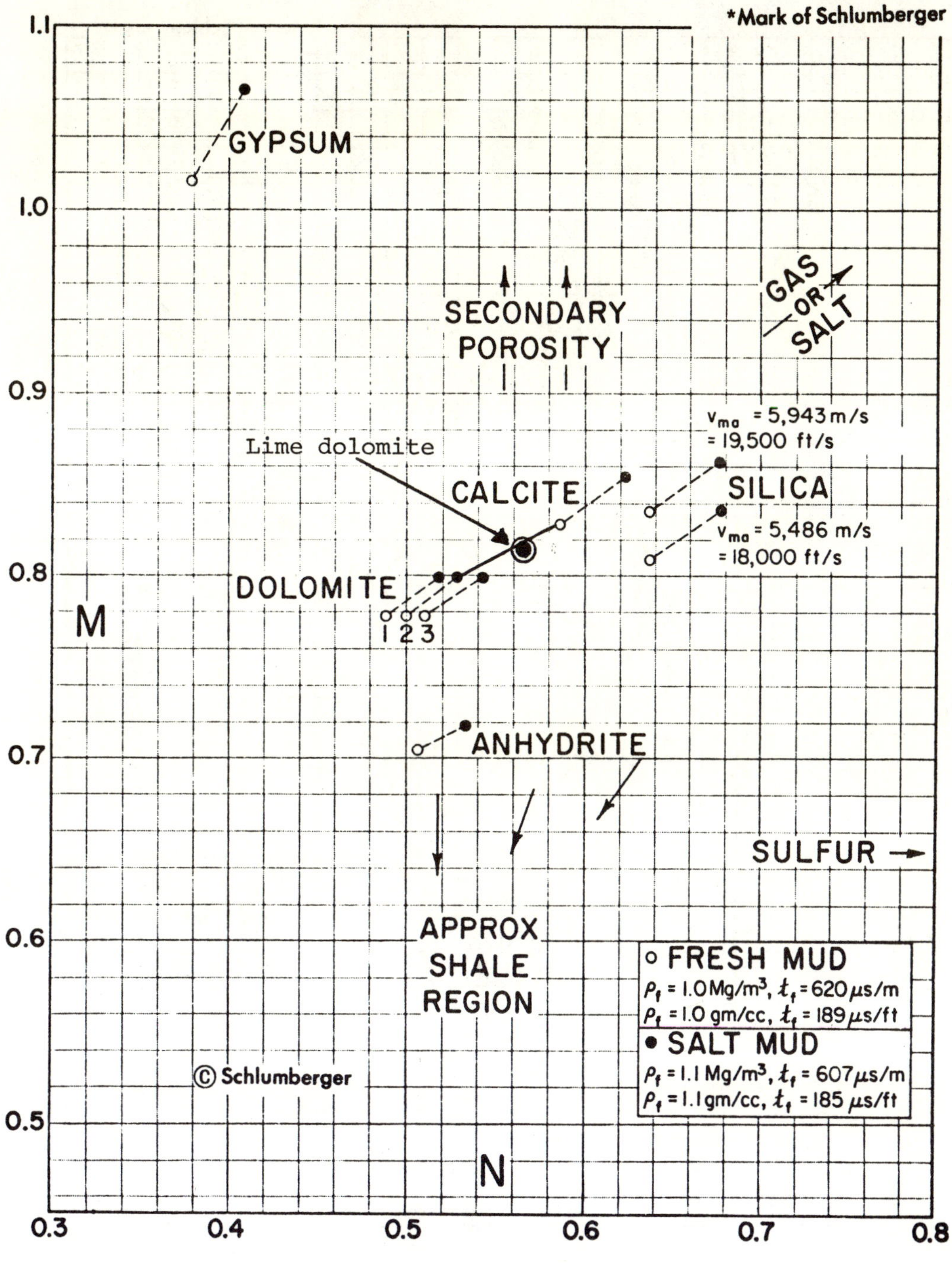

The dolomite lines are divided as to porosity as follows:

1) $\phi = 5.5$ to 30 p.u. 2) $\phi = 1.5$ to 5.5 p.u. and $\phi > 30$ p.u. 3) $\phi = 0$ to 1.5 p.u.

Figure 10-3. M-N plot showing a lime-dolomite point. (Courtesy of Schlumberger.)

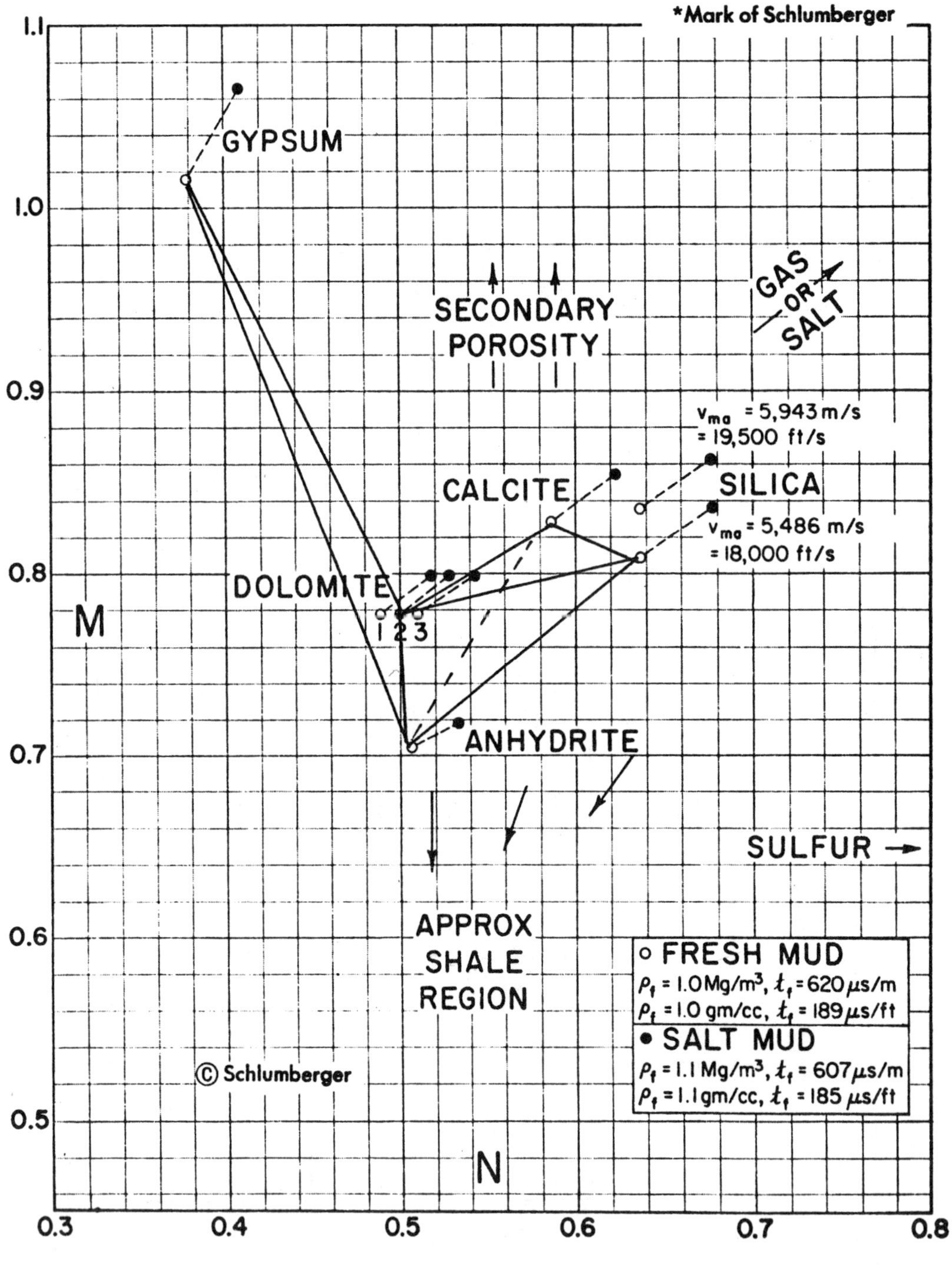

The dolomite lines are divided as to porosity as follows:

1) $\phi = 5.5$ to 30 p.u. 2) $\phi = 1.5$ to 5.5 p.u. and $\phi > 30$ p.u. 3) $\phi = 0$ to 1.5 p.u.

Figure 10-4. Illustration of the three possible combinations of minerals on the M-N plot. (Courtesy of Schlumberger.)

Calculating M:

$$M = \frac{\Delta t_f - \Delta t}{\rho_b - \rho_f} \times 0.01$$

Thus:

$$M = \frac{189 - 70}{2.48 - 1.0} \times 0.01 = 0.8$$

Calculating N:

$$N = \frac{\Phi_{nf} - \Phi_n}{\rho_b - \rho_f}$$

Thus:

$$N = \frac{1.0 - 0.15}{2.48 - 1.0} = 0.57$$

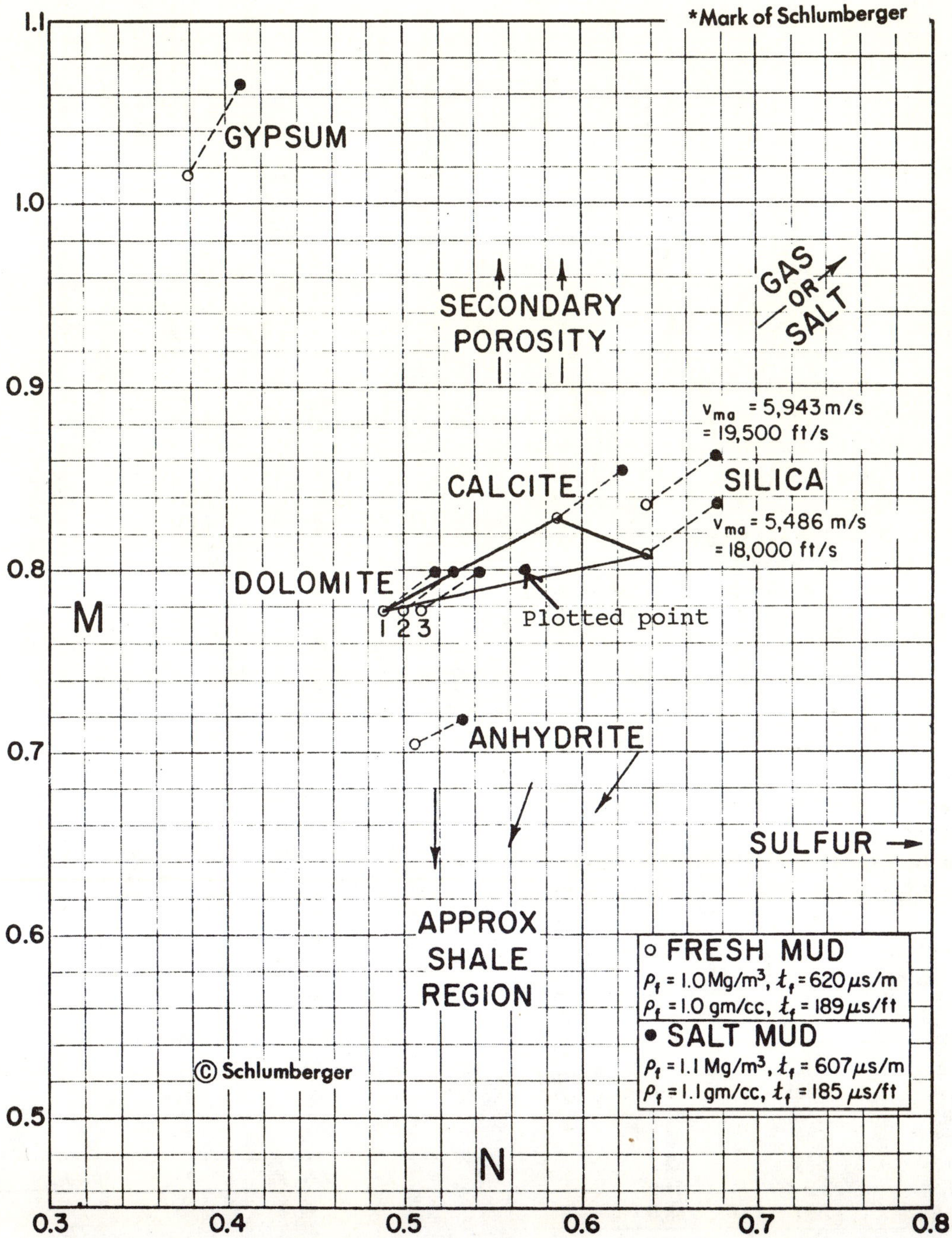

Figure 10-5. M-N plot showing a three mineral plot. (Courtesy of Schlumberger.)

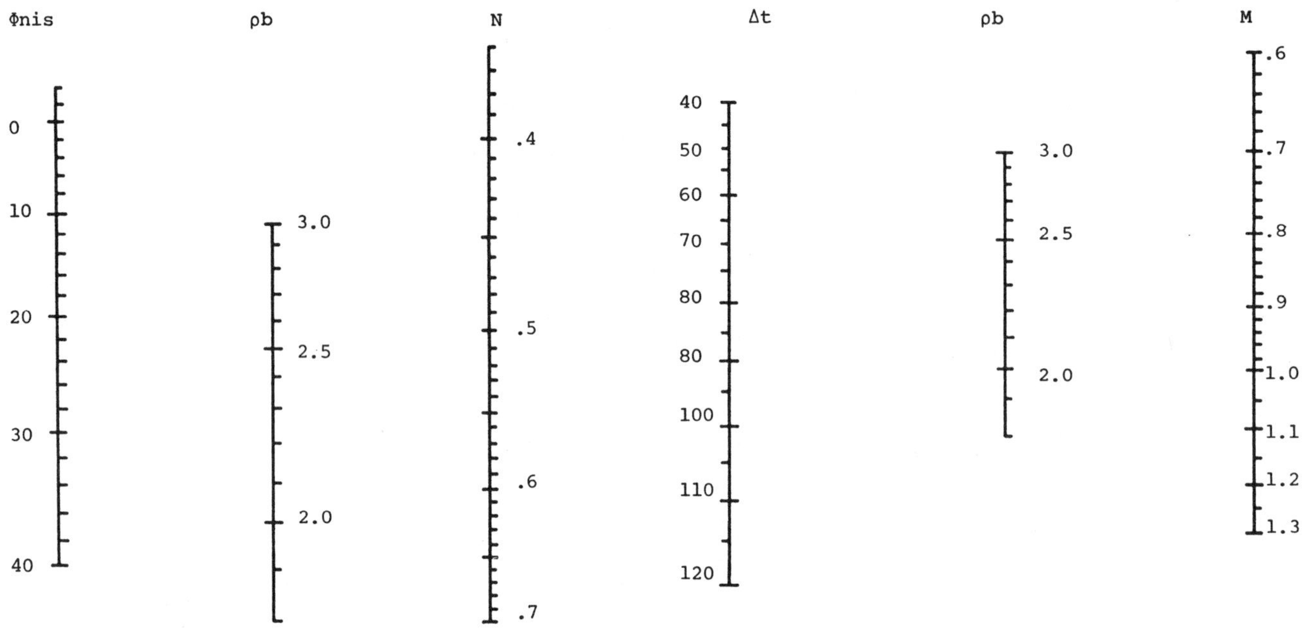

Figure 10-6. Nomograph used in calculating values for M and N. (Courtesy of Schlumberger.)

Figure 10-5 shows the point plots in the triangle defined as silica, calcite, and dolomite.

Before moving on to the next topic, we need to note that it is possible for a data point to plot outside a given line or triangle. If this is the case, arrows on the plot indicate why the plot is so. Of particular interest is the secondary porosity.

Figure 10-6 presents a nomograph for calculating the values of M & N.

Three-Porosity Matrix Plots

The three-porosity matrix crossplot is another matrix identification plot. This plot is trademarked *mid plot* by Schlumberger. This plot is not very popular, but I find that it is a good indicator of secondary porosity.

From the neutron, density, and sonic logs, we pick the following points in the zone of interest.

Φ_n	ρ_b	Δt
8.4	2.528	63.0
8.5	2.56	62.0
10.5	2.58	60.5
10.0	2.58	62.5
9.5	2.6	64.0
11.0	2.57	68.0

Now we use the chart in Figure 10-7 and crossplot the neutron porosity and Δt to obtain an apparent Δt matrix value $(\Delta t_{ma})_a$.

Φ_n	Δt	$\Delta(t_{ma})_a$
8.4	63.0	49.5
8.5	62.0	49.0
10.5	60.5	46.9
10.0	62.5	48.0
9.5	64.0	49.5
11.0	68.0	51.0

Now we go to the chart in Figure 10-8 and crossplot neutron porosity and bulk density to obtain an apparent matrix density $(\rho_{ma})_a$.

Φ_n	ρ_b	$(\rho_{ma})_a$
8.4	2.528	2.7
8.5	2.56	2.72
10.5	2.58	2.74
10.0	2.58	2.738
9.5	2.6	2.74
11.0	2.57	2.74

Combining the two apparent matrix values, they are plotted on the matrix identification plot of Figure 10-9.

$\Delta(t_{ma})_a$	$(\rho_{ma})_a$
49.5	2.7
49.0	2.72
46.9	2.74
48.0	2.738
49.5	2.74
51.0	2.74

(text continued on page 152)

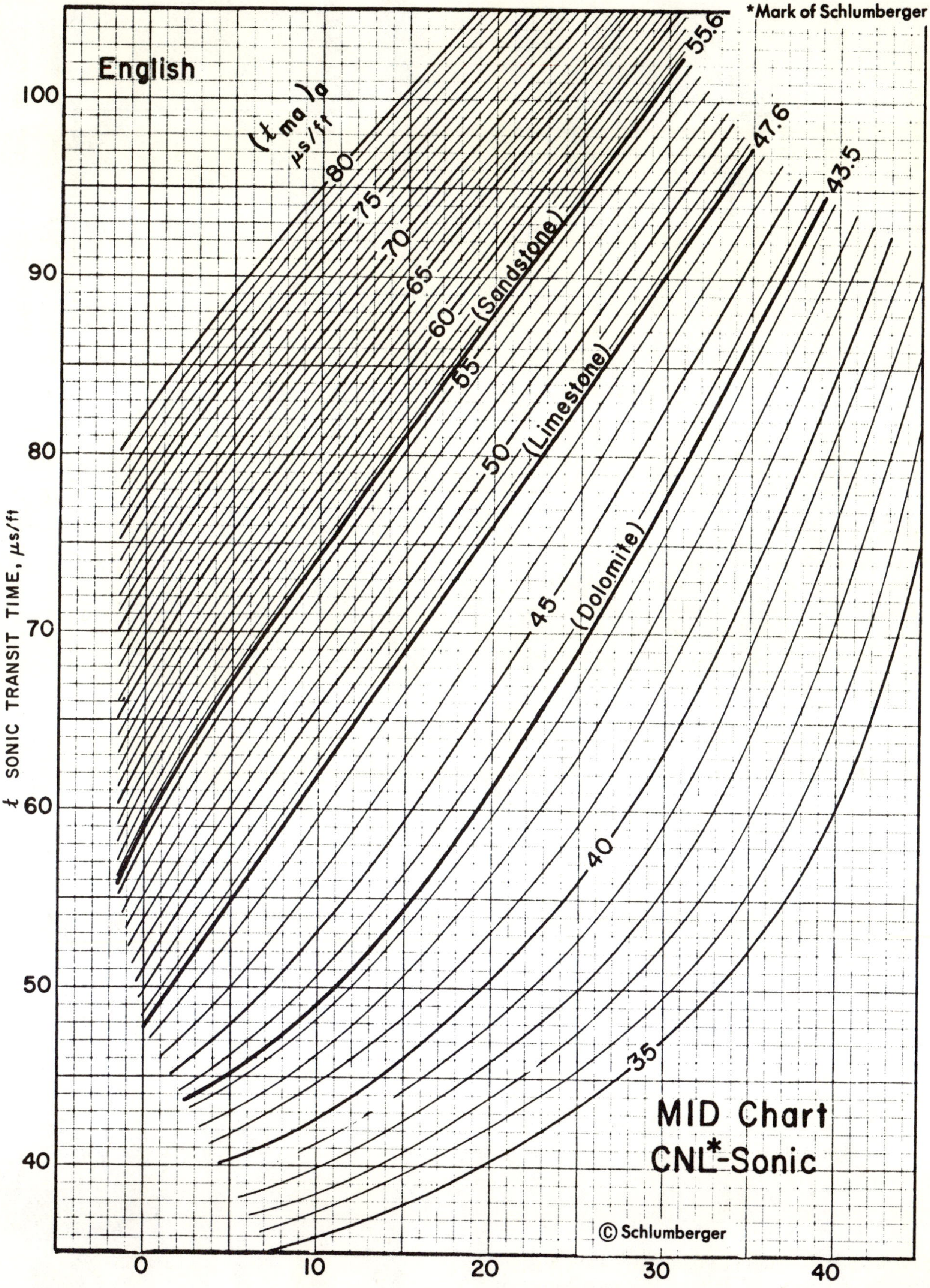

Figure 10-7. Determination of $(t_{ma})_a$ from sonic and neutron logs. (Courtesy of Schlumberger.)

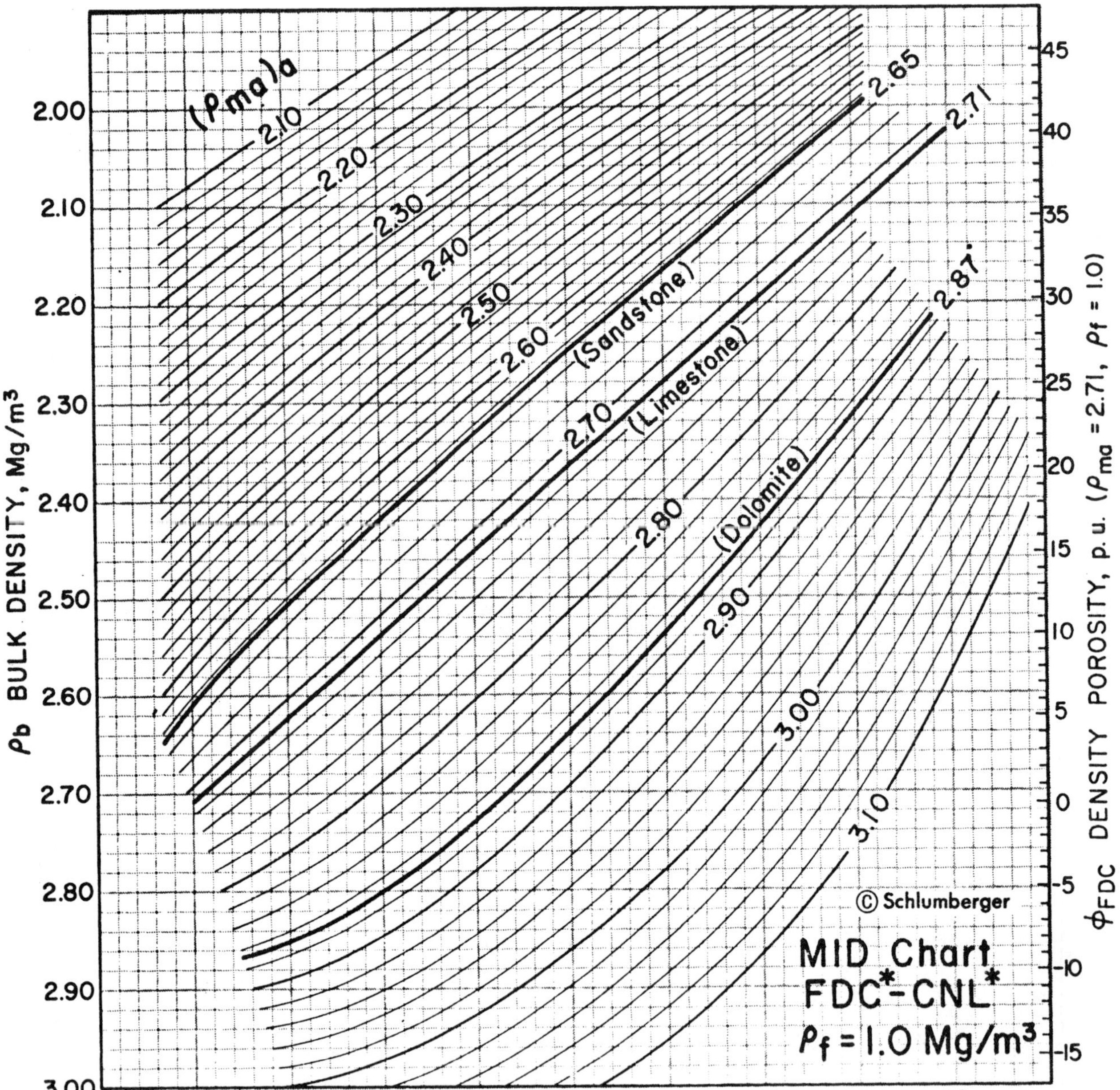

Figure 10-8. Determination of $(\rho_{ma})_a$ from density and neutron logs. (Courtesy of Schlumberger.)

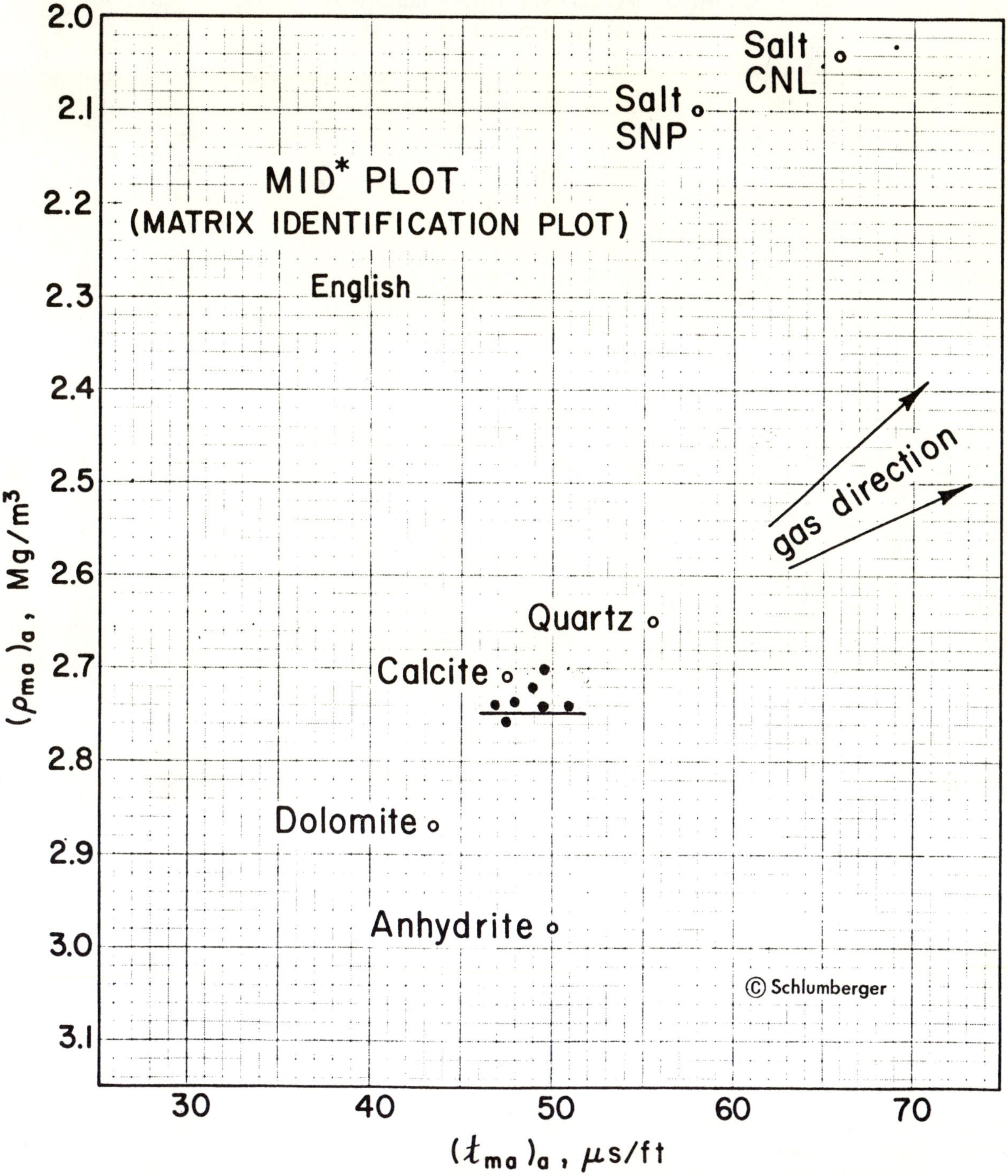

Figure 10-9. Illustration on the use of the mid plot. (Courtesy of Schlumberger.)

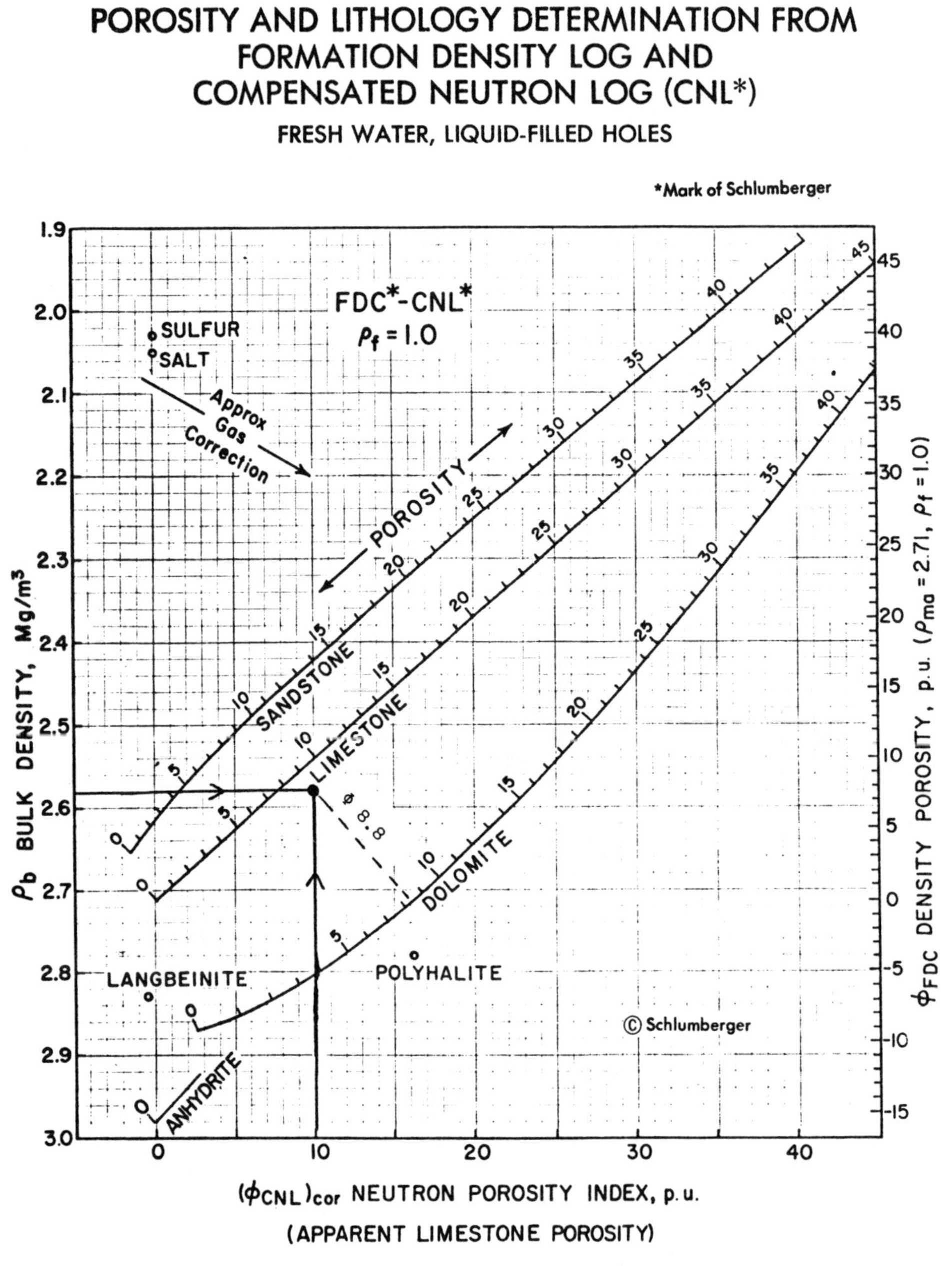

Figure 10-10. Crossplot illustrating crossplot porosity. (Courtesy of Schlumberger.)

The points congregate around the calcite lithology circle, indicating a limestone lithology. The points which are underlined represent secondary porosity. Secondary porosity (vugs or fractures) produces a displacement which is parallel to the sonic axis.

By comparison (and to crosscheck results), a two-porosity crossplot is made (Figure 10-10). As you can see, it is simpler to use the two-porosity crossplot than the three-porosity matrix identification plot.

Conclusion

It is often the case that mineral contents of a formation are overlooked. Simple analysis of a log can easily lead to improper conclusions. As much as the sonic log has fallen out of today's suite of logs, it is still a vital tool in log analysis.

11

Permeability

Permeability in some instances is not of great importance to the producibility of a well. In others, it is as crucial as water saturation. Unfortunately, there is no known tool that can measure permeability. Core-derived permeability is often calculated based on empirical rather than measured studies.

Well logs can predict relative indications of permeability, which are usually termed "permeability index," and are qualitative rather than quantitative. The quality of the index hinges on the quality of the data. The hardest parameter to pin down is that of the irreducible water saturation (S_{wirr}). Ideally, permeability is best determined at a zone which is at irreducible water saturation.

$$S_w = S_{wirr}$$

Irreducible Water Saturation

The simplest method of determining if a zone is at irreducible water saturation is to multiply the water saturation by the porosity. If the product is constant over an interval, then the zone is at irreducible water saturation.

$$C = \Phi S_w$$

If C is constant, then $S_{wirr} = S_w$

Another test for irreducible water has been proposed by G. R. Coates and J. L. Dumanoir. For their test to work, the reservoir must be homogeneous (constant lithology), contain the same type hydrocarbon, and have a fairly constant grain size and porosity. Zones with varying lithology or vugular (secondary) porosity could fail the test.

The test is as follows:

1. Over the zone of interest, list the ratios of R_w/R_t. This must be done for a large portion of the zone (i.e., every two feet). Find and mark the lowest ratio, and label it $R_w/R_{t\ min}$
2. Find a multiplier from Table 11-1 which corresponds to the R_w/R_t min. value.
3. Calculate $R_w/R_{t\ max}$ as follows: $R_w/R_{t\ max}$ = multiplier × $R_w/R_{t\ min}$
4. Take the average of all of the R_w/R_t values (Step 1) and label it $R_w/R_{t\ avg}$
5. If $R_w/R_{t\ avg} \leq R_w/R_{t\ max}$, then the zone is at irreducible water saturation.

Irreducible water saturation is best obtained through core samples.

Table 11-1
Multipliers for R_w/R_t Minimum Values

R_w/R_t Min	Multiplier
Less than 0.0001	6
0.0001 to 0.001	4
0.001 to 0.003	3.5
0.003 to 0.007	3.0
0.007 to 0.012	2.5
0.012 to 0.02	1.75
0.02 to 0.03	1.3
Greater than 0.03	1.1

Permeability Determination

By assuming a zone is at irreducible water by the relationship $C = \Phi S_w$, we can obtain the permeability index from the chart on Figure 11-1. Let us say that porosity is 15% and the constant "C" is 0.04. We plot the point on Figure 11-1 by entering the 15% porosity scale until it intersects the $\Phi S_w = C$ line of 0.04. S_{wirr} is 26%, and permeability is indicated by the semi-vertical lines. In this case, 10 millidarcies for oil, or 1 millidarcy for gas.

The chart is constructed from the relationship:

$$K = \sqrt{\frac{250\ \Phi^3}{S_{wirr}}} \quad \text{for medium gravity oils}$$

$$K = \sqrt{\frac{79\ \Phi^3}{S_{wirr}}} \quad \text{for dry gas}$$

If S_{wirr} is not known and the crude is not medium gravity, then the permeability is, of course, wrong.

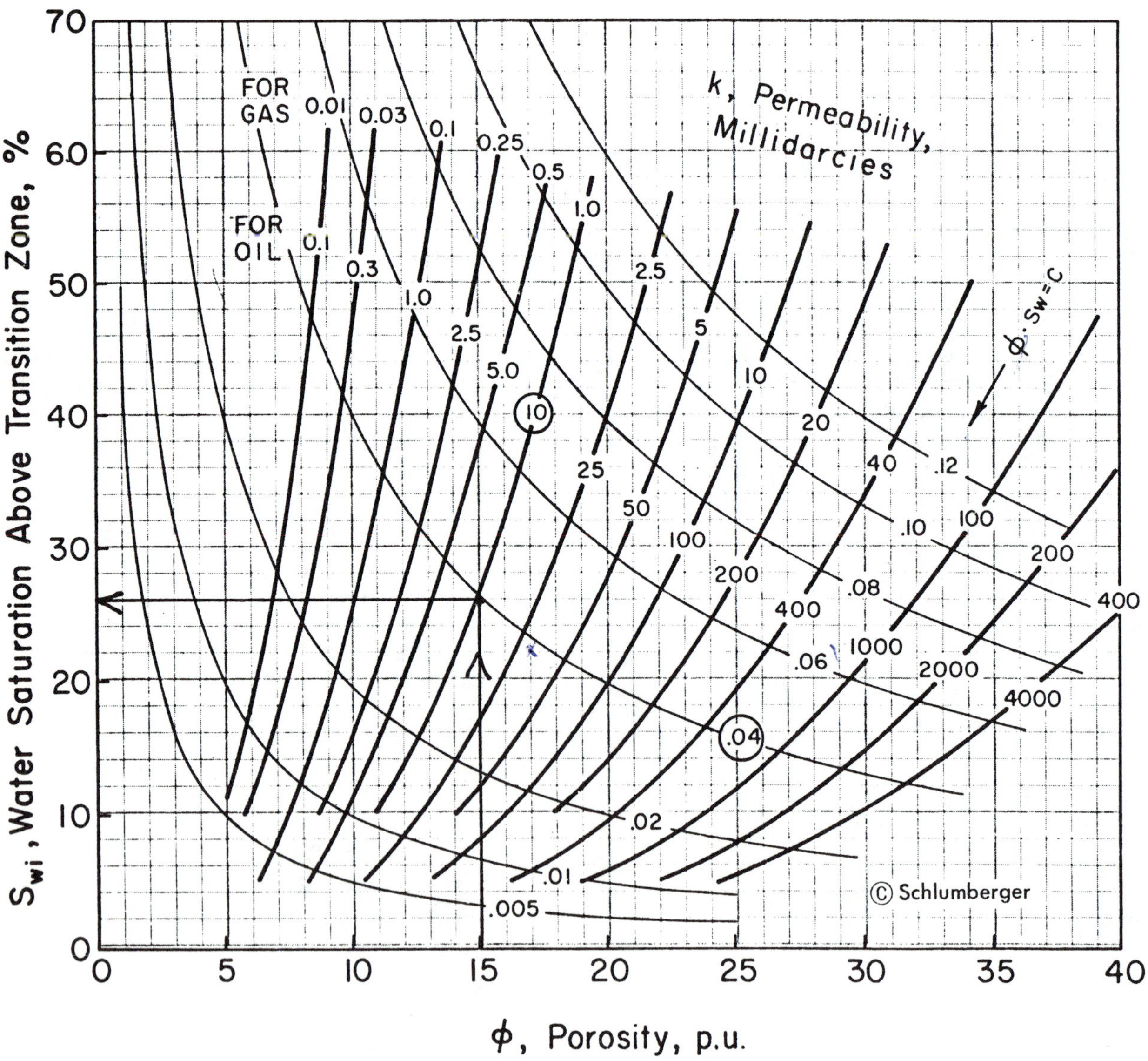

Figure 11-1. Permeability: sandstones, shaly sands. (Courtesy of Schlumberger.)

When irreducible water saturation and hydrocarbon density are not known, permeability can be obtained by using a resistivity gradient. Solving for permeability using the resistivity gradient requires the following parameters:

$$\begin{aligned} \Phi_n &= 16\% \\ \Phi_d &= 18\% \\ \Phi_{xp} &= 17\% \\ R_w &= 0.069 \ @ \ T_f \\ R_{mf} &= 0.4 \ @ \ T_f \\ R_{xo} &= 134 \text{ ohms} \\ R_t &= 23 \text{ ohms @ 6,330 ft} \\ R_t &= 6 \text{ ohms @ 6,370 ft} \end{aligned}$$

If hydrocarbon density is known then R_{xo} is not needed. The procedure for obtaining S_{hr} is outlined in Chapter 9.

The first step involves obtaining a hydrocarbon density, which means a residual hydrocarbon saturation is needed.

$$S_{hr} = 1 - \sqrt{\frac{F\ R_{mf}}{R_{xo}}}$$

Where F is in a limestone:

$$F = \frac{1}{.17^2} = 34.3$$

$$S_{hr} = 1 - \sqrt{\frac{34.3 \times 3 \times .4}{134}} = 0.68 \text{ or } 68\%$$

Taking the ratio Φ_n/Φ_d:

$$\frac{.16}{.18} = 0.89$$

We plot .89 and 68% on the CNL chart of Figure 11-2 to obtain a hydrocarbon density of 0.78 g/cc.

The API gravity of the oil is given by the equation:

$$API° = \frac{141.5}{\rho_{oil}} - 131.5$$

$$API° = \frac{141.5}{.78} - 131.5 = 49.9°$$

indicating a very light oil.

Next, we obtain the salinity level of the formation water. Figure 11-3 plots formation temperature and R_w to obtain 50 KPPM of sodium chloride. Alternatively, salinity may be calculated according to the following equation, using an R_w at 75°F.

$$\text{Salinity (PPM)} = 10^b$$

$$\text{where } b = \frac{3.562 - \log (R_w - 0.0123)}{0.955}$$

and R_w is at 75°F

The density of the water is determined from the following table:

KPPM NaCl	Density
0–49	1.0
50–99	1.034
100–149	1.071
150–199	1.109
200–249	1.148
250–299	1.189

Thus at 50 KPPM the water density is 1.034. The difference between the water and hydrocarbon density is:

$$\rho_w - \rho_h = 1.034 - 0.78 = 0.254$$

In order to solve the resistivity gradient chart of Figure 11-4, we solve for "a."

$$a = \frac{\Delta r}{\Delta d} \times \frac{1}{R_o}$$

where Δr is the resistivity difference in two points of the zone and Δd is the depth difference.
From the parameter table listed earlier:

$$a = \frac{23 - 6}{6{,}370 - 6{,}330} \times \frac{1}{R_o}$$

$$R_o = F\ R_w = 34.3 \times 0.065 = 2.4$$

$$a = 0.425 \times \frac{1}{2.4} = 0.18$$

The density difference and "a" are plotted to obtain a permeability of 150 md.

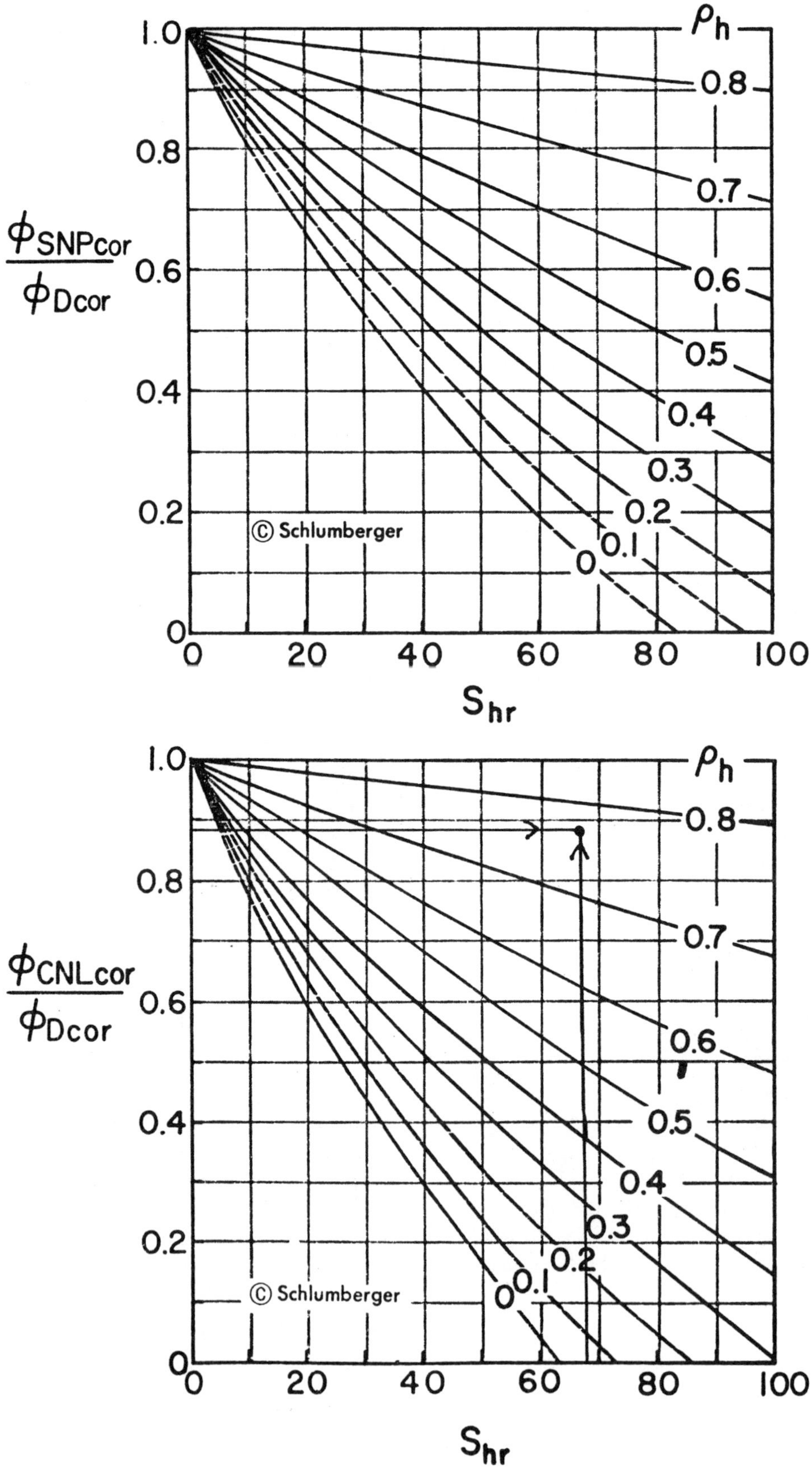

Figure 11-2. Estimation of hydrocarbon density. (Courtesy of Schlumberger.)

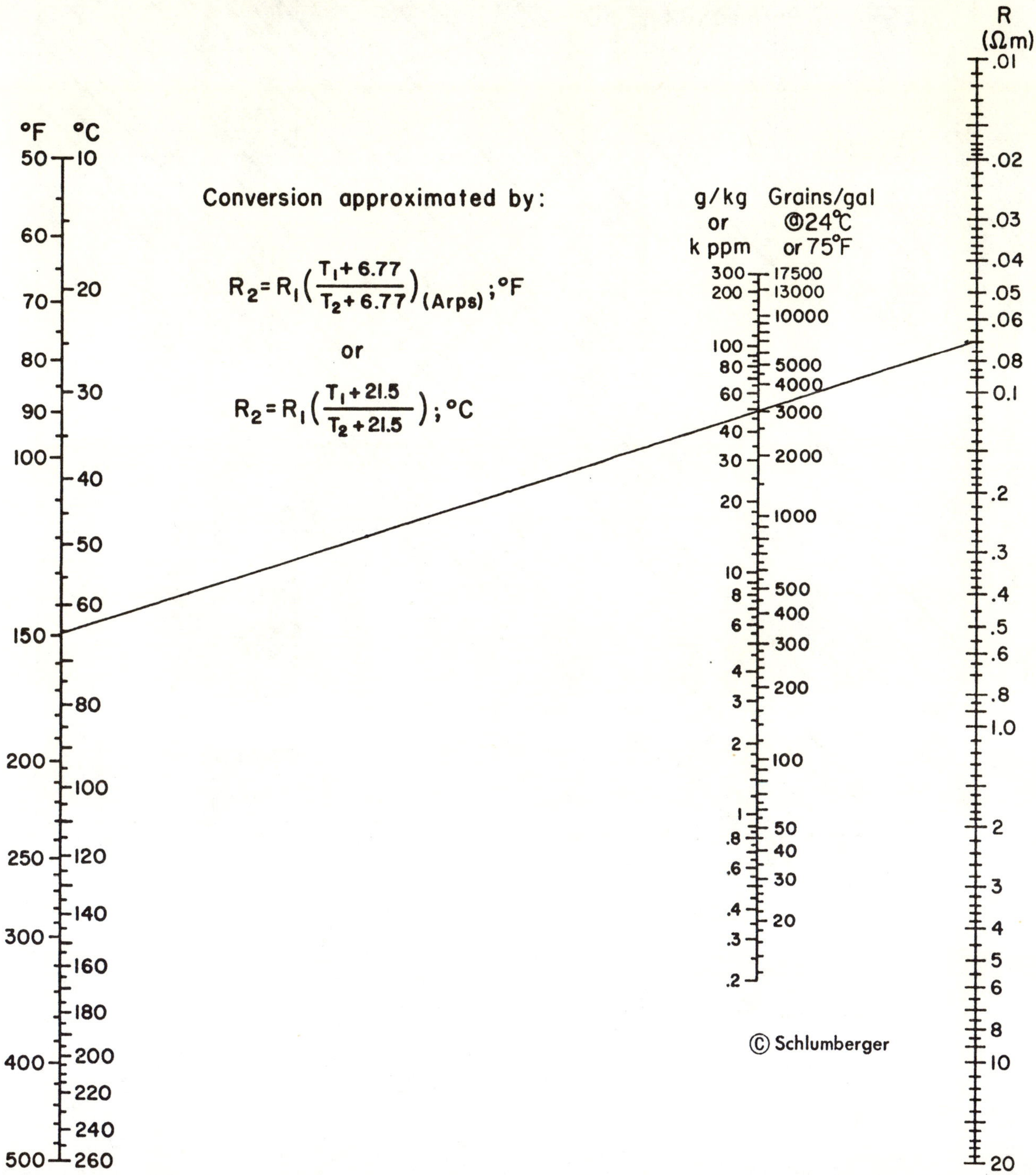

Figure 11-3. Resistivity nomograph for NaCl solutions. (Courtesy of Schlumberger.)

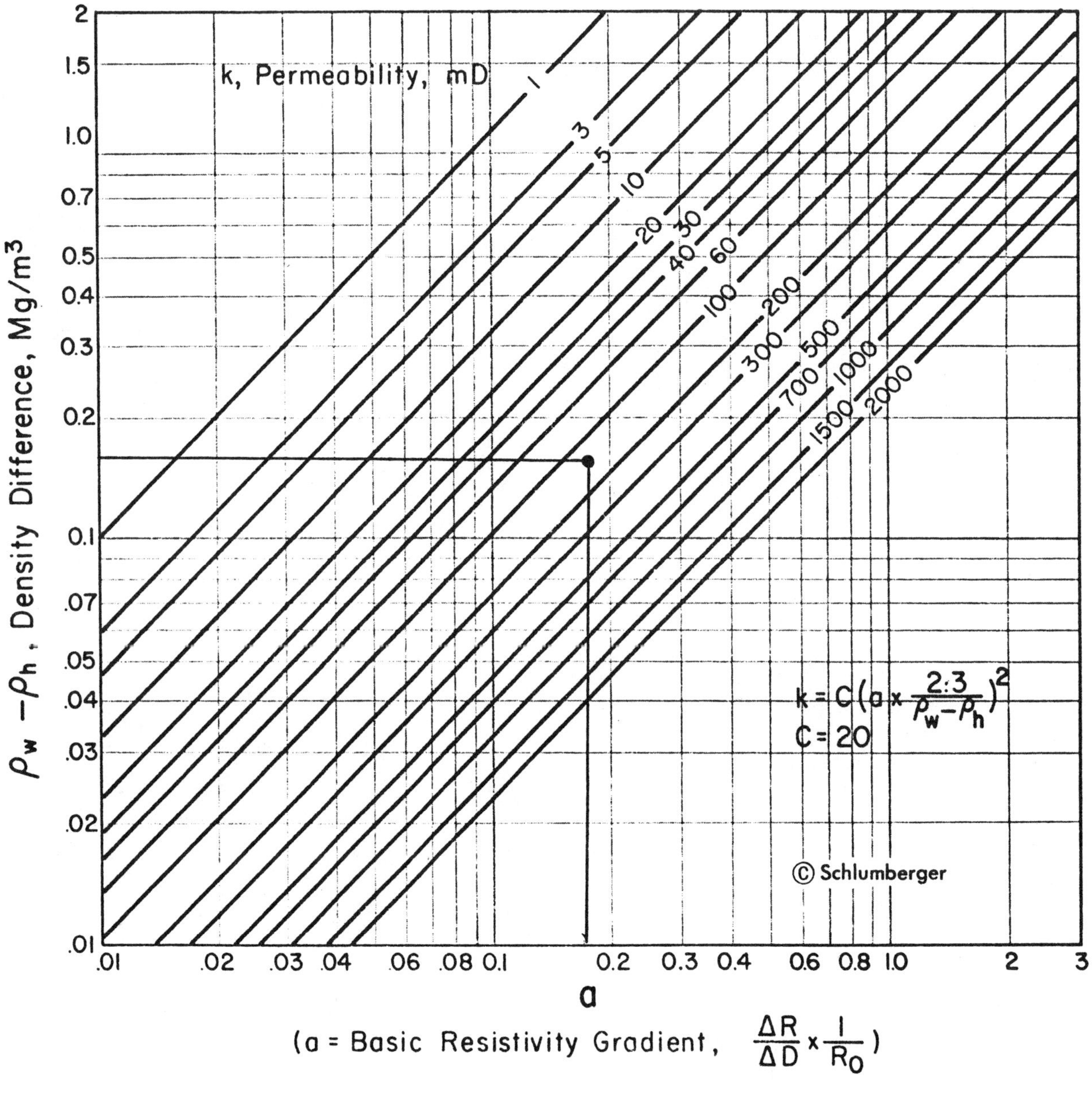

Figure 11-4. Permeability from resistivity gradient. (Courtesy of Schlumberger.)

Another method of obtaining a permeability indication follows the equation:

K (in millidarcies) =

$$\left[\frac{300\ (.077 + 1.55\ \rho_h - .627\ \rho_h^2)\ \Phi^{2w}}{W^4\ (R_w/R_{tirr})}\right]^2$$

where w =

$$\sqrt{(3.75 - \Phi) + 0.5\ (2.2 + \log R_w/R_{tcorr})^2}$$

and:

$$R_{tcorr} = R_{tirr}\ (0.077 + 1.55\ \rho_h - 0.627\ \rho_h)^2$$

ρ_h may be obtained from the chart, or:

$$\rho_h = \frac{141.5}{131.5 + API°}$$

or

$$\rho_h = \frac{.7 \times (1 - \Phi_n/\Phi_d)\, S_{hr} - (1 - \Phi_n/\Phi_d)}{(1 + .72\, \Phi_n/\Phi_d)\, S_{hr}}$$

R_{tirr} is the deep resistivity taken in a zone known to be at irreducible water saturation.

As is often the case, if no zone exists at irreducible water, the R_{tcorr} may be calculated by solving for R_{tirr} first:

$$R_{tirr} = \frac{R_w}{\left(\dfrac{10^{-6}}{G\,(\rho_{ma} - 2.6)^3}\right)}$$

Where G is approximately:

- For coarse grained limes, sands, dolomites—1
- For porous, low permeability, fine grained rocks —0.7
- For dense formations (less than 5% porosity) —2.5

R_{tcorr} essentially corrects for hydrocarbon density while W corrects for lithology.

With a reasonably good value of K, an estimation of irreducible water saturation is possible.

$$S_{wi} = 3.5 \times \left(\frac{\Phi^{1.26}}{K^{0.36}}\right) - 1$$

where K is in darcies

Water Cut

Estimates for water cut are possible with the knowledge of permeability and irreducible water saturation.

First, the relative permeability (permeability of one fluid relative to the other) is calculated.

The relative permeability of water is:

$$K_{rw} = \left(\frac{S_w - S_{wirr}}{1 - S_{wirr}}\right)^3$$

and for oil:

$$K_{ro} = \left(\frac{0.9 - S_w}{0.9 - S_{wirr}}\right)^2$$

given:

$S_w = 32\%$ and $S_{wirr} = 26\%$ (previous example):

$$K_{rw} = \left(\frac{0.32 - 0.26}{1 - 0.32}\right)^3$$

$$= 0.0007 \text{ darcy or } 0.7 \text{ md}$$

And for oil:

$$K_{ro} = \left(\frac{0.9 - 0.32}{0.9 - 0.26}\right)^2$$

$$= 0.821 \text{ darcy or } K_{ro} = 821 \text{ md}$$

The water oil ratio, WOR, is computed next.

$$WOR = B\,\frac{K_{rw}}{K_{ro}}$$

Where B is taken from the following table and is related to the viscosity of the fluids involved.

API° Gravity	B
14	50
19	20
27	10
35	5
45	2

Recall the gravity of the previous example was 50, therefore a value of 1.7 (extrapolated from the preceding table) will be used for B.

$$WOR = 1.7 \times \frac{0.0007}{0.821}$$

$$WOR = 0.0014$$

The water cut will be:

$$WC = \frac{WOR}{1 - WOR}$$

$$WC = \frac{0.0014}{1 - 0.0014}$$

$$WC = 0.0015$$

Multiplying by 100, the water cut is 0.15% water. We can expect 1.5 barrels of water per 1,000 barrels of oil.

Permeability from Microresistivity

The microlog is a two-short-spacing resistivity device with two shallow depths of investigation. One curve is representative of the mud cake reading, while the other measures formation.

The electrodes of the microlog are mounted on a rubber pad and by means of an arm, they are pushed against the borehole wall. The micro-inverse has a one-inch electrode spacing, while the normal has a two-inch spacing. Figure 11-5 shows the pad mounted against the borehole wall.

The microresistivity log is presented in Figure 11-6. Opposite hard impermeable beds, both curves exhibit high resistivities, and the curves usually swing wildly (see above zone A, Figure 11-6). Positive separation occurs opposite porous and permeable beds (zone A). Positive separation is the 2-in. normal, reading higher than the 1-in. inverse. Zone A is permeable except for a hard streak (gamma ray does not indicate shale) at 4,590 ft.

Since the microlog is a pad-mounted device, a caliper has to be included. In washed-out zones, no reading is possible because pad contact is lost. In the case of Figure 11-6, note that the caliper reads less than bit size. This indicates the presence of mud cake.

Opposite shales, the two curves will not separate, and will also show a low resistivity. Negative separation (1 in. higher than the 2-in.) may be seen in shales that have been washed out.

The 2-in. curve may be used as a measure of the flushed zone resistivity (R_{xo}) as long as the formation has a minimum of four inches of invading fluid. R_{xo} can be used to compute S_{xo} (saturation of the flushed zone).

While the microlog can indicate permeability, it cannot quantify it. This log is used strictly as a quality indicator. One other note is in order. In years past, the microlog was thought of as a porosity device. Experience has shown that porosity from the microresistivity tool is poor, at best. Figure 11-7 shows the neutron-density porosity curves opposite the microlog. It is easy to see that correlation is indeed poor.

The concepts presented in this chapter may be better understood by following these next two examples.

Example 1

Figures 11-8 and 11-9 show logs of the Cotton Valley Trend of East Texas. Zone A appears to be productive and relatively clean (i.e., no shale). The neutron-density logs show a porosity of 30% at the top of the interval.

First, we calculate a water saturation, given an R_w of 0.164 at T_f. R_{deep} is the same as R_t, having checked for possible corrections. The lithology is lime.

$$F = \frac{1}{.30^2} = 11.1$$

$$R_o = 11.1 \times 0.164 = 1.82$$

$$R_t = 12 \text{ ohms @ } 4{,}357 \text{ ft}$$

$$S_w = \sqrt{\frac{1.82}{12}} = .39 \text{ or } 39\%$$

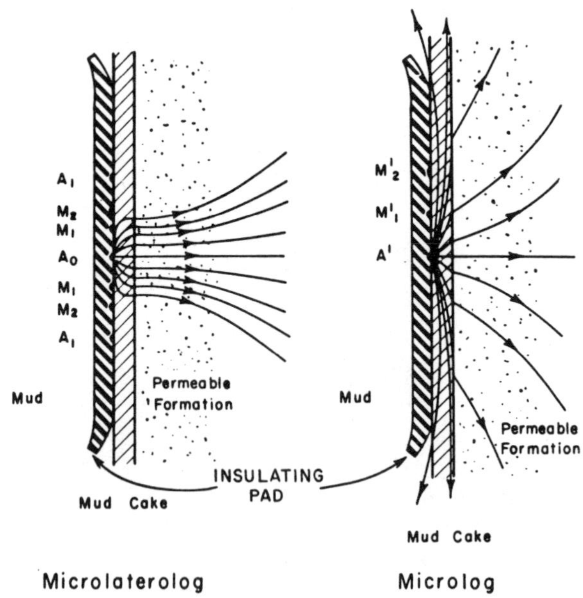

Figure 11-5. Illustration of the micro resistivity pads. (Courtesy of Schlumberger.)

(text continued on page 164)

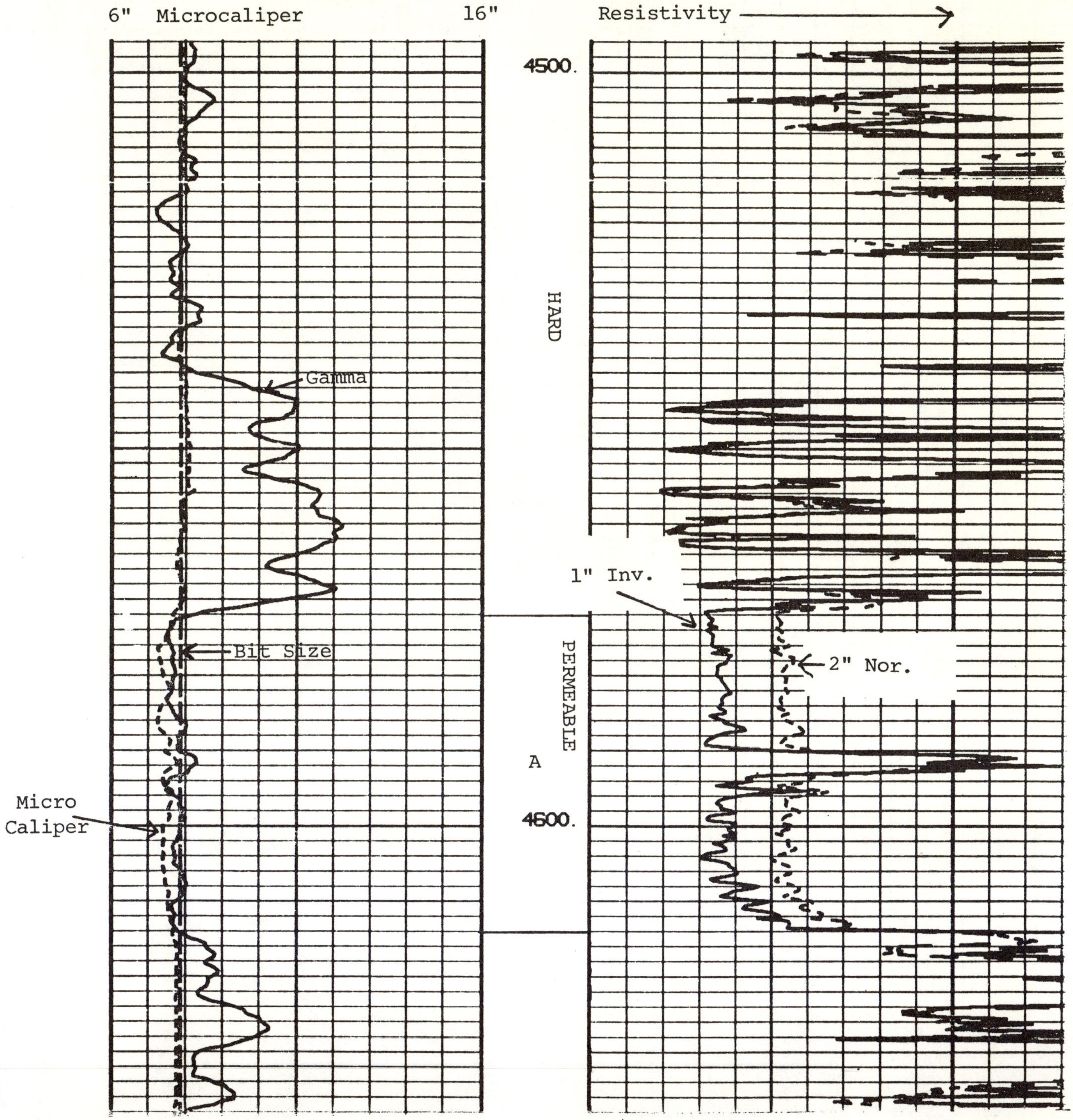

Figure 11-6. Microresistivity log showing qualitative permeability.

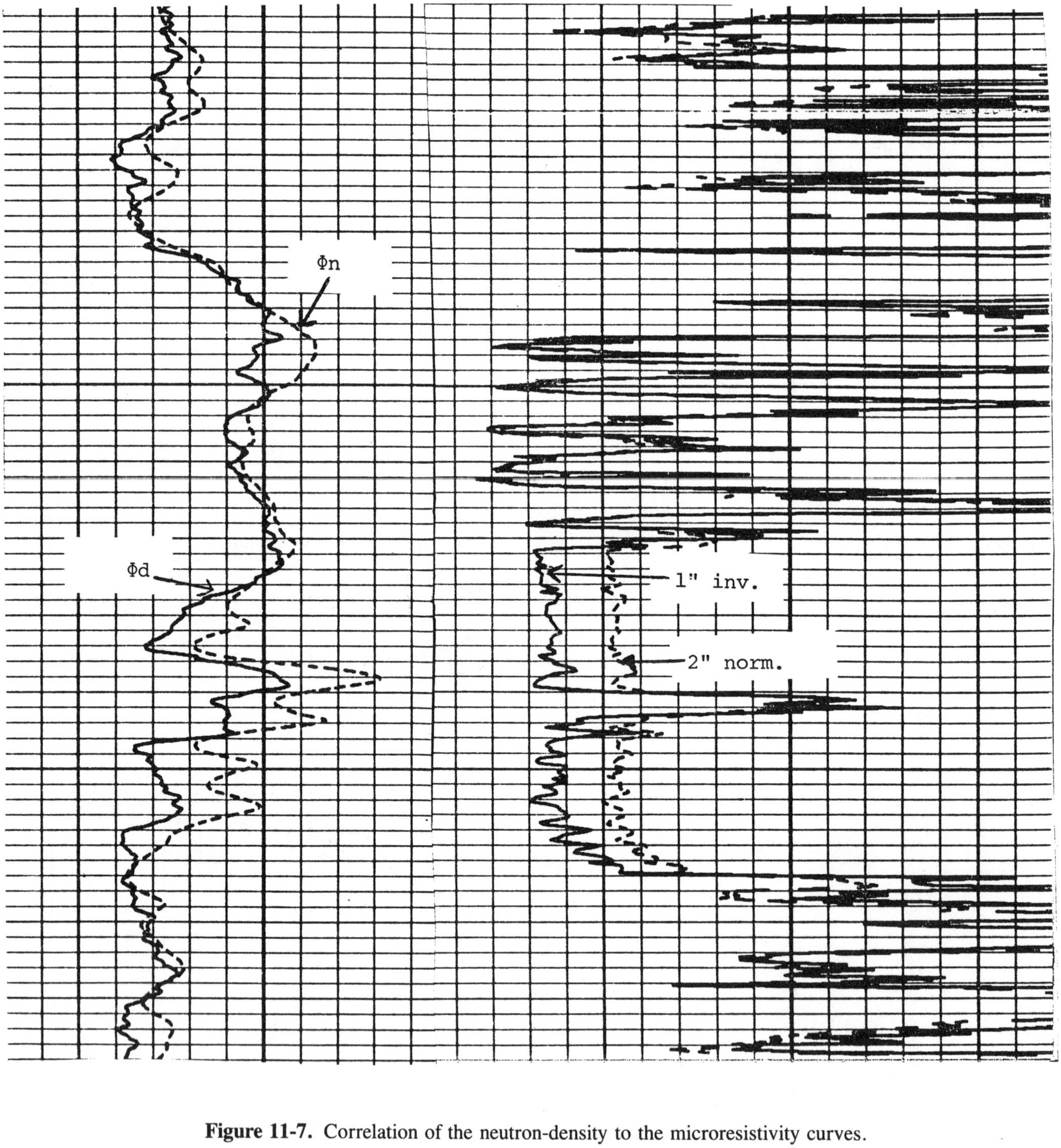

Figure 11-7. Correlation of the neutron-density to the microresistivity curves.

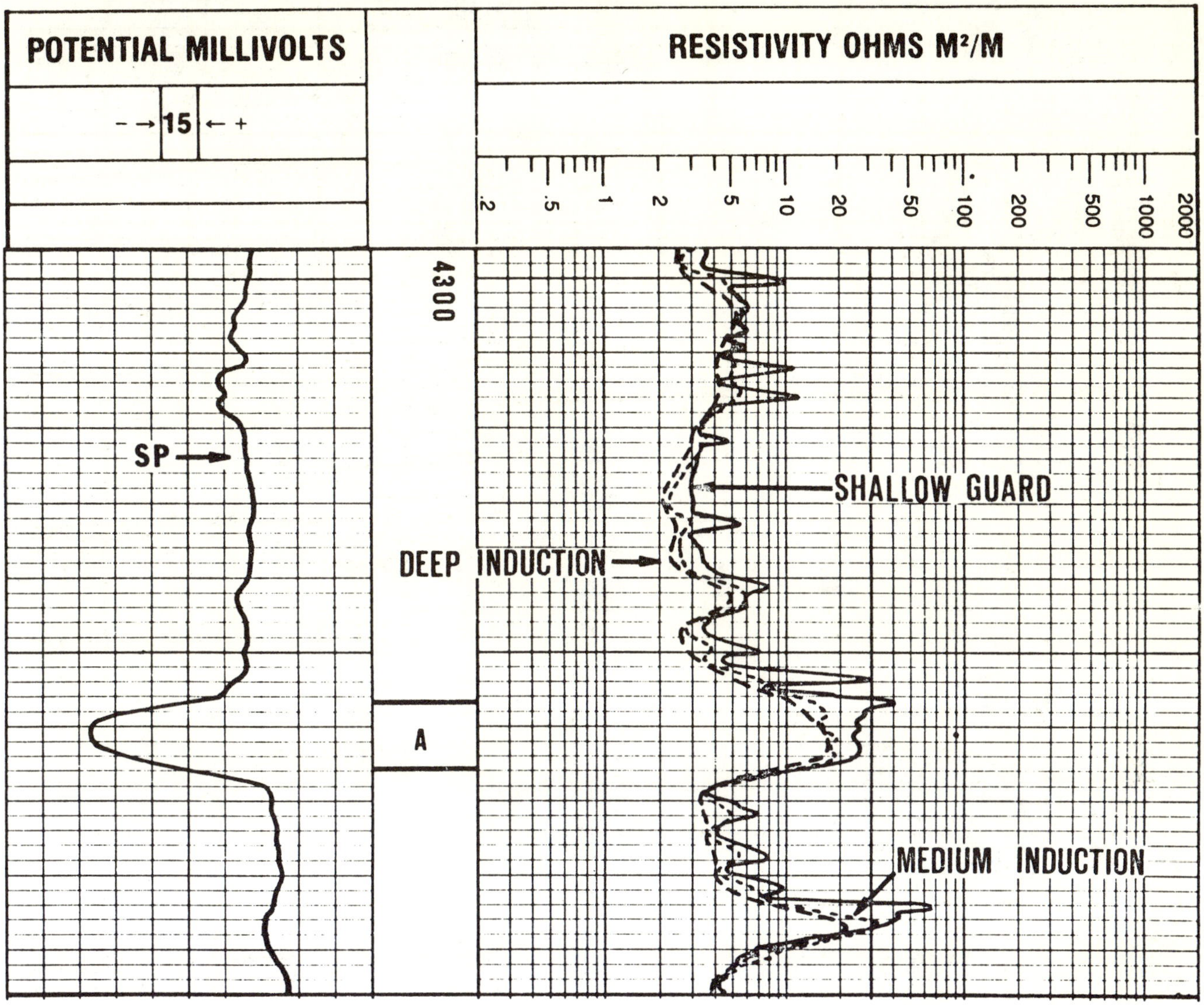

Location: East Texas

Formation: Cotton Valley Trend

Figure 11-8. Field example of a resistivity log. (Courtesy of Welex.)

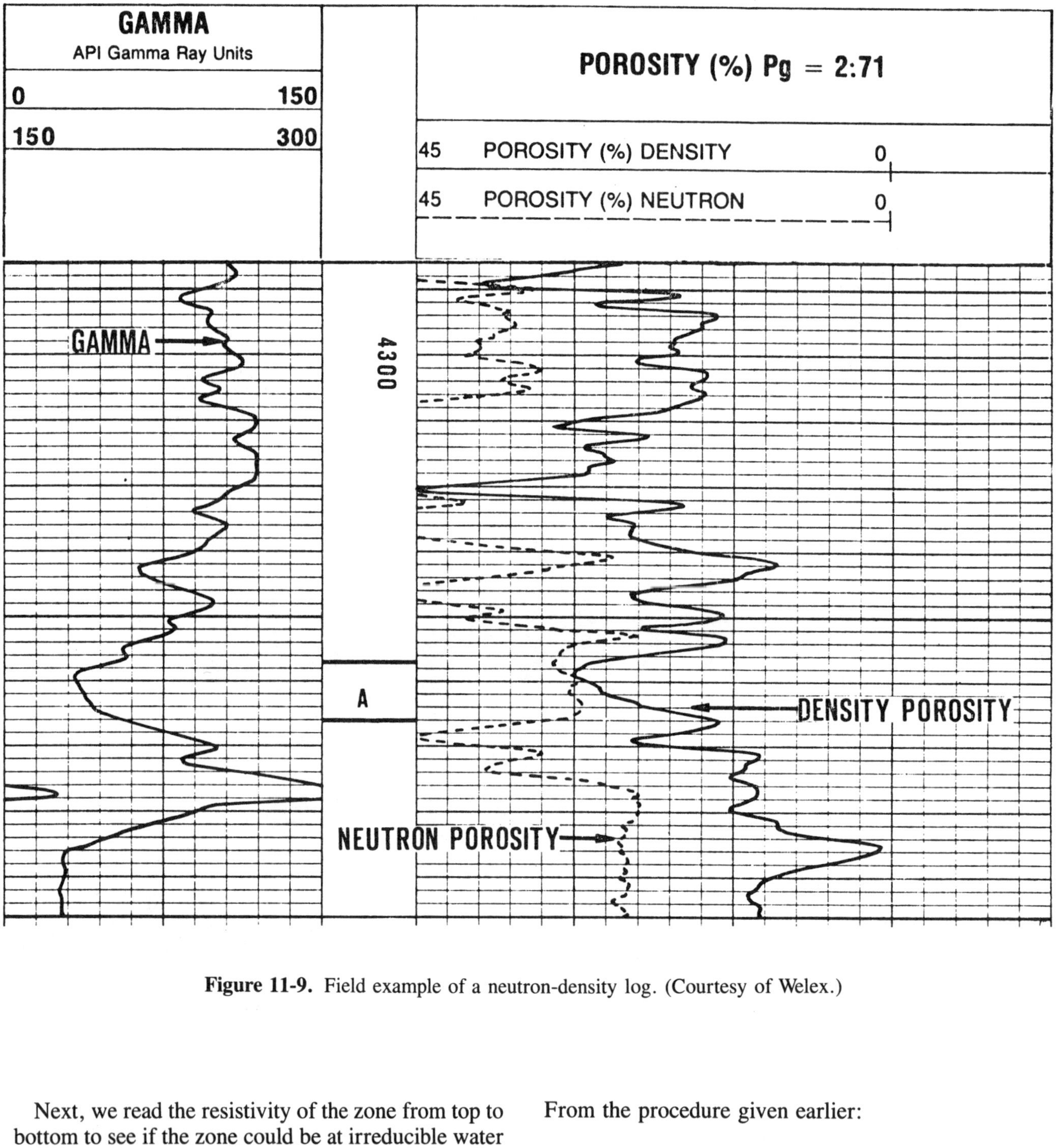

Figure 11-9. Field example of a neutron-density log. (Courtesy of Welex.)

Next, we read the resistivity of the zone from top to bottom to see if the zone could be at irreducible water and if not, then what oil/water cut we can expect.

	R_t	(R_w/R_t)
Top of zone	10	.0166
	12	.0137
	16	.0102
	18	.0091
	18	.0091
Bottom of zone	10	.0164

From the procedure given earlier:

$$(R_w/R_t)_{avg} = .0125$$

by adding the above ratios and dividing by 6.

$$(R_w/R_t)_{min} = .0091$$

lowest value.

From Table 11-1 (beginning of the chapter):

$$(R_w/R_t)_{max} = 2.5 \times .0091 = 0.0228$$

If $(R_w/R_t)_{avg} \leq (R_w/R_t)_{max}$, then the zone is at irreducible.

$$(.0125)_{avg} \leq (.0228)_{max}$$

This is certainly the case.

Further computation is not necessary, since the reservoir will produce water-free.

For the sake of completeness, the permeability of the zone is estimated.

Given: $R_d = 12$
$R_{med} = 16$
$R_s = 40$

we calculate their ratios and enter the tornado chart of Figure 11-10.

$$R_s/R_d = 40/12 = 3.3$$

$$R_{med}/R_d = 16/12 = 1.3$$

$$R_t = .975 \times 12 = 11.7$$

$$R_{xo} = 5 \times 11.7 = 58.5$$

R_{mf} @ T_f is .6 ohms

$$S_{hr} = 1 - \sqrt{\frac{R_{mf}\,F}{R_{xo}}}$$

$$S_{hr} = 1 - \sqrt{\frac{.6 \times 11.1}{58.5}}$$

$$S_{hr} = 66\%$$

Hydrocarbon density is given by the following equation *for oil*:

$$\rho h = \frac{.7 \times (1 - \Phi_n/\Phi_d)\,S_{hr} - (1 - \Phi_n/\Phi_d)}{(1 + .72\,\Phi_n/\Phi_d)\,S_{hr}}$$

Since $\Phi_n/\Phi_d = 1$

$$\rho_h = \frac{.7 \times 2 \times .66}{1.72 \times .66} = 0.814$$

Though not needed for this computation, it is of value to know the quality of the oil.

$$\text{API}^\circ\ (\text{oil}) = \frac{141.5}{.814} - 131.5$$

$$\text{API}^\circ = 42 \text{ or very light oil.}$$

The resistivity varies from 12 to 18 ohms over an interval of 8 ft. Thus the resistivity gradient is:

$$a = \frac{6}{8} \times \frac{1}{R_o} = \frac{6}{8} \times \frac{1}{1.82} = .41$$

The mud indicates less than 50 KPPM NaCl, so ρ_w is 1.0.

The density difference is $1 - .814 = .186$

Plotting these two values yields a permeability of about 600 md, according to the chart of Figure 11-11.

Example 2

Figures 11-12 and 11-13 show the logs for a well which contains gas in the upper interval. This is evident from the neutron density crossover on the porosity log.

Looking at the upper zone of the resistivity log (Figure 11-12) it is doubtful that it is at irreducible. Both the resistivity and porosity are not constant.

Plotting 4,050 ft on the chart found in Figure 11-14, we obtain a gas density of .1 g/cc.

Two points will be chosen for the initial analysis:

Data Point	Depth	Φ_n	Φ_d	R_{deep}
1	4,060	10%	23.8%	46 ohms
2	4,075	11%	30%	3 ohms

Solving for S_{hr} using the gas saturation chart on Figure 11-15:

At 4,060 ft $S_{hr} = 52\%$ and $\Phi_{xp} = 17\%$
At 4,075 ft $S_{hr} = 60\%$ and $\Phi_{xp} = 22\%$

One point should be made here before continuing. The Welex neutron-density log found on Figure 11-13 uses a nonstandard curve coding which can be confusing. The dashed curve should represent the neutron response, whereas it is represented by the solid curve here.

(text continued on page 172)

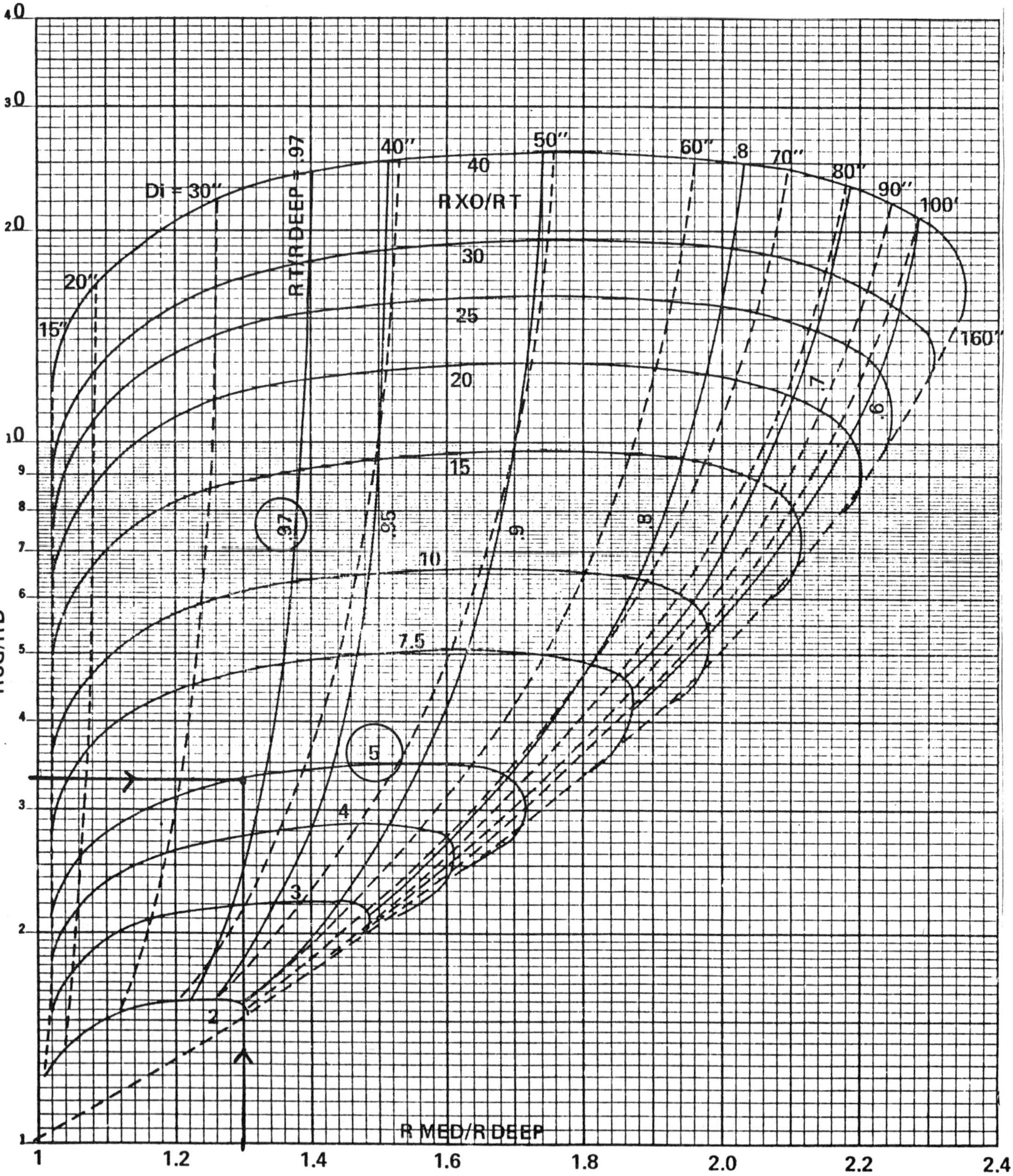

Figure 11-10. Tornado chart showing resistivity plot of a field example. (Courtesy of Welex.)

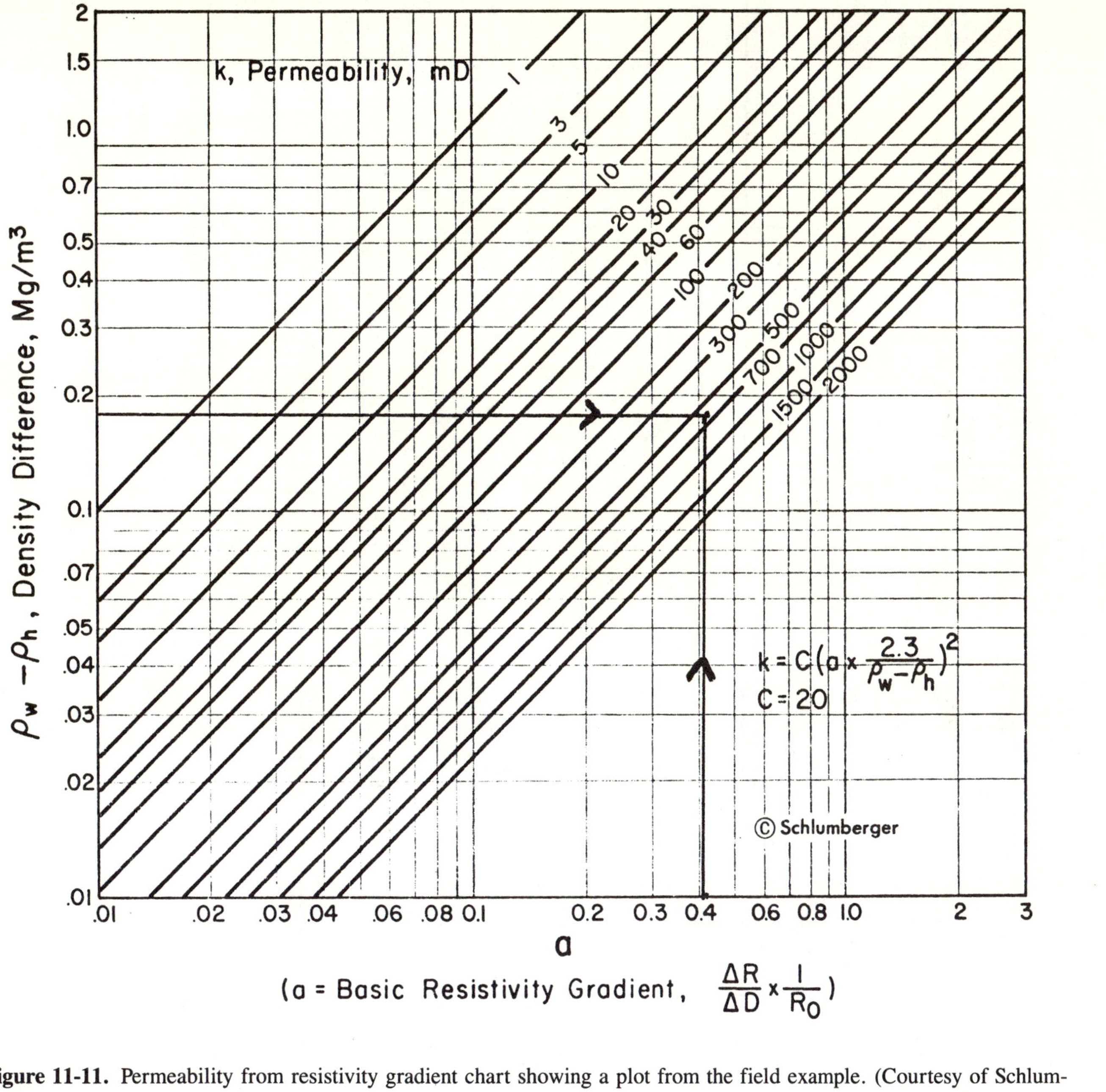

Figure 11-11. Permeability from resistivity gradient chart showing a plot from the field example. (Courtesy of Schlumberger.)

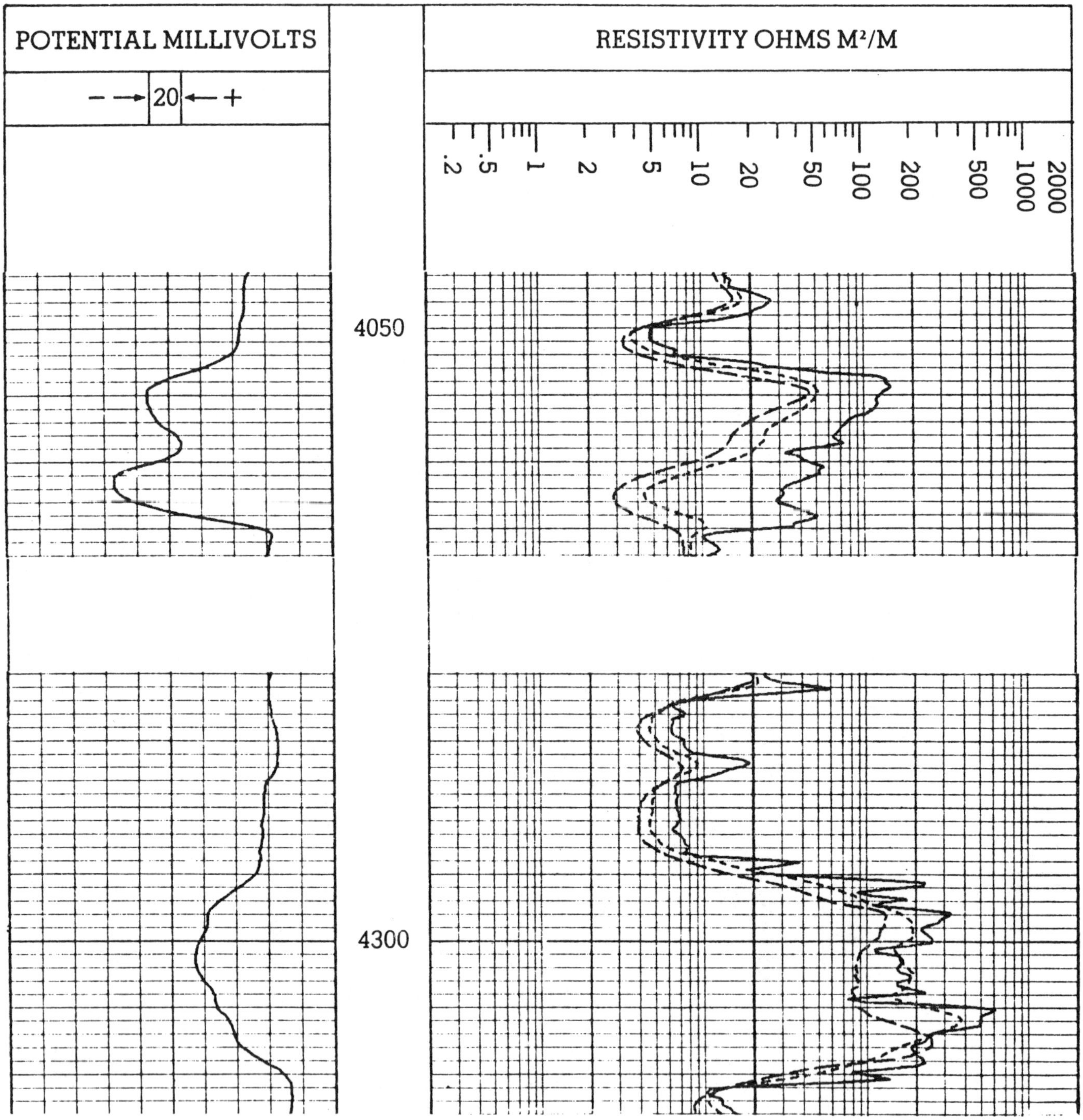

Figure 11-12. Field examples of a formation containing gas. (Courtesy of Welex.)

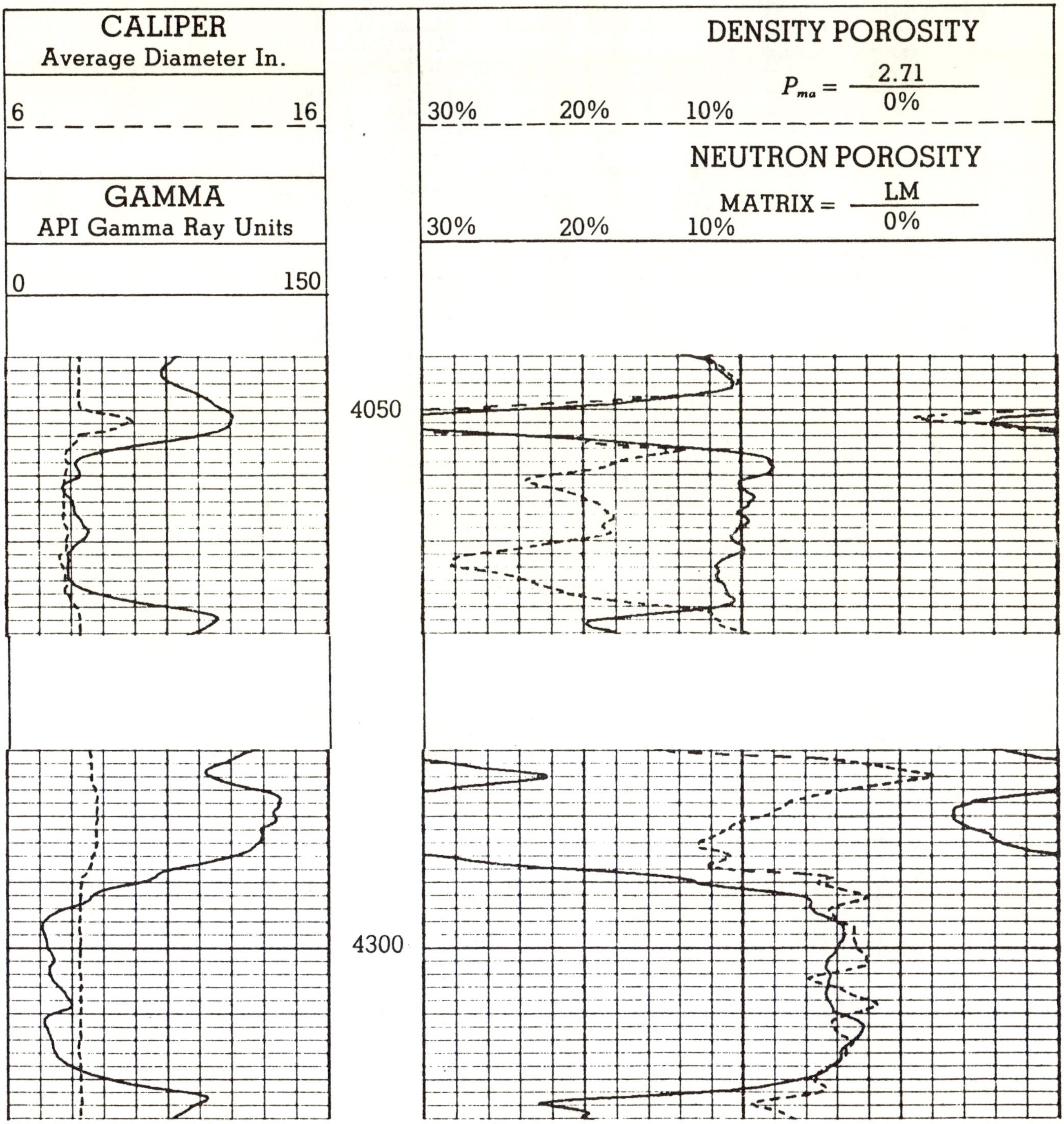

Figure 11-13. Nonstandard coding of the neutron density curves. (Courtesy of Welex.)

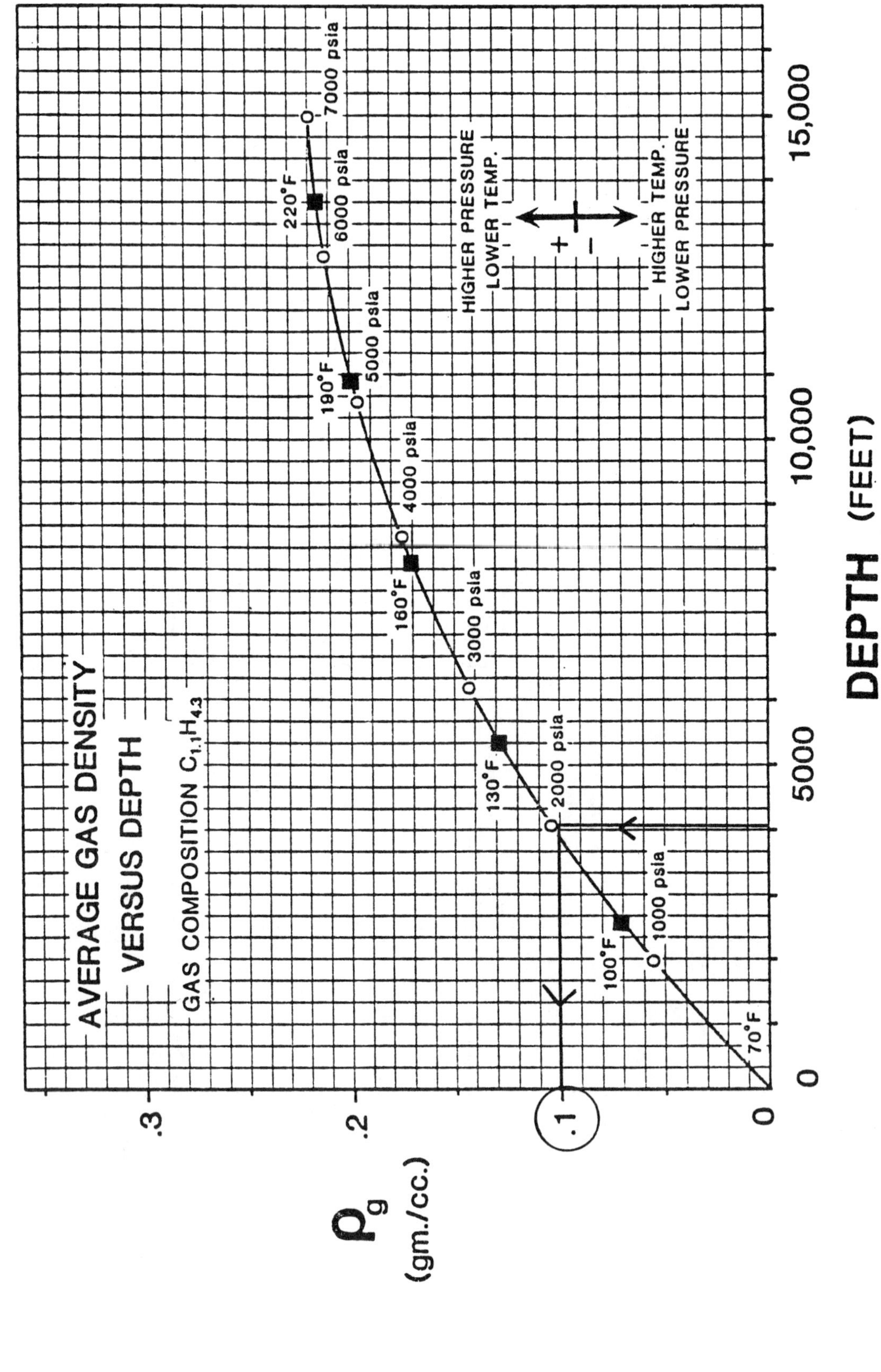

Figure 11-14. Estimating gas density as a function of depth. (Courtesy of Welex.)

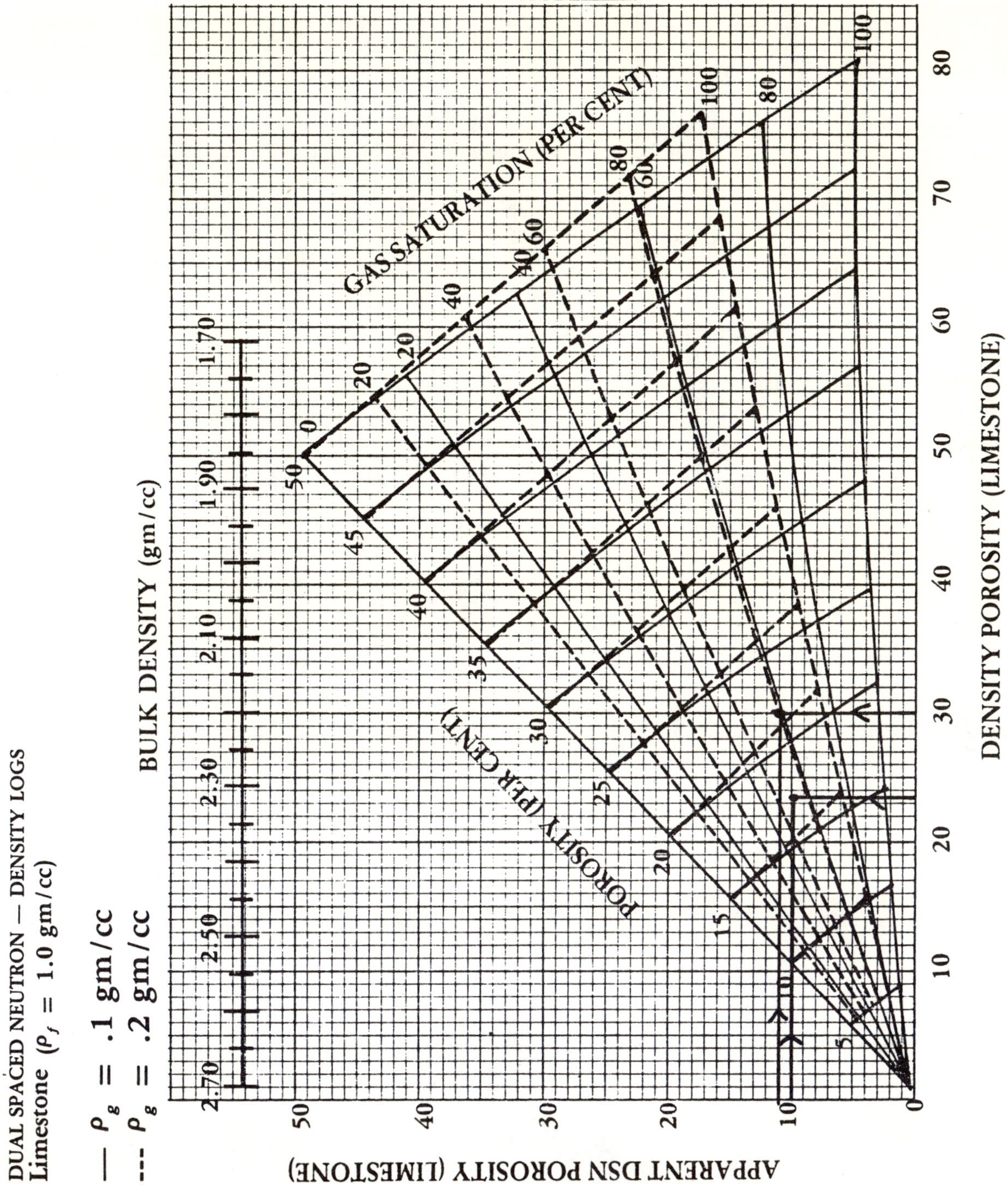

Figure 11-15. Porosity and gas saturation crossplot showing a plot of the field example. (Courtesy of Welex.)

From the discussion of Chapter 9:

$$\Phi_{dcorr} = \Phi_d - \Delta\Phi_d \text{ (4,060 ft)}$$

$$\Delta\Phi_d = \frac{1.07 \times .238 \times .52\ (1.11\ (1 - .1) - .03)}{2.71 - 1}$$

$$\Delta\Phi_d = 0.075$$

$$\Phi_{dcorr} = .238 - .075 = .163 \text{ or } 16.3\%$$

$$\Phi_{dcorr} = \Phi_d - \Delta\Phi_d \text{ (4,075 ft)}$$

$$\Delta\Phi_d = \frac{1.07 \times .3 \times .6\ (1.11\ (1 - .1) - .03)}{2.71 - 1}$$

$$\Phi_{dcorr} = .30 - .109 = .191 \text{ or } 19.1\%$$

$$\Phi_{ncorr} = \Phi_n - \Delta\Phi_n \text{ (4,060 ft)}$$

$$\Delta\Phi_n = .1 \times .52\ (2.2 \times .1 - 1) = -.0406$$

$$\Phi_{ncorr} = .1 - (-.0406) = .1406 \text{ or } 14\%$$

$$\Phi_{ncorr} = \Phi_n - \Delta\Phi_n \text{ (4,075 ft)}$$

$$\Delta\Phi_n = .11 \times .6\ (2.2 \times .1 - 1) = -.0515$$

$$\Phi_{ncorr} = .11 - (-.0515) = .1615 \text{ or } 16.2\%$$

Thus we crossplot Φ_n 14%, Φ_d 16.3% for 4,060 ft, and Φ_n 16.2%, Φ_d 19.1% for 4,075 ft on Figure 11-16.

At 4,060 ft Φ_{xp} = 15% and matrix value mix is 85%/15% which gives us a matrix value of 2.69.

At 4,075 ft Φ_{xp} = 17.9% and the matrix mix is also 85%/15%, indicating the matrix value also to be 2.69.

Using an R_w of .04 ohms:
At 4,060 ft

$$F = \frac{1}{.15^2} = 44.4$$

$$R_o = 44.4 \times .04 = 1.78$$

$$S_w = \sqrt{\frac{1.78}{48}} = .193 \text{ or } 19.3\%$$

At 4,075 ft

$$F = \frac{1}{.191^2} = 27.4$$

$$R_o = 27.4 \times .04 = 1.09$$

$$S_w = \sqrt{\frac{1.09}{3}} = .604 \text{ or } 60.4\%$$

We have the two extremes of the reservoir. To make the analysis thorough, one should calculate a statistically significant sum of points in between. This should yield the point at which too much water will be produced.

To continue, we calculate R_{tirr} at the bottom of the zone.

At 4,075 ft

$$R_{tirr} = \frac{R_w}{\left(\dfrac{10^{-6}}{G\ (\rho_{ma} - 2.6)^3}\right)}$$

where G = 1

$$R_{tirr} = 29.16$$

This suggests that the value of G = 1 is not appropriate, because the resistivity at the top of the zone is 46 ohms, indicating, with great confidence, that the very top is irreducible. Therefore, an R_{tirr} of 46 ohms is used.

$$R_{tcorr} = R_{tirr}\ (.077 + 1.55\ \rho_h - .627\ \rho_h^2)$$

$$R_{tcorr} = 10.4$$

and

$$W = \sqrt{3.75 - + .5 \log R_w/R_{tcorr} + 2.2)^2}$$

$$W = 1.89$$

Now for the same bottom zone we calculate the permeability:

$$K = \left[\frac{300\ (.077 + 1.55\ \rho_h - .627\ \rho_h^2)\ \Phi^{2w}}{W^4\ (R_w/R_{tirr})}\right]^2$$

where K is in md.

$$K = 23.6 \text{ md.}$$

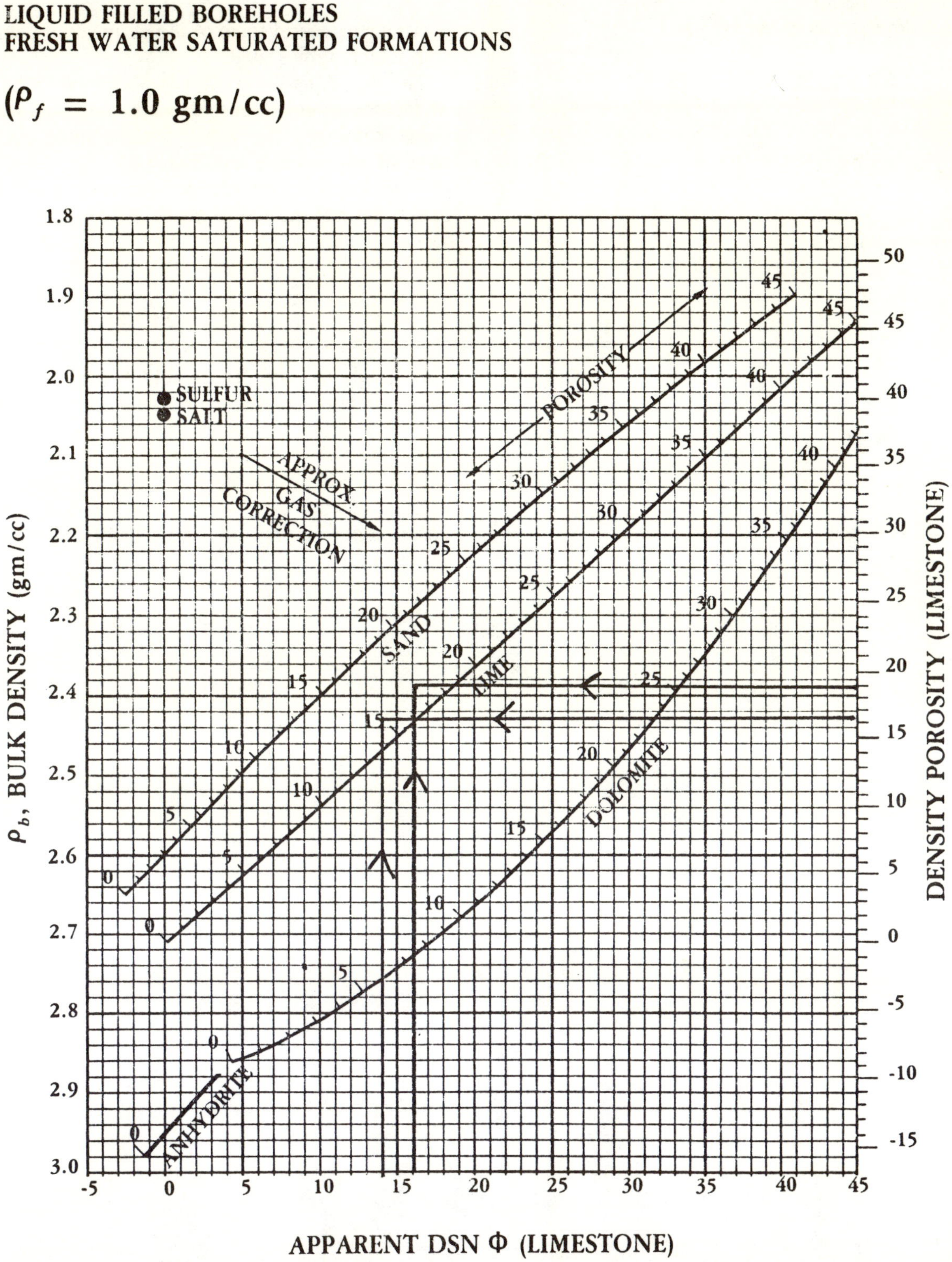

Figure 11-16. Neutron density crossplot showing a plot of the field example. (Courtesy of Welex.)

This figure agrees quite well with the average permeability of the area.
Irreducible water is given by:

$$S_{wirr} = \left(\frac{(R_w/R_{tirr})}{\Phi}\right)^{1/w}$$

$$S_{wirr} = 12.6\%$$

To further confirm:

$$K^{\frac{1}{2}} = \frac{79\ \Phi^3}{S_{wirr}}$$

Transposing:

$$S_{wirr} = \frac{79\ \Phi^3}{K^{\frac{1}{2}}}$$

where K is in md.

$$S_{wirr} = 11.5\%$$

Another general form of the permeability computation which agrees well.

The S_{wirr} equation used was preferable because the W factor accounted for hydrocarbon density as well as some lithology. The zone contains shale but in the zone of interest it is less than 10%.

Water cut for a gas reservoir involves many factors which are beyond the scope of this text.

Conclusion

Permeability and irreducible water computations are extremely helpful to the analyst in determining oil/water ratios, hydrocarbon/water contacts, and the economic factors of production.

One should always keep in mind that permeability calculations are merely estimates based on empirical observations and not on pure scientific fact.

12

New Technology

As technology advances at a geometrical rate, the ability of making more sophisticated measurements is greatly enhanced. A new generation of tools has come about in recent years that hold promise. It is too early to tell if any or all of them will prove to be as tried and true as our familiar logs. Perhaps some will go the way of the Redox Potential tool (oxidation reduction) or become as popular as the neutron-density tools. Only time and experience will tell.

Gamma Ray Spectrometry

The late 1930s saw the development of the first natural gamma ray detector tool which we are all familiar with today. Recent advances in technology allow us to look at the same natural gamma ray radiation in a different way. When we see an ordinary gamma ray curve on a log, we see the total energy emission of the series uranium, thorium, and potassium. The spectral gamma ray tool separates the energy levels of each component so that we may be able to view each separately.

There are three types of natural potassium isotopes:

K 39
K 40
K 41

K 40 is the only radioactive isotope of the three. The ratio of potassium 40 to total potassium has been observed to be a constant on the earth. Since this ratio is constant, the measurement of K 40 provides a direct indication of a formation's total potassium content.

In the alteration process, micas and feldspars are largely destroyed, yielding the clay minerals illite, interstratified illite montmorillonite, montmorillonite, chlorite, and kaolinite. A small part of the total potassium concentration enters into the formation of those clay minerals, but the major part is dissolved by water.

Montmorillonite disappears progressively by illitization and chloritization as the shales are buried and pressure and temperature increase. Kaolinite disappears in the process. During the transformation, the amount of potassium increases in the clay.

Thorium has only one long-lived isotope: Th 232. The average concentration of thorium in the earth's crust is 12 PPM.

The rock origin of thorium is silicic igneous rock in which it exists as a number of minerals, such as cheralite, huttonite, thorite, etc. Thorium compounds are insoluble. During alteration, thorium is easily hydrolyzed, giving it a tendency to concentrate in residual minerals such as clay and bauxite. Due to its insolubility, thorium is mostly transported in suspension. As a result, thorium concentrates in the silty fraction of shales.

Uranium can be either soluble or insoluble. Uranium can be oxidized to form insoluble uraninite. When combined with organic acids, uranium forms soluble uranyl salts. Uranium is chiefly transported in solution. Some is transported by suspension in the form of heavy metals, such as phosphate, schoepite, or as uraninite. Uranium can precipitate carbonate complexes, sulfate complexes, and organic and phosphate complexes. It can also be absorbed by amorphous silica, alum, aluminum, coal, and lignite.

Thermal diagenesis of organic matter, which produces hydrocarbon, does not affect the uranium concentration. The uranium remains with the organic matter and not the hydrocarbon.

With a brief study of each component behind us, one or two applications are obvious. There are cases when perhaps the gamma ray log may indicate streaks of shale. Since uranium (a component of the total gamma ray log) can exist in sandstones, as well as carbonates, it is possible to mistake uranium streaks for shale. The gamma spectrometry tool can differentiate between the two cases.

The other obvious application is that of a better shale volume indicator. In the past, we have used the gamma ray tool as an indicator of shale volume by the equation:

$$V_{sh} = \frac{GR - GR_{clean}}{GR_{sh} - GR_{clean}}$$

It is now possible to examine the shale volume of each component of the gamma spectrum.

$$V_{sh} = \frac{Th - Th_{clean}}{Th_{sh} - Th_{clean}} \quad \text{for thorium}$$

$$V_{sh} = \frac{U - U_{clean}}{U_{sh} - U_{clean}} \quad \text{for uranium}$$

$$V_{sh} = \frac{K - K_{clean}}{K_{sh} - K_{clean}} \quad \text{for potassium}$$

Where clean refers to the minimum reading in a nonshaly (clean) formation. Sh is the reading in 100% shale, and the unsubscripted variable is in the zone of interest.

Uranium, of the three components, is perhaps not a good indicator of shale because it is found in other organic materials. It is possible that we may mistake shale for uranium-rich formation matrix. If uranium is excluded as an indicator of shale, then we can define a fourth curve, combining thorium and potassium, as follows:

$$GRS = A\,Th + CK$$

where A and C are calibration constants, and GRS is the gamma ray shale curve.

We may now refine the shale volume by the substitution of GRS in the shale volume equation.

$$V_{sh} = \frac{GRS - GRS_{clean}}{GRS_{sh} - GRS_{clean}}$$

All four curves (GRS, U, Th, and K) are recorded on the log. Further interpretation of this log is lacking predictability at this time, and may require the use of sophisticated computer power to extract the information. Figure 12-1 shows the log configuration.

Photoelectric Absorption Index

The new litho-density tool is an improved version of the compensated density tool, with the addition of a curve called Pe. Pe is the index of the effective photoelectric absorption cross section of the formation.

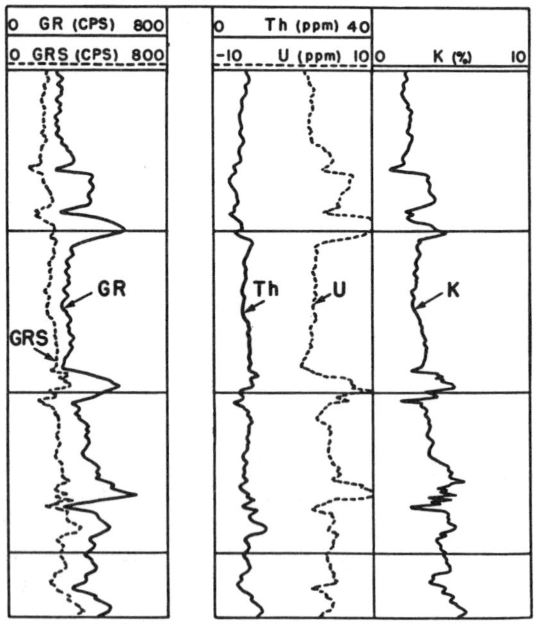

Figure 12-1. Gamma ray spectometry log. (Courtesy of Schlumberger.)

When matter is bombarded by gamma radiation, three interactions take place. The three interactions are as follows:

1. Compton scattering—when a gamma ray collides with an atom and dislodges an electron, and is not absorbed in the process, it is called compton scattering.
2. Pair production—this is when the gamma ray produces an electron-positron pair conforming to Einstein's equation of $E = MC^2$.
3. Photoelectric absorption—the gamma ray is lost as a result of collision, and its energy is given to an orbiting electron of one of the formation's nuclei.

The bulk density measurement uses the compton scattering principle because it is not sensitive to the specific material of the formation. The measurement relates electron density to bulk density of the ratio of the atomic number to its mass. This relationship is nearly constant for formation materials, thus isolating bulk density.

$$\rho_E = 2\,\rho_b \frac{Z_1}{M}$$

where ρ_E = electron density
ρ_b = bulk density
Z_1/M = 0.5 for most formation elements

At low gamma ray energies, photoelectric absorption takes place. Photoelectric absorption is highly dependent on the atomic number of the nuclei in the formation. The atomic number refers to the number of protons in the nucleus of the atom. As this number increases, so does the electron count. Photoelectric absorption is increased in probability as the number of electrons increase. If each element has a different atomic number, the rate of absorption will vary accordingly.

Compton scattering occurs at an energy level which is higher than the level required for photoelectric absorption. A density tool is designed to see compton scattering which yields bulk density. If the same tool is allowed to sample the gamma ray count for a slightly longer time, it would measure the rate of photoelectric absorption.

Pair production occurs at high energy levels. By using nuclear radioactive sources below its level, the possibility of pair production is eliminated. The source is chosen to produce compton scattering and as time elapses and gamma rays lose (transfer) their energy, photoelectric absorption occurs.

This gives two useful parameters. One measures the total bulk of material which is relatively insensitive to the material, while the other is sensitive to the material itself. ρ_b is, in essence, a porosity reading insensitive to the material, and Pe is sensitive to the material but insensitive to the porosity.

The lithology-density tool looks at both parameters and presents ρ_b and Pe (for lithology).

One other parameter of interest is the effective photoelectric absorption cross section index per unit volume, which is designated by the letter U. By definition:

$$U = Pe\,\rho_e$$

$$\text{where } \rho_e = \frac{\rho_b + 0.1883}{1.0704}$$

ρ_e is a measure of the number of electrons per unit volume.

In more traditional terms, we may express:

$$\rho_b = \Phi\,\rho_f + (1 - \Phi)\,\rho_{ma}$$

Transposing:

$$\Phi = \frac{\rho_b - \rho_{ma}}{\rho_{ma} - \rho_f}$$

Accordingly, we express U in similar terms:

$$U = \Phi\,U_f + (1 - \Phi)\,U_{ma}$$

or

$$\Phi = \frac{U - U_{ma}}{U_{ma} - U_f}$$

where U, U_f, and U_{ma} are in barns/volume (barn = 10^{-24} CM^2) and $U_f = 0.398$

Table 12-1 shows the values of interest for each parameter.

ρ_b for oil = 1.222 oil − 0.188

ρ_b for gas = 1.33 gas − 0.188

U for oil = 0.136 ρ oil

Table 12-1
Values of Interest for Each Parameter

Material	Pe	Matrix	ρ_b	U
Quartz	1.81	2.65	2.64	4.78
Calcite	5.08	2.71	2.71	13.8
Dolomite	3.14	2.87	2.88	9.0
Anhydrite	5.05	2.96	2.98	14.9
Halite	4.65	2.17	2.04	9.68
Siderite	14.7	3.94	3.89	55.9
Pyrite	17.0	5.0	4.99	82.1
Barite	267.0	4.48	4.09	1065.0
Water	0.358	1.0	1.0	0.398
Water (100 KPPM) NaCl	0.734	1.06	1.05	0.850
Water (200 KPPM) NaCl	1.12	1.12	1.11	1.36
Oil	0.119	—	—	—
Gas	0.095	—	—	—

U for gas = 0.119 ρ gas

As can be seen from the table, use of barite may forbid the use of the Pe curve.

Lithology Density Determination

We have seen that the Pe curve is related to the nature of the matrix and relatively unaffected by the porosity. The Pe measurement may be used in the following ways:

1. By itself, as a matrix indicator.
2. In combination with bulk density, to analyze a binary matrix and determine porosity.
3. In combination with density-neutron, for complex lithology determination and porosity (three-mineral matrix).

The first case simply involves reading the Pe curve from the log (Figure 12-2) and comparing the value to Table 12-1. The Pe curve is not entirely unaffected by porosity. Figure 12-3 shows an approximation of its effect.

The second case uses Pe and ρ_b to derive a crossplot porosity and the mineral pair involved. The use of this analysis is somewhat misleading because in order for it to be useful, one must know what the exact mineral pair is. If the quantity is not known, then a density neutron crossplot must be performed first.

Assuming that the mineral pair is unknown, ρ_b and Φ_n are plotted on the left-hand crossplot of Figure 12-4 (this is a conventional neutron-density crossplot). Apparent true porosity (Φ_{ta}) is noted, and the mineral is also noted. For instance, $\rho_b = 2.62$ and $\Phi_n = 14\%$; this plots between calcite (lime) and dolomite, with an apparent true porosity of 10%.

Next we plot ρ_b and Pe on the right-hand chart of Figure 12-4. Since the mineral pair calcium-dolomite is indicated, the ribs (lines) connecting the points C & D are used. If the mineral pair quartz (sand)-dolomite were indicated, then the ribs connecting the Q-D points are used. The C-Q ribs are similarly used for the calcium-quartz pair. Using the example for the neutron-density (left half of chart), we plot $\rho_b = 2.62$ and Pe = 3.8 to obtain an apparent true porosity of 10%. If both values of apparent true porosity are equal, then the right mineral pair is chosen, and the pore space is liquid filled (oil and/or water). If the values do not agree, another pair of minerals are chosen.

Where gas is concerned, the values may not be reconcilable. Assuming that the matrix value pair is

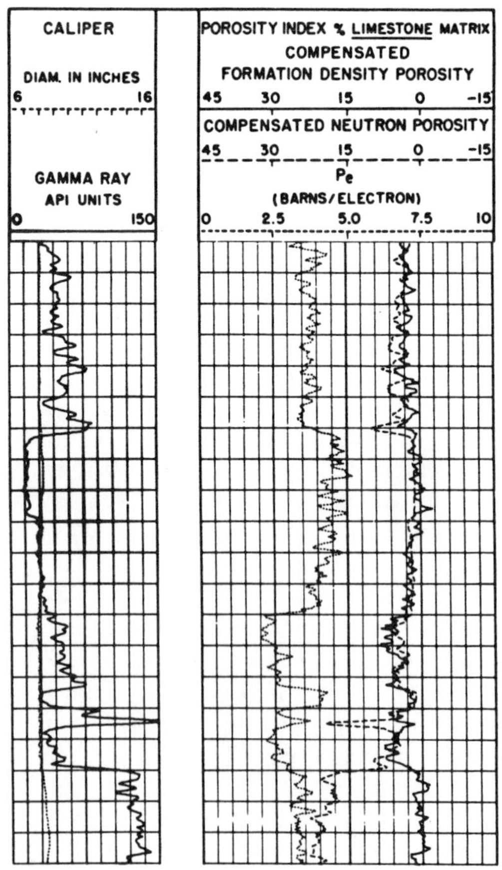

Figure 12-2. Litho density log showing the photoelectric absorption index. (Courtesy of Schlumberger.)

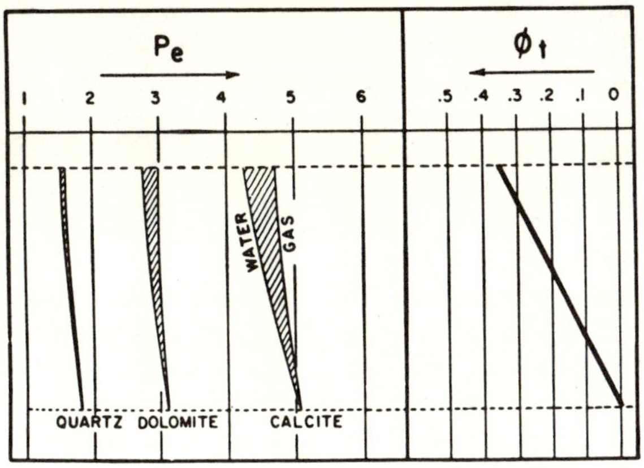

Porosity Effect on P_e

Matrix	ϕ_t	100 percent H_2O	100 percent CH_4
Quartz	.00	1.81	1.81
	.35	1.54	1.76
Calcite	.00	5.08	5.08
	.35	4.23	4.96
Dolomite	.00	3.14	3.14
	.35	2.66	3.07

Figure 12-3. General patterns of the P_e curve. (Courtesy of Schlumberger.)

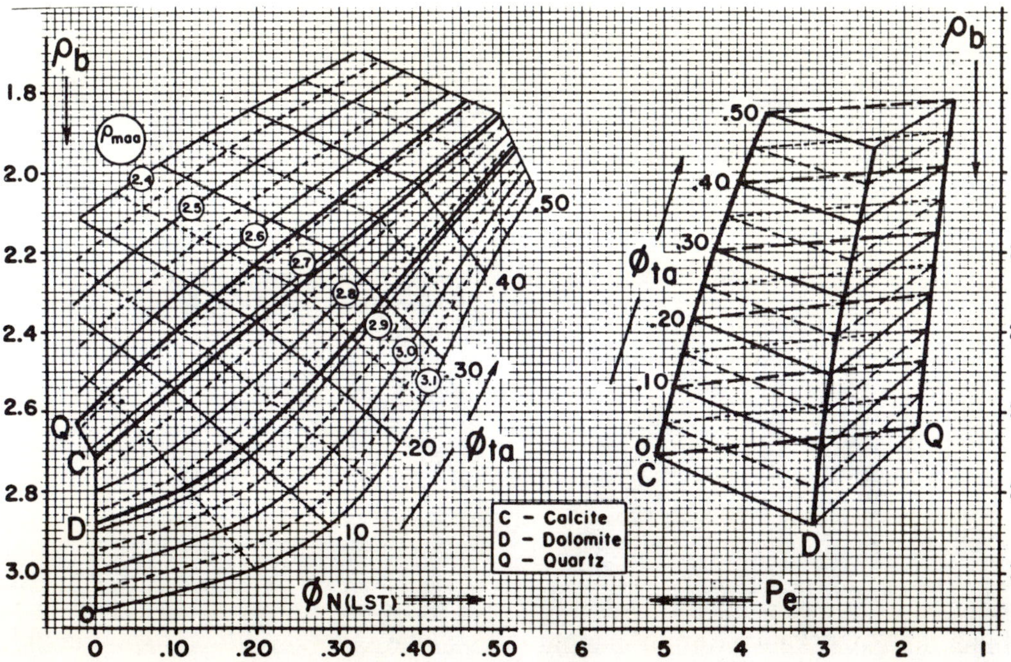

Figure 12-4. Neutron-density-P_e crossplot. (Courtesy of Schlumberger.)

known and gas is suspected, the Φ_n Φ_d crossplot porosity will be less than the ρ_b – Pe crossplot porosity.

The third case solves for a three-mineral matrix. In order to determine the relative proportions of the three minerals present, we need to know the apparent U of the matrix U_{maa}. The chart of Figure 12-5 may be used, or it may be solved by the following equation:

$$U_{maa} = \frac{Pe\ \rho_e - \Phi_{ta}\ U_f}{1 - \Phi_{ta}}$$

where $$\rho_e = \frac{\rho_b + 0.1883}{1.0704}$$

$U_f = 0.398$
Φ_{ta} = apparent true porosity (crossplot)

Pe is from the log

Working the example of Figure 12-5:

$Pe = 3.65$

$\rho_b = 2.52$

$\Phi_{ta} = 0.16$

$$\rho_e = \frac{2.52 + 0.1883}{1.0704} = 2.53$$

$$U_{maa} = \frac{3.65 \times 2.53 - 0.16 \times 0.398}{1 - 0.16}$$

$U_{maa} = 10.9$

Apparent matrix density is obtained by crossplotting neutron-density on the left half of the chart of

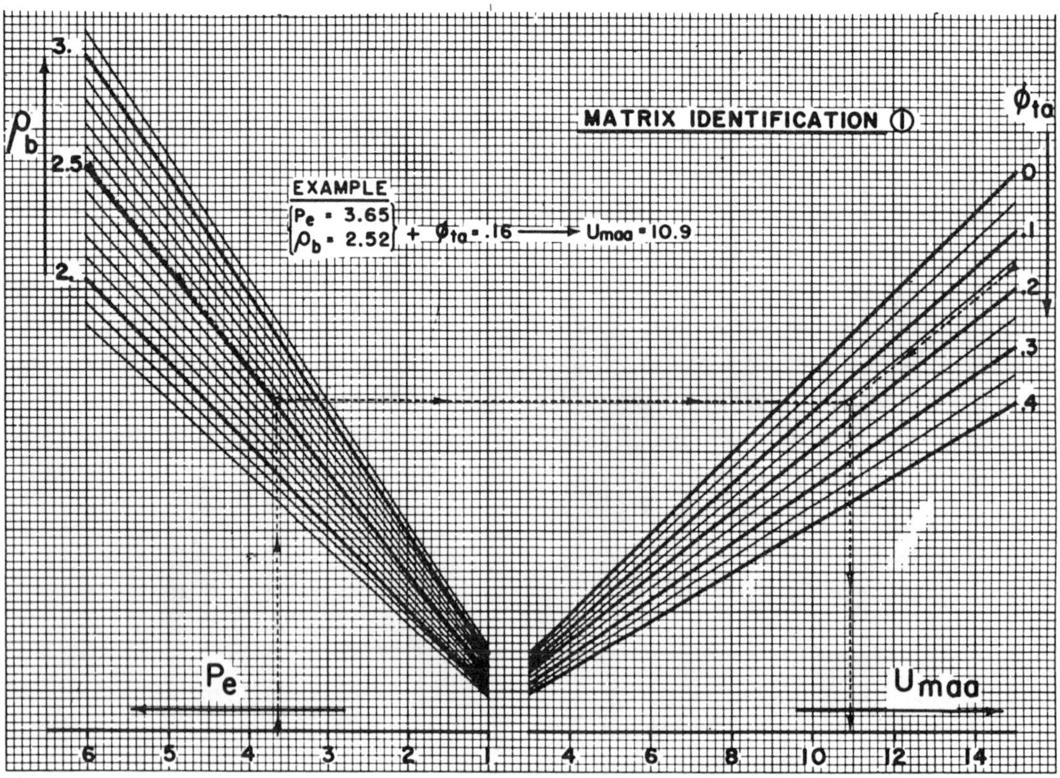

Figure 12-5. P_e matrix identification crossplot. (Courtesy of Schlumberger.)

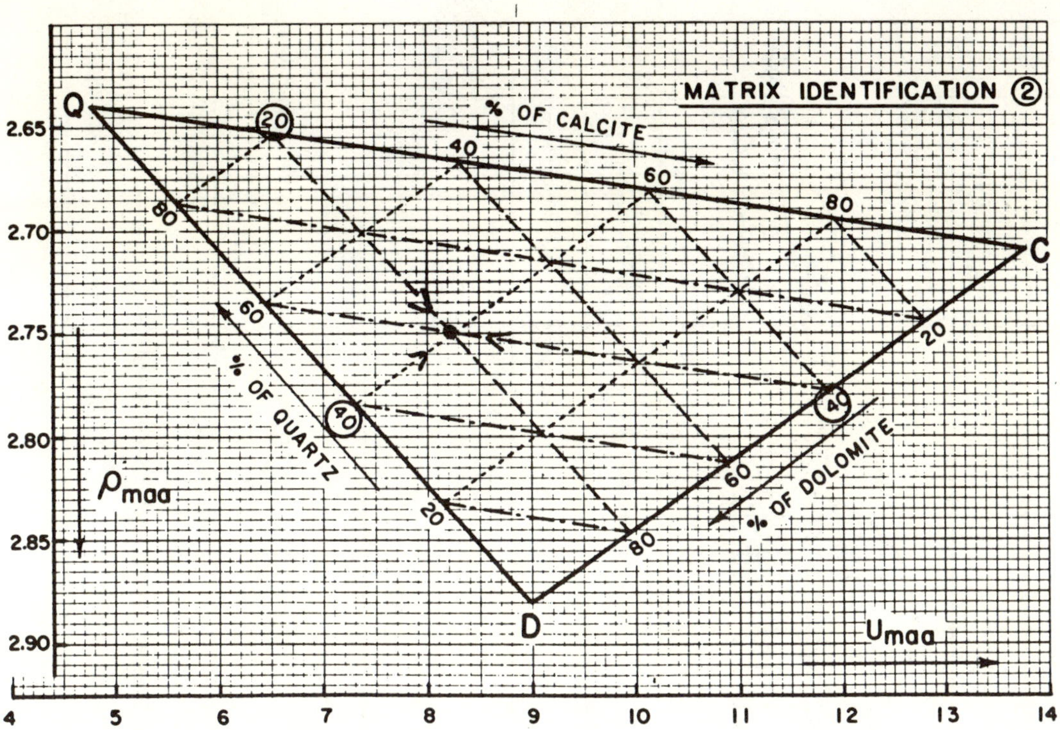

Figure 12-6. Matrix identification crossplot showing a point containing three minerals. (Courtesy of Schlumberger.)

Figure 12-4. U_{maa} and ρ_{maa} are plotted on the matrix identification chart of Figure 12-6. Given U_{maa} = 8.25 and ρ_{maa} = 2.75, the points plot on Figure 12-6 as follows:

- Quartz = 40% (sand)
- Calcite = 20% (lime)
- Dolomite = 40%

Figure 12-7 shows the general position of various minerals on the plot.

Electromagnetic Propagation

If you can think of the sonic tool operating at microwave frequencies, then you have the electromagnetic propagation tool. The tool is borehole compensated and just as the sonic, it contains two transmitters and two receivers. The basic principle of operation is electromagnetic propagation through a dielectric material. In short, just as sound passes through different materials at different rates of speed, so do electromagnetic waves. In the case of this tool, the importance of measuring water saturation requires that the transmitter frequency is slightly over one gigahertz (10^9 Hz). At this frequency, the dielectric constant of water as compared to other minerals or fluids in the formation is very obvious. As a result of the dielectric, the tool measures the propagation of the wave and thus presents an output "travel time" called T_{pl} (in nanoseconds per meter). Table 12-2 shows minerals with their dielectrics and associated propagation travel time.

The contrast between water and other minerals is very clear cut. How then, can T_{pl} be converted to useful information? Simple. If the time-average equation works for the sonic log, it should work for propagation time. Thus, ΦEPT* is calculated as follows:

$$\Phi EPT = \frac{T_{pl} - T_{pma}}{T_{pw} - T_{pma}}$$

where T_{pma} = matrix propagation time
T_{pw} = water propagation time
T_{pl} = log propagation time

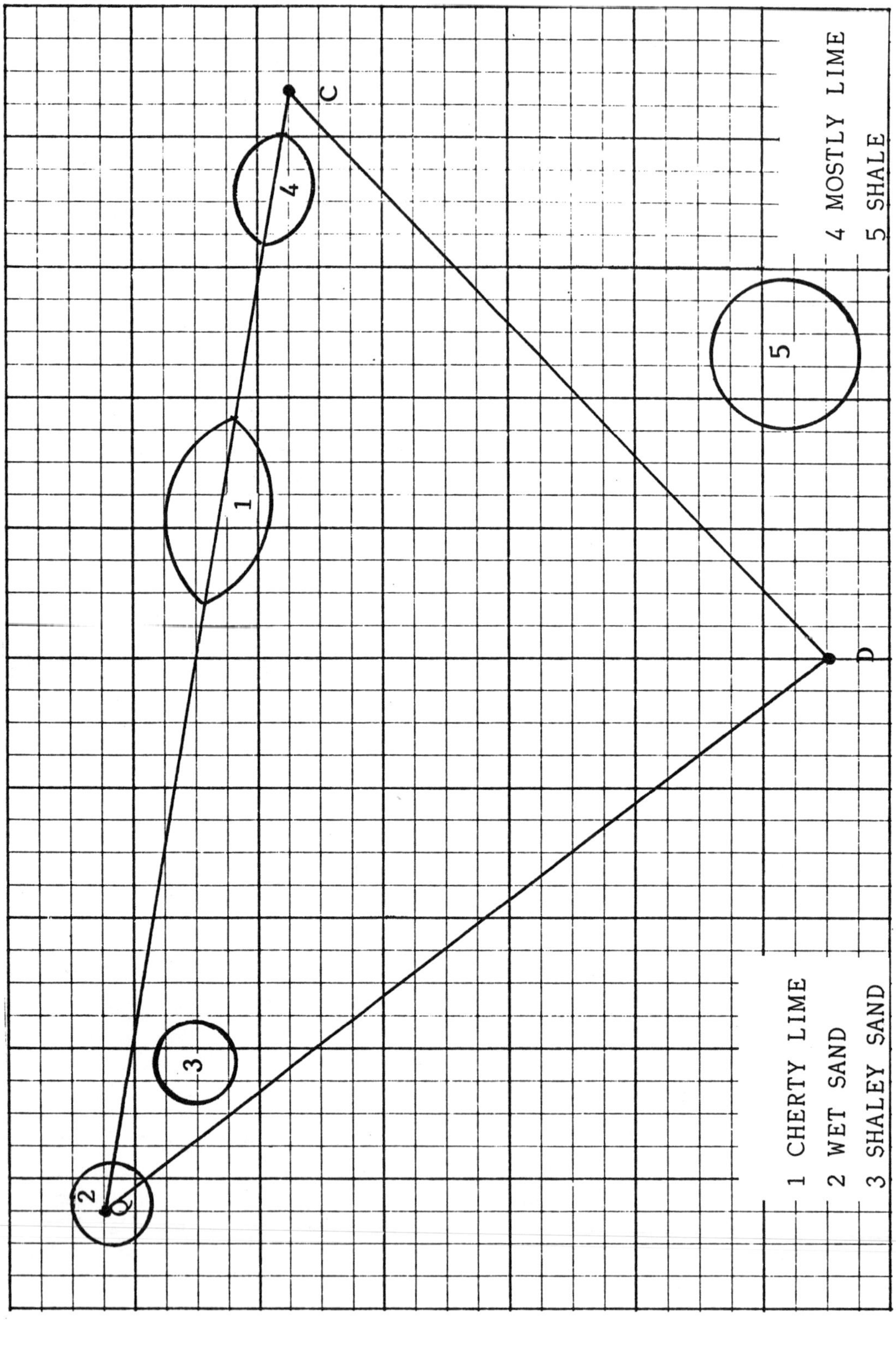

Figure 12-7. Illustration of plot areas for various minerals.

Table 12-2
Dielectrics and Associated Propagation Travel Time of Various Materials

Material	Dielectric Constant	Tpl nsec/m
Sandstone	4.65	7.2
Dolomite	6.8	8.7
Limestone	7.5–9.2	9.1–10.2
Anhydrite	6.35	8.4
Oil	2.0–2.4	4.7–5.2
Gas/Air	1.0	3.3
Shale	5–25	7.45–16.6
Water (Fresh)	78.3	29.5

and

$$T_{pw} = 20 \times \frac{710 - T_{f/3}}{444 - T_{f/3}}$$

where T_f is formation temperature in °F

Electromagnetic propagation porosity differs from conventional porosity in that it is measured very close to the borehole wall (1 in. to 4 in. depending on formation conductivity), and thus reacts to formation fluid and mud filtrate. If the EPT* porosity is equal to the neutron-density porosity, then the flushed zone water saturation is 100%. A good approximation is:

$$S_{xo} = \frac{\Phi EPT^*}{\Phi_{nd}}$$

A better value of S_{xo} can be obtained if the propagation time of the hydrocarbon is known.

$$S_{xo} = \frac{(T_{pl} - T_{pma}) + \Phi_{nd}(T_{pma} - T_{ph})}{\Phi_{nd}(T_{pw} - T_{ph})}$$

The bulk volume of water is then stated as:

$$BVW = S_{xo} \times \Phi_{nd}$$

The tool has a four-centimeter spacing between transmitter and receiver antennas. This spacing is necessary because of the microwave frequency used. On the tool's behalf, this spacing gives very fine vertical resolution making the oil/water contact easy to locate. Working against the tool is the shallow depth of investigation (1 in. to 4 in.).

* EPT is a registered trademark of Schlumberger

Advantages

- Analysis of heavy oil (movable oils)
- Fresh water detection
- Evaluation of S_{xo}

Disadvantages

- Shallow investigation
- Cannot tolerate more than 3/4 in. mud cake
- Unreliable in rugose holes
- Operable in fresh muds only

Electromagnetic propagation is an analyst's dream. It is unfortunate that its disadvantages limit its usefulness severely. For the tool to operate successfully, the hole must be as smooth as possible, and the mud must be balanced with great care, so as not to create excessive mud cake. Overall, for most cases, the expense involved in maintaining these conditions does not justify the data obtained. This tool has limited applications.

Conclusion

Of the three new services introduced in the past few years, spectral gamma ray logs and photoelectric effect logs are real and practical. While they do not add to the analysis, they do refine the art towards science. These two services are very likely to become as dear to us as neutron density.

Propagation tools, as noble as the idea seems, have a long road ahead to prove that their advantages are worth the cost of their disadvantages.

As of this writing the price difference of a compensated density and that of the lithology density is little or none in comparison. Any potential investor, engineer, geologist, or analyst should insist on the litho-density tool.

13

Shaly Sand Analysis

Shaly sand interpretation presents to the analyst an exercise not only in science, ability, and experience, but also intuition, insight, and talent. The interpretation of a shaly matrix requires the analyst to choose from several models available, evaluating each one answer and deciding which is correct.

This type of analysis has been reserved for the last chapter of this volume because one needs every tool available for successful analysis. Shales vary in characteristics and composition, making each one an individual case. What can be generalized about shales is at best broad and thus leaves room for interpretation (or misinterpretation).

This chapter presents the various methods and techniques available to the analyst.

Part I

Shale

Shale is a sedimentary rock composed of clay minerals, precipitates, and fine clastic particles. The chemical composition and physical nature of a shale cannot be uniquely defined. In various proportions, shales are composed of clay minerals, silt, carbonates, and other non-clay minerals. Silt is mostly fine-grained material that is predominantly quartz, possibly containing (in small quantities) feldspar and calcite. The clay minerals are kaolinite, illite, and montmorillonite.

If shale is composed of clay and silt, and silt is predominantly quartz, then our porosity tools should see the silt portion of the shale as sand matrix (assuming a sandstone matrix) and clay volume. That is to say, what we are really concerned with is clay and not shale. Log analysis has a tendency to place emphasis on lithological terms, rather than in terms of the mineral involved. Logging tools see the mineral rather than the lithology. We frequently use limestone rather than calcite and sandstone in place of quartz. Shale, being no exception, should be called clay. Not to be confusing, the term shale will be used throughout but it should be taken as clay.

Shaly Sands

Shale can be distributed in a sand in one of three ways. Laminated shale (Figure 13-1A) exists as layers of shale (thin shale beds or streaks) between sand beds. This type of shale has no effect on the intermatrix porosity or permeability of each individual sand bed. Note that while each sand bed is not affected by the shale laminae, the zone as a whole has less porosity. Figure 13-1A shows an equation which relates the effective porosity to the amount of shale present.

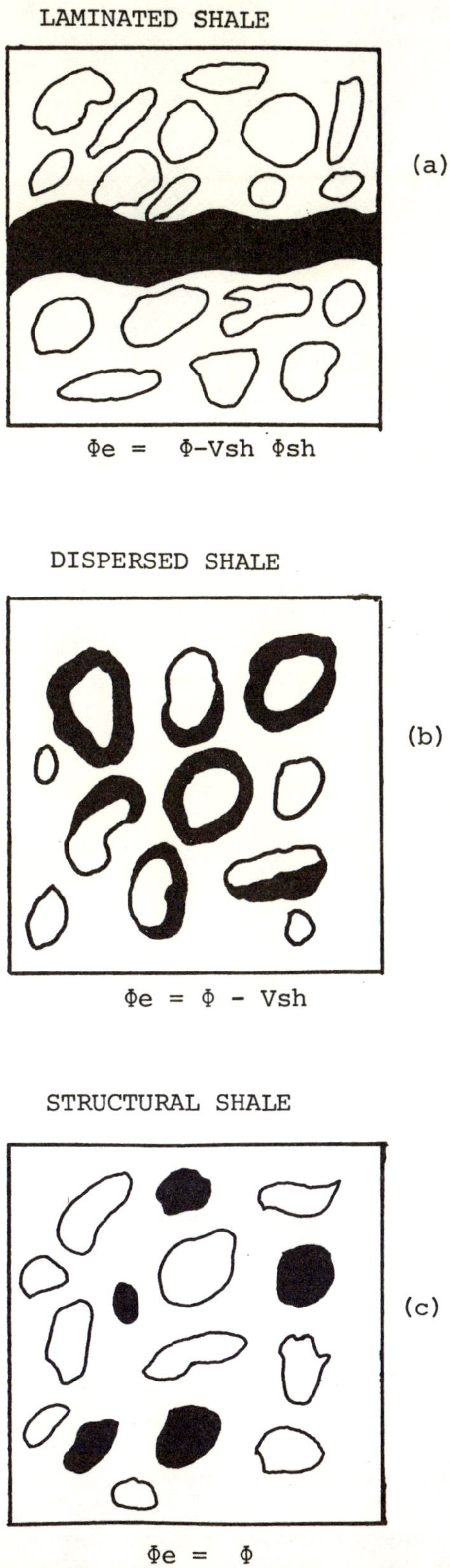

Figure 13-1. Illustration of three basic shale models: (A) laminated shale ($\phi_e = \phi - V_{sh} - \phi_{sh}$); (B) dispersed shale ($\phi_e = \phi - V_{sh}$); (C) structural shale ($\phi_e = \phi$).

Dispersed shale (Figure 13-1B) is shale that either accumulates and adheres to the grains of the sand, or acts as a cementing factor between two grains. Dispersed shale reduces porosity and permeability to a great degree. Effective porosity is reduced proportionately by the amount of shale present. An 18% porosity is thus reduced to 8% from a shale volume of 10%.

Structural shale, unlike the last two models, has no effect on porosity or permeability. Structural shale occurs in small granules just as the sand grains, and is considered part of the matrix. The shale replaces matrix rather than adds to it. Figure 13-1C shows graphically how the shale appears and its porosity relationship.

Porosity

The term porosity can be (and often is) confusing in terms of shale. While the previously mentioned porosities are actual porosities, it is not to say that log porosity will be unaffected. Tool response will be affected because the mineral makeup of the matrix has changed. Thus in the case of structural shale, the response of the particular porosity tool will represent a different porosity than what is actual. The sonic porosity, for instance, will not exhibit the true porosity even if the relationship for this model states that $\Phi_e = \Phi$.

The sonic porosity is affected by the type of shale present. Neutron and density porosity are affected by the shale and not the shale type (or model).

True Resistivity

In the case of laminated shale, the resistivity of the sand is affected by the resistivity of the shale by a parallel conductivity.

$$\frac{1}{R_t} = \frac{1 - V_{lam}}{R_{sd}} + \frac{V_{lam}}{R_{sh}}$$

where V_{lam} = the percent volume of laminated shale.
R_{sd} = the clean sand resistivity.
R_{sh} = the shale resistivity.

Thus, a clean sand with a resistivity of 8 ohms, a shale resistivity of 1 ohm, and a shale volume of 40% would show R_t as:

$$\frac{1}{R_t} = \frac{1 - 0.4}{8.0} + \frac{0.4}{1.0} = 0.475$$

$$R_t = \frac{1}{0.475} = 2.1 \text{ ohms}$$

As can be seen, the laminae can make R_t pessimistic and possibly mask the presence of hydrocarbons.

Dispersed shale acts as an electrolyte which conducts electrical current through a network composed of the pore, water, and dispersed shale. R_t is expressed as:

$$\frac{1}{R_t} = \frac{\Phi_{im}^2 S_{im}}{a}\left[\frac{q}{R_{dis}} + \frac{S_{im} - q}{R_w}\right]$$

where Φ_{im} = the intermatrix porosity which includes all of the space occupied by fluids and dispersed shale.
S_{im} = the fraction of the intermatrix porosity, Φ_{im}, occupied by the formation water, dispersed shale mixture.
q = the fraction of the intermatrix porosity occupied by the shale.

$$q = \frac{\Phi_{im} - \Phi}{\Phi_{im}}$$

or approximated by:

$$q \simeq \frac{\Phi_s - \Phi_d}{\Phi_s}$$

where d = density
s = sonic

Intermatrix porosity is taken from the sonic porosity. S_{im} constitutes a "slurry" of water and shale. If this is the case, then $S_{im} - q$ (q represents the shale portion) will yield only the water portion in the pore space. With this principle in mind, we can state that water saturation is related as:

$$S_w = \frac{(S_{im} - q)}{(1 - q)}$$

It is evident that in order to derive S_{im}, we need S_w. S_w is the parameter sought, therefore, a practical approach must be taken. Combining all of these relations, we may write the following equation:

$$S_w = \sqrt{\frac{\dfrac{aR_w}{\Phi_{im}^2 R_t} + \dfrac{(qR_{dis} - R_w)^2}{2R_{dis}} - \dfrac{q(R_{dis} + R_w)}{2R_{dis}}}{1 - q}}$$

where Φ_{im} = sonic porosity
R_{dis} = dispersed shale resistivity

R_{dis} is extremely difficult to evaluate since it may not have the same value as nearby bedded shales. Nearby bedded shales, as well as structural and laminated shales, are load-bearing. The pressure tends to squeeze water out of them, thus altering their resistivity and porosity response. Dispersed shale is not load bearing, therefore will most likely exhibit different characteristics than nearby shales.

The value of R_{dis} is several times greater than that of R_w. R_{dis} is not critical, thus allowing S_w to be simplified.

$$S_w = \sqrt{\frac{\dfrac{aR_w}{\Phi_{im}^2 R_t} + \dfrac{8^2}{4} - \dfrac{q}{2}}{1 - q}}$$

Laboratory investigations and extensive field experience indicate that a simple relationship can be used for any clay model. The relationship works well in most shaly sands.

$$\frac{1}{R_t} = \frac{\Phi^2 S_w^2}{aR_w(1 - V_{sh})} + \frac{V_{sh}S_w}{R_{sh}}$$

where R_{sh} is a nearby shale resistivity.

In the case of laminated shale, after R_t is corrected for shale and porosity is corrected for shale, then S_w may be computed in the usual manner. Dispersed shale needs the mathematical formula just presented.

Shale Effects on Porosity

The following discussion is for the purpose of illustrating the effect of shale on porosity. It does not follow that each computation be made in order to establish a shale-corrected porosity.

The neutron tool responds to the total amount of hydrogen in the formation. Shale has a large amount of bound water, thus neutron porosity will be higher in shaly formations. All factors considered (including hydrocarbons), the neutron porosity response is:

$$\Phi_n = \Phi S_{xo} \Phi n_{mf} + V_{sh} \Phi_{nsh} + \Phi S_{hr} \Phi_{nh} - \Delta\Phi_{nex} + (1 - \Phi - V_{sh}) \Phi_{nma}$$

where Φ = the actual or true porosity
S_{xo} = the flushed zone saturation
S_{hr} = residual hydrocarbon saturation
Φ_{nmf} = 1 for fresh mud filtrate
Φ_{nsh} = nearby shale porosity
Φ_{nh} = hydrocarbon response (for average oil = 1)
$\Delta\Phi_{nex}$ = the excavation effect (1.0 for sandstone, 1.046 for limestone, and 1.173 for dolomite)
Φ_{nma} = the matrix response. If the tool is calibrated in the lithology log, then term is zero.
V_{sh} = shale volume

For the density log:

$$\rho_b = \Phi(S_{xo} \rho_{mf} + S_{hr} \rho_{ha}) + V_{sh} \rho_{sh} + (1 - \Phi - V_{sh}) \rho_{ma}$$

You may note two similarities between the neutron and density response. One is that neither makes a special case for the shale model type. The other is (with the exception of $\Delta\Phi_{nex}$) that each has identical terms. $\Delta\Phi_{nex}$ can be eliminated since it is compensated for by the computer gear. If the reading is taken in a zone 100% saturated with water, then hydrocarbon effects are also eliminated. In the case of a hydrocarbon-bearing interval, the porosities may be corrected for hydrocarbon first. We can state the log porosity as:

$$\Phi_n = \Phi_e + V_{sh} \Phi_{nsh}$$

$$\Phi_d = \Phi_e + V_{sh} \Phi_{dsh}$$

where Φ_n = neutron log porosity
Φ_d = density log porosity
Φ_e = effective porosity

Transposing the equations for effective positity;

$$\Phi_e = \Phi_d - V_{sh} \Phi_{dsh}$$

$$\Phi_e = \Phi_n - V_{sh} \Phi_{dsh}$$

Thus, we have two identical and workable shale-corrected porosities that are independent of shale type. Φ_{dsh}, it should be noted, is computed using a sandstone ρ_{ma} for shaly sands and a limestone ρ_{ma} for shaly limes.

For the sonic porosity, all three types of shale must be considered.

$$\Phi_s = \Phi_e + V_{dis} + (V_{lam} + V_{str}) \Phi_{slam}$$

where V_{lam} = laminated shale volume
V_{dis} = dispersed shale volume
V_{str} = structural shale volume

Isolating each case:

$\Phi_s = \Phi_e + V_{dis}$ dispersed

$\Phi_s = \Phi_e + V_{lam} \Phi_{slam}$ laminated

$\Phi_s = \Phi_e$ structural

Solving for effective porosity:

$$\Phi_e = \Phi_s - V_{dis}$$

$$\Phi_e = \Phi_s - V_{lam} \Phi_{slam}$$

$$\Phi_e = \Phi_{s,\ structural}$$

As stated earlier, the equations only serve to account for the shale material in the bed and do not account for the tool response in the combined matrix.

Three types of porosities are thus required:

Φ_a, which is the raw porosity from the log.

Φ_{corr}, corrected for shale response.

Φ_e, taking out the net effect of the shale.

Making this point clear, the structural case suggests no correction is needed in porosity.

$$\Phi_e = \Phi_{corr}$$

However, the porosity Φ_{corr} is not the same as Φ_a because the matrix is not pure, but rather a binary mixture of two minerals, one of which is clay.

A crossplot is indicated so that a correct value of matrix can be used to correct for the mixture, then the appropriate shale model can be applied.

Many analysts are confused on this point and consequently compute erroneous porosities.

Shale Volume

Shale volume can be determined from one of three sources (other sources will be developed later): the gamma ray curve, the spectral gamma ray curve, or the SP curve.

The SP does not reach static condition in a shaly sand. Thus, in a thick shaly sand, the SP will reach a pseudostatic SP state. The ratio of SSP to PSP is taken to be the difference caused by shale. Therefore, shale volume from the SP is stated as:

$$V_{sh} = 1 - \frac{PSP}{SSP}$$

where PSP is the pseudostatic SP and taken from the shaly zone of interest. SP is taken from a clean zone (i.e., no shale).

$$SSP = -60 \text{ mv}$$

$$SP = -120 \text{ mv}$$

$$V_{sh} = 1 - \frac{-60}{-120} = 0.5 \text{ or } 50\%$$

Shale volume from the SP is best used in water-bearing sands. In hydrocarbon-bearing sands, it will tend to give an upper limit of shale content.

The gamma ray log may also be used. Figure 13-2 shows the gamma ray curve and the parameter picked for the shale volume computation. GR_{clean} is taken to be the point where the gamma ray reads the lowest (minimum). The reasoning is that the cleanest sand will have the least amount of radioactivity. GR_{shale} is chosen where the gamma ray reads the highest (maximum), again reasoning that it is pure shale, uncontaminated by sand. GR is, of course, in the zone of interest. Sometimes the relation of gamma ray values to V_{sh} is stated in terms of a gamma ray index (GI), then the index is transformed into a shale volume.

$$GI = \frac{GR - GR_{clean}}{GR_{sh} - GR_{clean}}$$

$$V_{sh} = 100 \times GI^{1.4} \quad \text{for older formations, Gulf Coast}$$

V_{sh} relation in this case is for older formations in the Gulf Coast region. If the rock is just beyond the Miocene period, then the following relationship exists:

$$V_{sh} = 0.21\ [2^{(2.9GI)} - 1]$$

Again, this is for the Gulf Coast region.

If such a relationship exists in one area or field, then use it. If no known relationship exists then assume that V_{sh} is the same as GI. Perhaps the better method is to use the GRS uranium-free curve from the spectral gamma tool (see Chapter 12).

$$V_{sh} = \frac{GRS - GRS_{min}}{GRS_{max} - GRS_{min}}$$

Page 193 shows the relationship index to the shale for three types of rock. An individual plot for each field is easily accomplished using the techniques presented later in this chapter.

Water Saturation

Chapter 9 (Figure 9-5) discussed a crossplot method for determining the type of shale present in the zone of interest. Assuming the crossplot indicated, a laminated shale, we might solve for water saturation as follows:

Laminated Model.

1. Read R_{sh} in nearby shale bed.
2. Calculate V_{sh}.

$$V_{sh} = \frac{GRS - GRS_{min}}{GRS_{max} - GRS_{min}}$$

or

$$V_{sh} = 1 - \frac{PSP}{SSP}$$

3. Read clean sand streak R_{sd}.
4. Correct R_t

$$\frac{1}{R_t} = \frac{1 - V_{sh}}{R_{sd}} + \frac{V_{sh}}{R_{sh}}$$

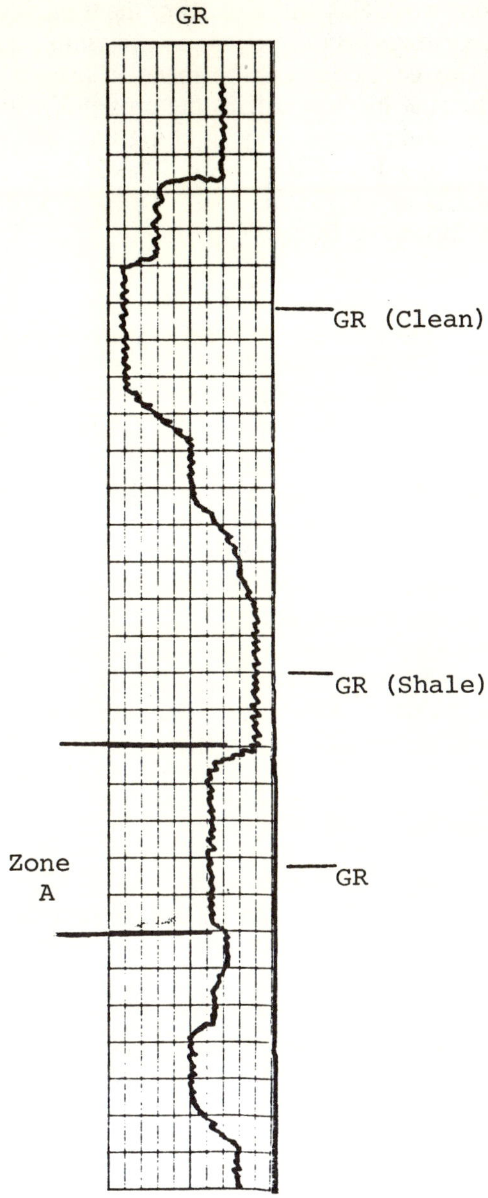

$$Vsh = \frac{GR - GRclean}{GRshale - GRclean}$$

Figure 13-2. Gamma ray curve showing picks for the gamma ray index.

5. Calculate or read shale porosity: For sonic and density matrix use a sandstone value for sands, and a limestone matrix for limes in the computation of shale porosity.
6. Correct porosity.

$$\Phi_e = \Phi\log - (V_{sh} \times \Phi_{sh})$$

for neutron, density, or sonic

Note that the term ($V_{sh} \times \Phi_{sh}$) corrects for the curve response. If $\Phi_e = \Phi_a$, as in the case of structural shale, the porosity matrix must represent both shale and sand (or lime).

7. Correct porosity for hydrocarbon type.
8. Calculate S_w as usual.

$$F = \frac{a}{\Phi^m} \quad \text{(corrected)}$$

$$R_o = FR_w$$

$$S_w = \sqrt{\frac{R_o}{R_t}} \quad \text{(corrected)}$$

Dispersed Model.

1. Using the sonic curve, calculate Φ_s in the shaly zone of interest.

$$\Phi_s = \Phi_e + V_{dis} = \Phi_{im}$$

2. Sonic porosity can be taken as the sum of effective porosity and shale volume.
3. Calculate q.

$$q = \frac{V_{dis}}{\Phi_{im}}$$

4. Read R_t from the zone of interest.
5. Calculate S_w

$$S_w = \sqrt{\frac{\frac{aR_w}{\Phi_{im}{}^2R_t} + \frac{q^2}{4} - \frac{q}{2}}{1 - q}}$$

Part II

Part I presented a system of shaly sand analysis which is based on sound theoretical foundation. Part II presents a more advanced and perhaps more accurate method of shaly sand interpretation. The method presented earlier suffers from the drawbacks of using one porosity device and a shale value which renders very little assurance of its validity.

Two-Porosity Shale Crossplot

The neutron-density curves can be combined to form a two-porosity shale crossplot. The two following equations represent the neutron-density response in a liquid-filled shaly formation.

$$\Phi_d = \Phi_e + V_{sh}\,\Phi_{dsh}$$

$$\Phi_n = \Phi_e + V_{sh}\,\Phi_{dsh}$$

The relationship can be used to construct the plot of Figure 13-3A. Zero percent neutron density porosity represents the pure quartz point, 100% porosity represents 100% water. The −5% density, 20% neutron porosities represent a dry clay (no water). Shale is composed of silt (mostly very fine quartz), dry clays, and water. Figure 13-3B shows the shale line connecting the three points. It is assumed that the density and neutron porosities were calculated in a sandstone matrix and no matrix error exists.

The dry clay point is of little concern to the analyst, therefore the point of interest is the wet clay point. Figure 13-4 shows the crossplot constructed. It should be noted that the 100% shale point is really 100% clay. The shale point is chosen from the log by finding a point where the neutron-density curves are separated the farthest apart in a shale. Always use the SP or gamma ray curves to confirm the shale readings.

The object of picking this value is to find a shale that is as silt-free as possible. Shales that are completely silt-free will not be found. In essence, this fact will introduce some error in the plot. Point A on Figure 13-4 shows the actual silt-free clay point. The error in this case is not very significant. It will cause shale volume to be a little on the high side. Effective porosity is not affected. The accuracy of the data obtained from the plot is a function of how well the shale point represents the silt-free clay point. Chapter 7 gives the step-by-step instructions on the actual construction of the neutron-density shale crossplot.

Although for this part, it is not essential that the type of shale is known, it is presented for study in Figure 13-5. A significant number of points must be plotted in order to establish a trend.

Heavy- to medium-density hydrocarbons have about the same effect on the plot as water. Light to very light hydrocarbons will displace the point to the northwest. Gas will plot off of the triangle and plot in a northwesterly direction, as shown in Figure 13-6.

Gas correction can be made in one of two ways. The first method utilizes the crossplot with the addition of the gamma ray. Plotting the data point Φ_d = 24% and Φ_n = 10%, we see that it is gas filled (circled point of Figure 13-7). Correcting the point is performed as follows:

1. Find on the log the cleanest sand possible. This will be the lowest gamma ray API units. Note the gamma reading and label it γ_{clean}.
2. At the point where the neutron-density was chosen for the shale point (100% shale), note the gamma reading and label it γ_{shale}.
3. Place the γ_{clean} reading at the 50% neutron-density porosity scale.
4. Place the γ_{shale} reading by the 100% shale point.
5. Scale the graph (by the porosity lines) linearly in gamma API units.

For Figure 13-7, the γ_{clean} was 32 API units, and the γ_{shale} was 92 API units. There are five divisions, so each incrcment of the linear scale is:

$$\text{Increment} = \frac{92 - 32}{5} = 12$$

Thus from 32, we add 12 and the same for each succeeding number.

To correct the circled point of Figure 13-7, we read the gamma ray curve where the point was taken from. A line is drawn from the indicated gamma ray reading parallel to the shale volume lines. In this case 56 API units are indicated. Another line is drawn from the gas-filled point (circled point) parallel to the average gas correction line until it intersects the gamma ray line. The intersection represents the point, had it been liquid filled. Figure 13-7 shows the process completed. Porosity is slightly less than 10% and V_{sh} is about 38%.

A more accurate method of correcting for hydrocarbons is to correct the neutron-density porosities before they are plotted.

For the neutron porosity:

$$\Phi_{ncorr} = \Phi_n - \Delta\Phi_n$$

$$\text{where } \Delta\Phi_n = \Phi S_{hr}\,\frac{\rho_h - 0.7 + 0.4P}{1 - 0.4P} \qquad \text{for oil}$$

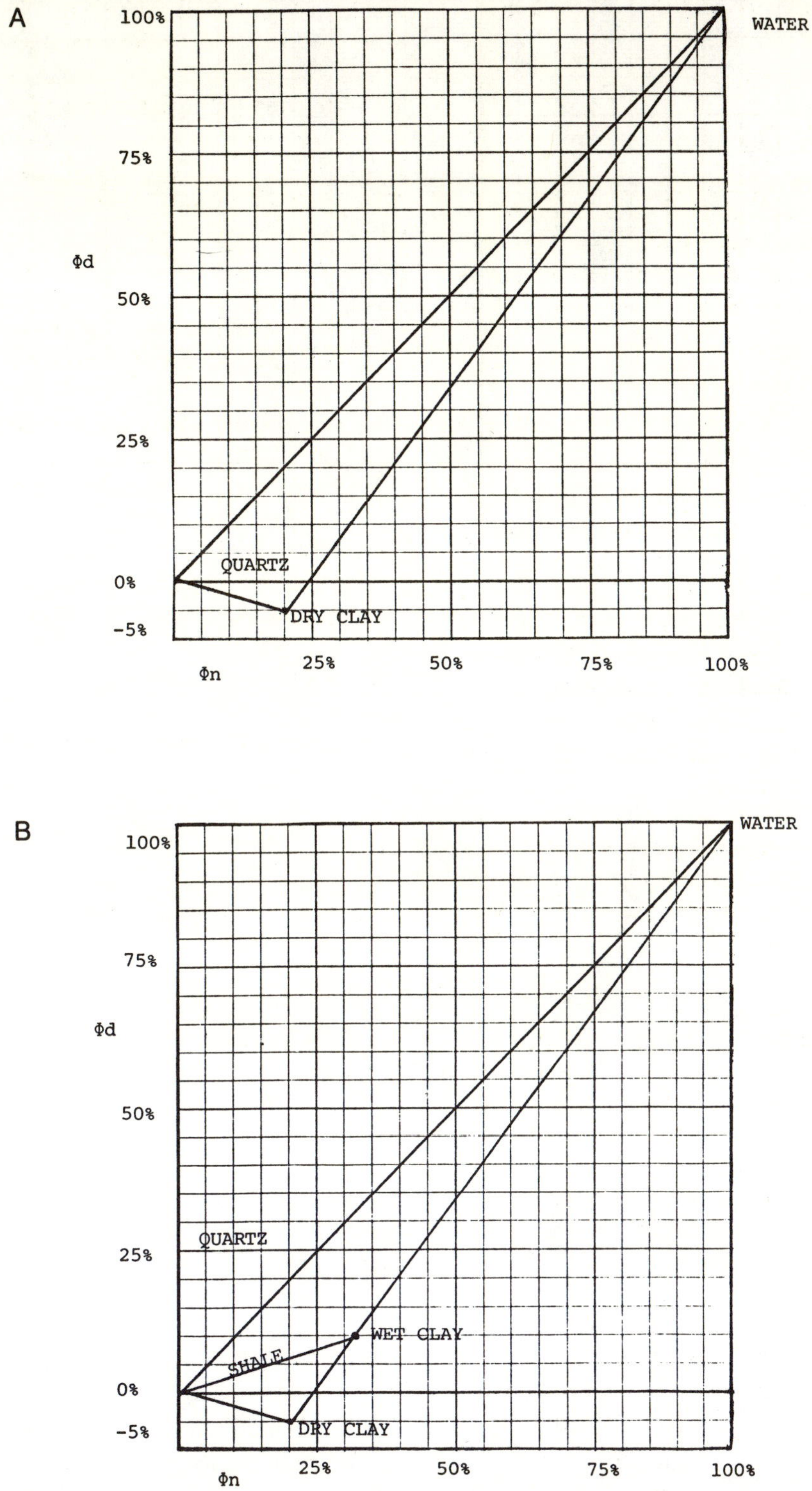

Figure 13-3. Neutron density shale crossplots.

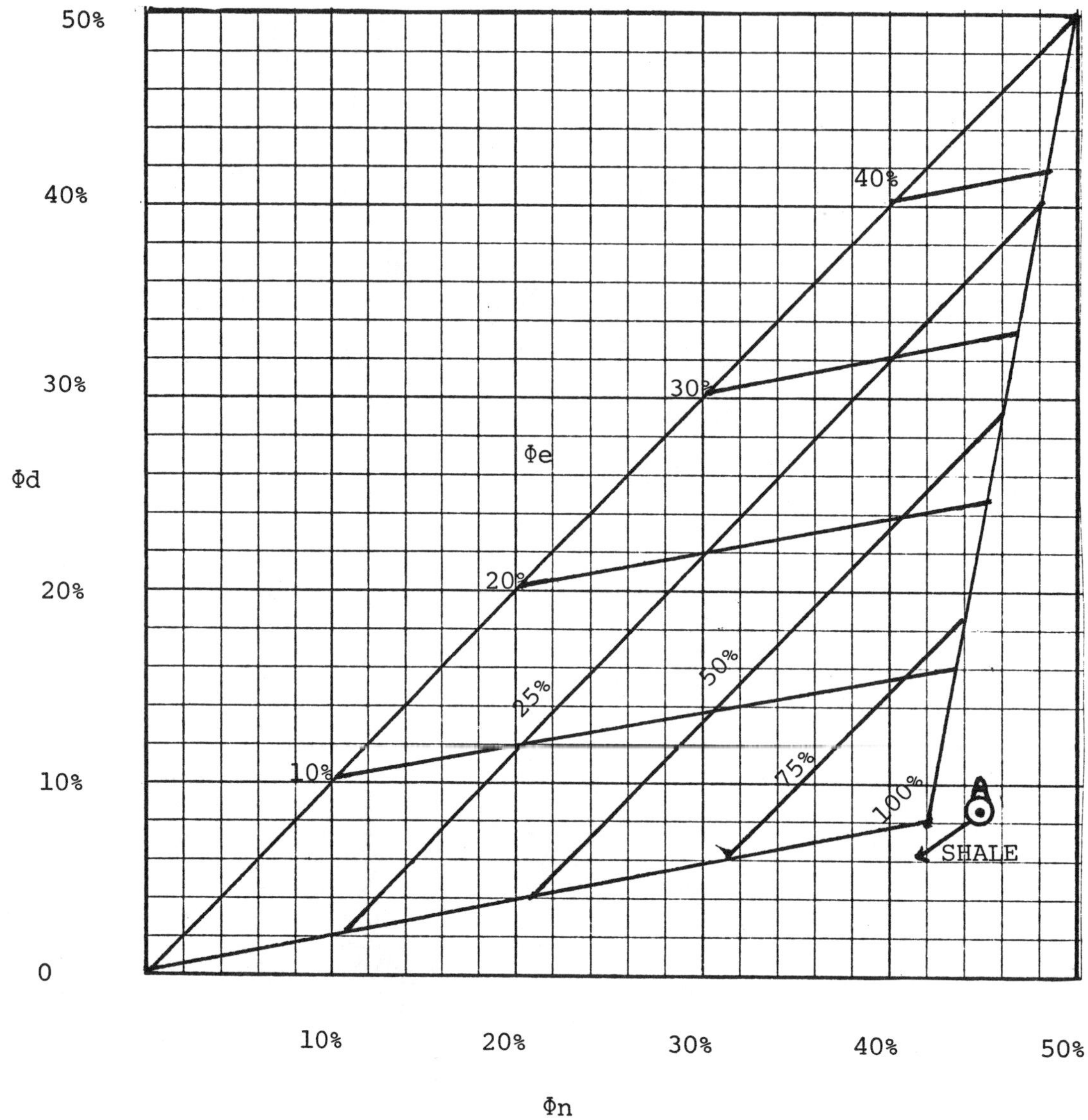

Figure 13-4. Neutron density shale crossplot chart showing scales.

$$\Delta\Phi_n = \Phi S_{hr} \frac{2.2\,h - 1.0 + 0.4P}{1 - 0.4P}$$

for gas

and P = salinity of the mud in PPM $\times 10^{-6}$
ρ_h = hydrocarbon density
S_{hr} = residual oil saturation

In fresh muds (less than 50,000 PPM) the equations can be reduced to:

$\Delta\Phi_n = \Phi_{Shr}\,(\rho_h - 0.7)$ for oil
$\Delta\Phi_n = \Phi_{Shr}\,(2.2\,\rho_h - 1.0)$ for gas

For the density curve:

$$\Phi_{dcorr} = \Phi_d - \Delta\Phi_d$$

(See Chapter 9 for detailed discussion on these corrections.)

and:

$$\Delta\Phi_d = \frac{1.07 \times \Phi_{Shr} (1.11 + 0.65P - 1.24\ \rho h)}{\rho_{ma} - 1 - 0.7P}$$

for gas

$$\Delta\Phi_d = \frac{1.07 \times \Phi_{Shr} (1.11(1 - \rho_h) + 0.65P - 0.03)}{\rho_{ma} - 1 - 0.7P}$$

for oil

In salinities of less than 50,000 PPM (fresh muds), P can be considered 0.

A simpler approach is:

$$\Phi_{dcorr} = \Phi_e + \left(\frac{1 - \rho_h}{\rho_{ma} - 1}\right) \times \Phi_{e\ Shr}$$

where $\Phi_e = \Phi_d$ in 100% water.

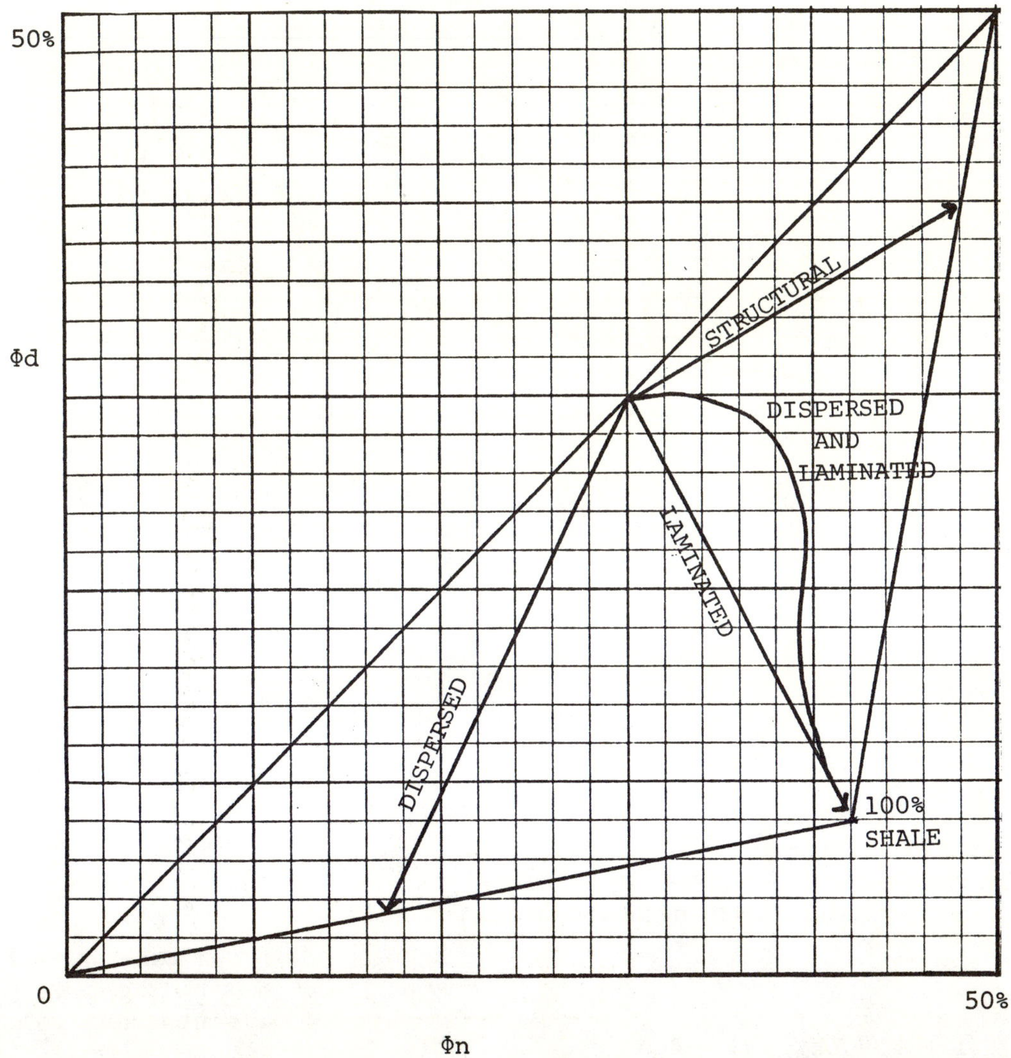

Figure 13-5. Neutron density shale crossplot showing the plot lines for the three basic types of shale.

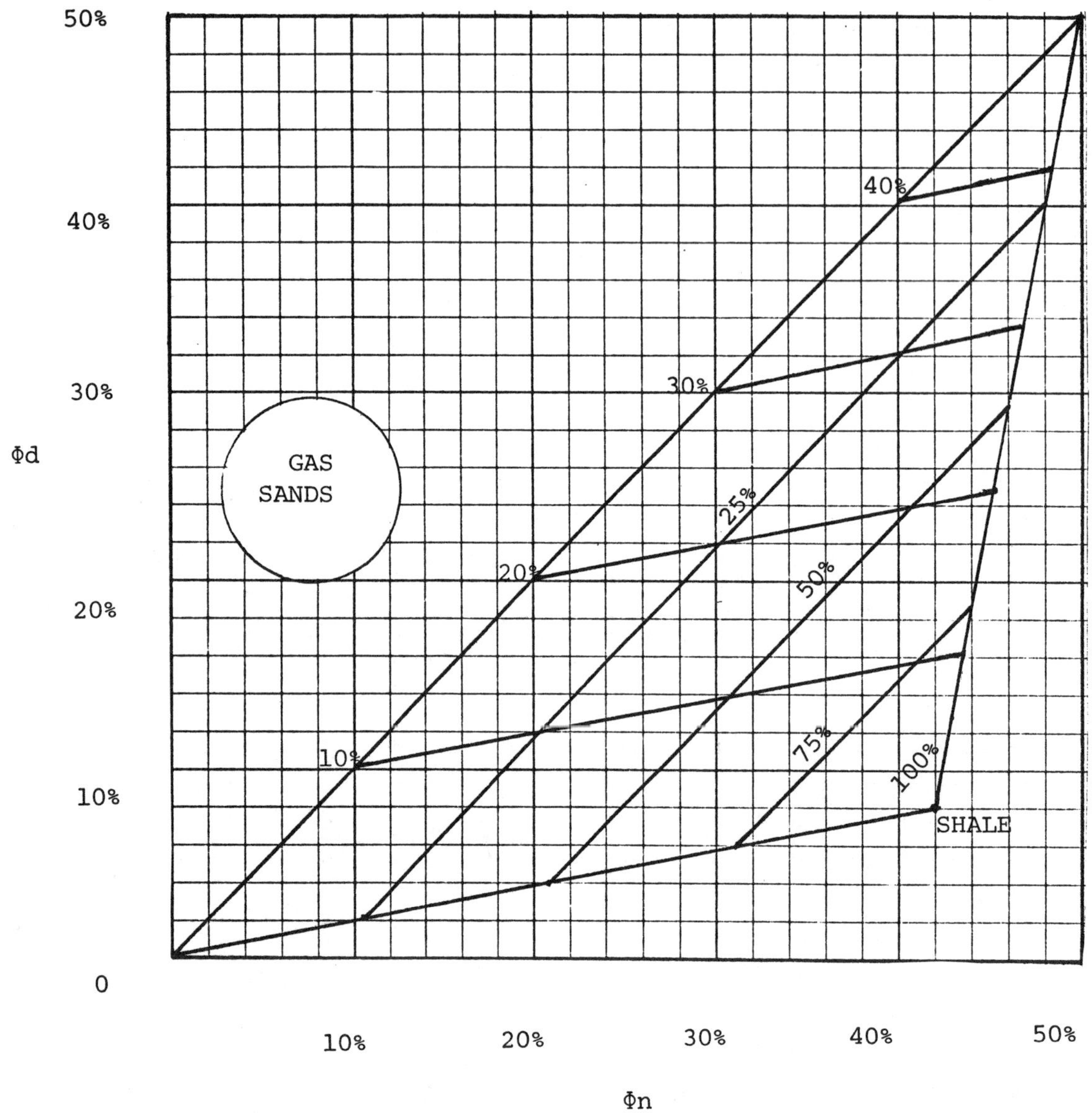

Figure 13-6. Neutron density shale crossplot showing the general area of gas plots.

An estimate of hydrocarbon density ρ_h can be obtained by the use of the chart in Chapter 10, Figure 10-14, or it may be calculated directly by the following equations:

For oil

$$\rho_h = \frac{0.7 \times (1 + \Phi_n/\Phi_d)\ S_{hr} - (1 - \Phi_n/\Phi_d)}{(1 + 0.72\ \Phi_n/\Phi_d)\ S_{hr}}$$

For gas

$$\rho_h = \frac{(1 + 0.72\ \Phi_n/\Phi_d)\ S_{hr} - (1 - \Phi_n/\Phi_d)}{(2.2 + 0.8\ \Phi_n/\Phi_d)\ S_{hr}}$$

Residual hydrocarbon saturation is used because the neutron-density tools investigate the flushed zone. The portion of hydrocarbon in the zone is given by:

$$S_{hr} = 1 - \sqrt{\frac{a}{\Phi^m} \times \frac{R_{mf}}{R_{xo}}} = 1 - \sqrt{\frac{FR_{mf}}{R_{xo}}}$$

The value of R_{xo} can be obtained from an R_{xo} resistivity tool, or the dual induction tornado chart combination.

S_{hr} can also be estimated from the gas-saturation charts discussed in Chapter 9.

Given $\Phi_n = 10\%$
$\Phi_d = 24\%$
$S_{hr} = 80\%$

$$\rho_h = \frac{(1 + .72 \times (.1/.24))\ .8 - (1 - (.1/.24))}{(2.2 + .8\ (.1/.24))\ .8}$$

The gas density vs. depth chart can also be used.

$$\rho_h = 0.23$$

$$\Delta\Phi = \frac{1.07 \times .24 \times .8 - (1.11 - 1.24 \times .23)}{1 - 2.65}$$

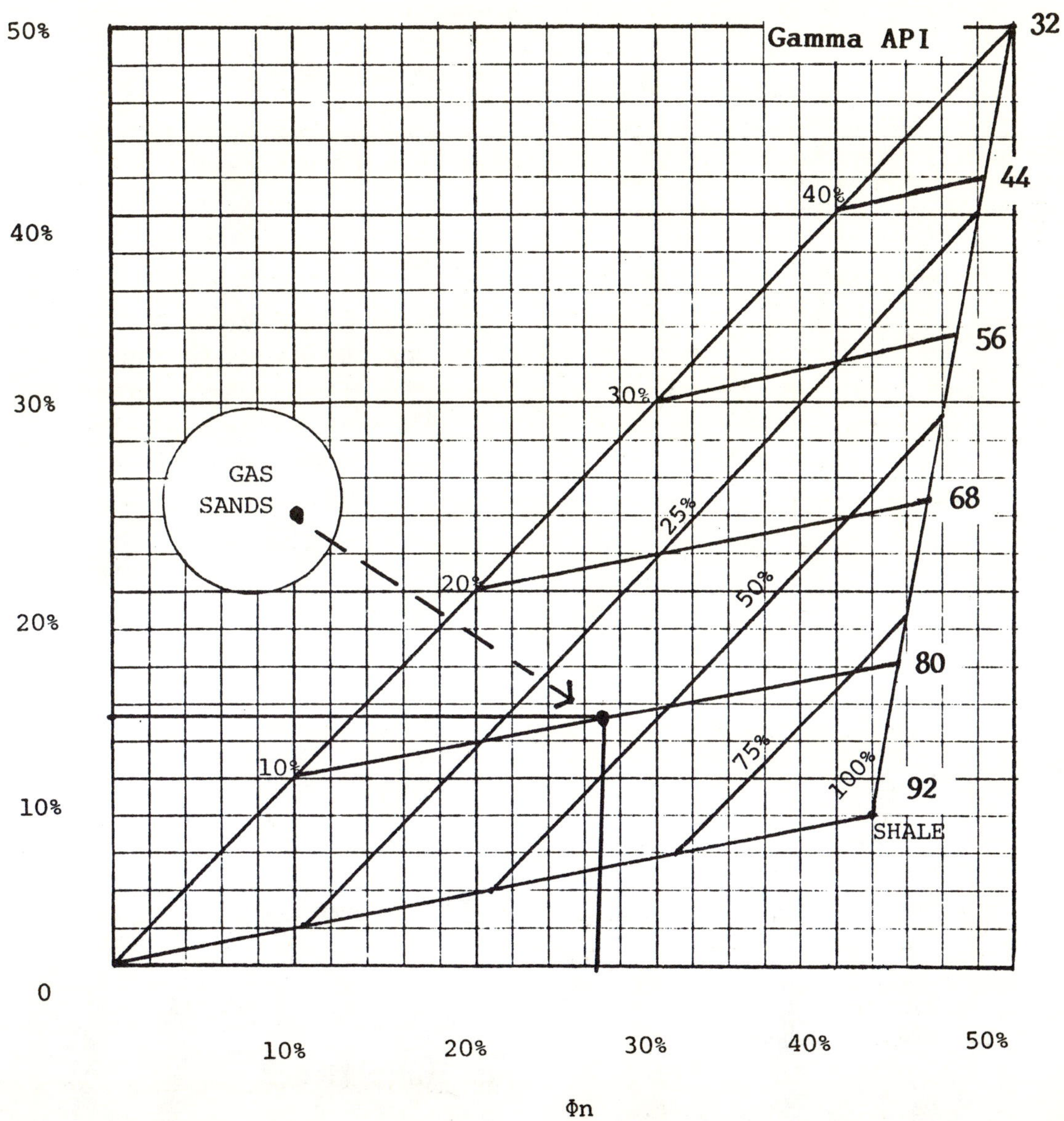

Figure 13-7. Illustration of a chart correction for gas.

$\Delta\Phi_d$ = .103 or 10.3%

$\Phi_{dcorr} = \Phi_d - \Delta\Phi_d$

Φ_{dcorr} = 24 − 10.3 = 13.7%

$\Delta\Phi_n$ = .1 × .8 (2.2 × .23 − 1)

$\Delta\Phi_n$ = −0.136 or −13.6%

$\Phi_{ncorr} = \Phi_n - (-\Delta\Phi_n)$

Φ_{ncorr} = 10 + 13.6 = 26.6%

Figure 13-7 shows the crossplot of the uncorrected gas and the corrected gas. Shale volume for this case is 36%, and the effective porosity is 10%.

To check the validity of the plot and the computations, the density porosity is calculated independently.

Φ_{dcorr} = 13.6% (gas corrected)

Φ_{dshale} = 8% (used as the shale point for the crossplot)

V_{sh} = 36%

$\Phi_e = \Phi_d - (\Phi_{dsh} V_{sh})$

Φ_e = .136 − (.08 × .35) = .108 or 10.8%

and thus proves to be correct. Further proof is that the density curve is the least affected by shale.

Water Saturation

There are a number of water saturations for the solutions of shaly formations. One which yields consistent results over broad areas is the *Simandoux Equation*.

$$S_w = \sqrt{\frac{aR_w}{\Phi^m R_t} + \frac{aR_w V_{sh}{}^2}{2R_{sh}\Phi^m}} - \frac{aR_w V_{sh}}{2R_{sh}\Phi^m}$$

where R_{sh} = nearby shale and Φ is corrected (i.e., crossplot Φ)

Fine Tuning

The saturation of the flushed zone was calculated, initially, with the assumption that the formation was clean. Now that a reasonably accurate value of porosity exists, the process is repeated, using:

$$S_{xo} = \sqrt{\frac{aR_w}{\Phi^m R_{xo}} + \left(\frac{aR_w V_{sh}}{2R_{sh}\Phi^m}\right)^2} - \frac{aR_w V_{sh}}{2R_{sh}\Phi^m}$$

$$S_{hr} = 1 - S_{xo}$$

As you can see, the S_{xo} value is now adjusted for gas and shale.

The new S_{xo} value is used to recalculate the hydrocarbon density and, in turn, new corrected neutron-density porosities. Several "loops" may be required until an answer is convergent.

Procedure

1. Choose neutron-density in the shaliest point. Plot as 100% V_{sh} (shale point).
2. Construct shale crossplot (see Chapter 7).
3. Obtain R_{xo}.
   a. Dual induction tornado
   b. R_{xo} tool
4. Calculate S_{xo} and S_{hr}.
   a. Use gas chart
   b. Use the mean average (Chapter 9)
5. Compute the hydrocarbon density.
6. Correct neutron-density porosities.
7. Crossplot for shale volume and effective porosity.
8. Recompute S_{xo} and R_{xo} using V_{sh} and porosity crossplot.
9. Repeat Steps 4–8 as many times as necessary.
10. Calculate S_w using a shale equation such as the Simandoux.

Part III

Dual Water Model

The dual water model is a relatively young system of shaly sand analysis. It is based on the theory that there are two types of water in the shaly formation. The first type of water is said to be "bound" water. Bound water is the water which adheres or is contained by the clay. Figure 13-8 shows schematically how the water is bound to the clay. A certain amount of water is directly absorbed by the clay crystal, followed by a layer of water at its surface surrounding a sodium cation. This thin sheet of salt-free water is quite significant. Clays can have up to 6,300 acres of surface per cubic foot (900 km^2/m^3) as compared to one tenth of an acre per cubic foot (0.014 km^2/m^3) for an average sand. One can easily see that clay water contributes a significant amount of water, as compared to the total pore spaces.

The second type of water is termed "free" water. This by no means implies that the water is producible. It simply states that it is water not bound to clay. It includes irreducible water as well as water that can move.

Porosity is given as Φ_t or total porosity. It is the sum of the free water porosity plus the bound water porosity and the hydrocarbon porosity.

$$\Phi_t = \Phi_f + \Phi_b + \Phi_h$$

While shales contain salt-free water, their cationic membrane properties allow ionic currents to conduct. In the dual water system, conductivities rather than resistivities are used. The F-Φ relationship used is $1/\Phi^2$.

The first step in the process is the computation of an apparent water resistivity (R_{wa} is usually obtained from the log, otherwise it may be manually computed).

$$\frac{R_t\Phi^m}{a} = R_{wa}$$

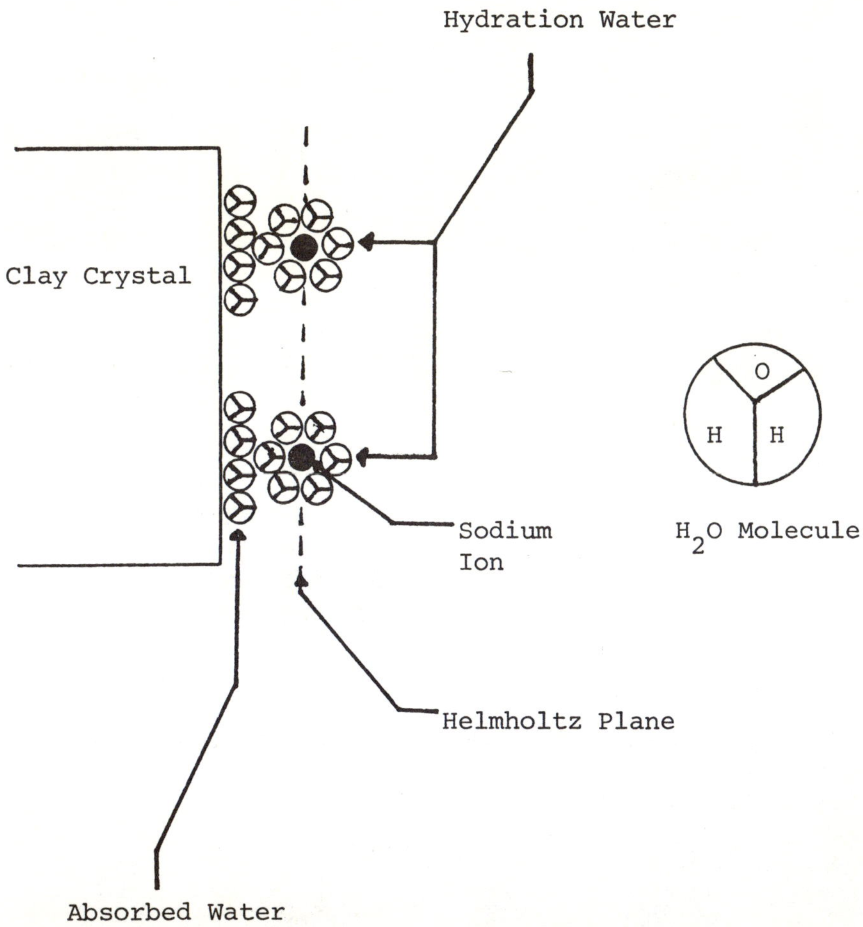

Figure 13-8. Illustration of the dual water model.

The conductivity of the free water is:

$$C_{wf} = \frac{1,000}{R_{wa}}$$

in a zone 100% saturated with water.

The conductivity of the bound water is:

$$C_{wb} = \frac{1,000}{R_{wa}} \text{ (in a nearby shale)}$$

The water saturation of the shale is taken to be the same as the shale volume.

$$S_{wb} = V_{sh}$$

This quantity may be obtained from conventional methods already discussed, or simply:

$$S_{wb} = \left(\frac{GR_{shaly\ sand}}{GR_{shale}}\right)^2$$

Fluid conductivity is determined next:

$$C_{fl} = S_{wb} \times C_{wb} + (1 - S_{wb})\, C_{wf}$$

Total porosity takes the bound water into account, and uses the neutron-density porosity.

$$\Phi_t = \frac{1 + \Phi_n}{1 + \left(\frac{1}{\Phi_d}\right)} + \frac{S_{wb}}{3 + 20\,\Phi_n}$$

Conductivity of the fluid:

$$C_o = \Phi_t^2 \times C_{fl}$$

$$C_t = \frac{1,000}{R_t} \quad \text{gives true conductivity}$$

If the laminated shale is present, take R_t from the shallow reading curve.

Water saturation is finally arrived at by:

$$S_w = \sqrt{\frac{C_t}{C_o}}$$

Effective porosity is:

$$\Phi_e = (1 - S_{wb})\, \Phi_t$$

The whole process seems quite new. Actually, if you stop and think about it, there is great similarity between the dual water model and conventional analysis.

$$C_{wf} = R_{wa} \text{ in a 100\% water wet sand}$$

$$C_{wb} = R_{wa} \text{ in a shale}$$

$$S_{wb} = V_{sh}$$

$$C_o \cong FXR_w \text{ or } R_o$$

$$C_t = R_t$$

$$S_w = \sqrt{\frac{C_o}{C_t}} \text{ inverted is } \sqrt{\frac{R_o}{R_t}}$$

Φ_t is a shale-corrected porosity.

About the only real difference is the parameter C_{fl} or conductivity of the fluid.

$$C_{fl} = S_{wb}\, C_{wb} + (1 - S_{wb})\, C_{wf}$$

or translated:

$C_{fl} = V_{sh}\, R_{wa}$ (shale) $+ (1 - V_{sh})\, R_{wa}$ (100% water)

All things taken into consideration, the dual water model works well. It has the plus of not needing a value of R_w, and it is relatively easy to use.

The equations used are listed in sequence below:

1. $R_{wa} = \Phi \times R_t/a$
2. $R_{wf} = R_{wa}$ in 100% water
3. $C_{wf} = \frac{100}{R_{wf}}$
4. $R_{wb} = R_{wa}$ in nearby shale

5. $C_{wb} = \dfrac{1{,}000}{R_{wb}}$

6. $S_{wb} = V_{sh}$

or

$$S_{wb} = \left(\frac{GR_{shaly\ sand}}{GR_{shale}}\right)^2$$

7. $C_{fl} = S_{wb} \times C_{wb} + (1 - S_{wb})\, C_{wf}$

8. $\Phi_t = \dfrac{1 + \Phi_n}{1 + \left(\dfrac{1}{\Phi_d}\right)} + \dfrac{S_{wb}}{3 + 20\ \Phi_n}$

9. $C_o = \Phi_t^2 \times C_{fl}$

10. $S_w = \sqrt{\dfrac{C_t}{C_o}}$

11. $\Phi_e = (1 - S_{wb})\, \Phi_t$

Example

Given in the zone of interest:

$R_{wa} = 11.4$ (hydrocarbon zone)

$R_t = 118$ (hydrocarbon zone)

$R_{wa} = .042$ (shale R_{wb})

$GR_{sh} = 118$ (shale)

$GR_{shsd} = 55$ (shaly sand)

$\Phi_n = 33.5\%$, $\Phi_d = 28\%$ (hydrocarbon zone)

Conductivity of the free water:

$$C_{wf} = \frac{1{,}000}{11.4} = 87.7$$

Conductivity of the bound water:

$$C_{wb} = \frac{1{,}000}{.042} = 23{,}800$$

Bound water saturation:

$$S_{wb} = \left(\frac{55}{118}\right)^2 = .22\ (V_{sh})$$

Fluid conductivity:

$$C_{fl} = .22 \times 23{,}800 + (1 - .22)\ 87.7$$

$$C_{fl} = 5{,}304$$

$$\Phi_t = \frac{1 + .335}{1 + \left(\dfrac{1}{.28}\right)} + \frac{.22}{3 + 30 \times .22}$$

$$\Phi_t = .309 \text{ or } 30.9\%$$

100% water conductivity:

$$C_o = .309^2 \times 5{,}304$$

$$C_o = 1{,}639$$

True conductivity:

$$C_t = \frac{1{,}000}{118} = 8.47$$

$$S_w = \sqrt{\frac{8.47}{1{,}639}} = 0.72 \text{ or } 7.2\%$$

The water saturation of 7.2% is rather incredible in that it seems low. The log data was taken from the logs of Figures 13-9 and 13-10. R_w for the zone is .09 ohms.

As a quick check, by conventional analysis:

$$F = \frac{1}{.309^2} = 10.5$$

$$R_o = 10.5 \times .09 = .95$$

$$S_w = \sqrt{\frac{.95}{118}} = 0.897 \text{ or } 8.97\%$$

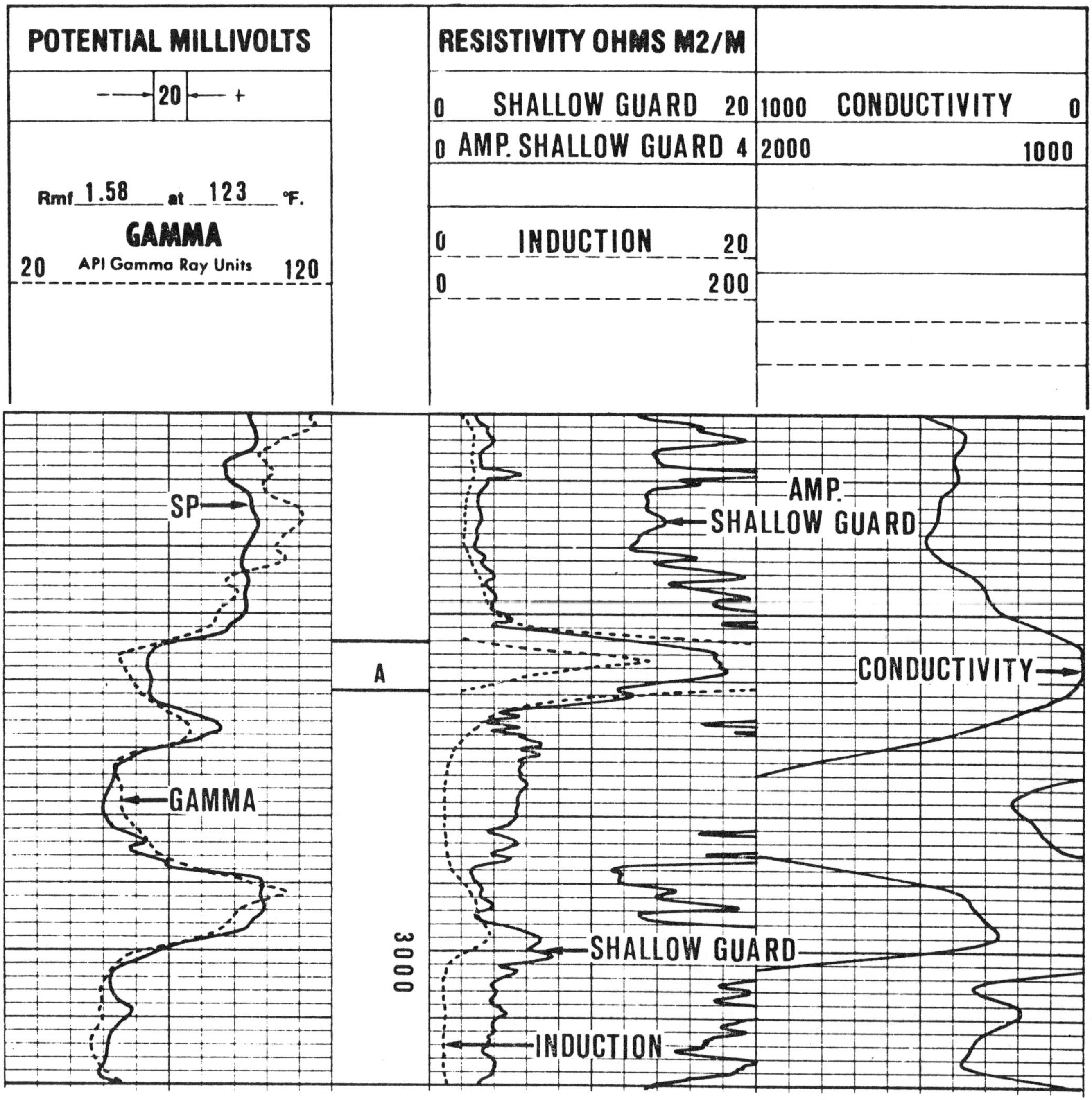

Location: Van Zandt Co. Tx.

Formation: Createous Age Sand

Figure 13-9. Field example of a resistivity log. (Courtesy of Welex.)

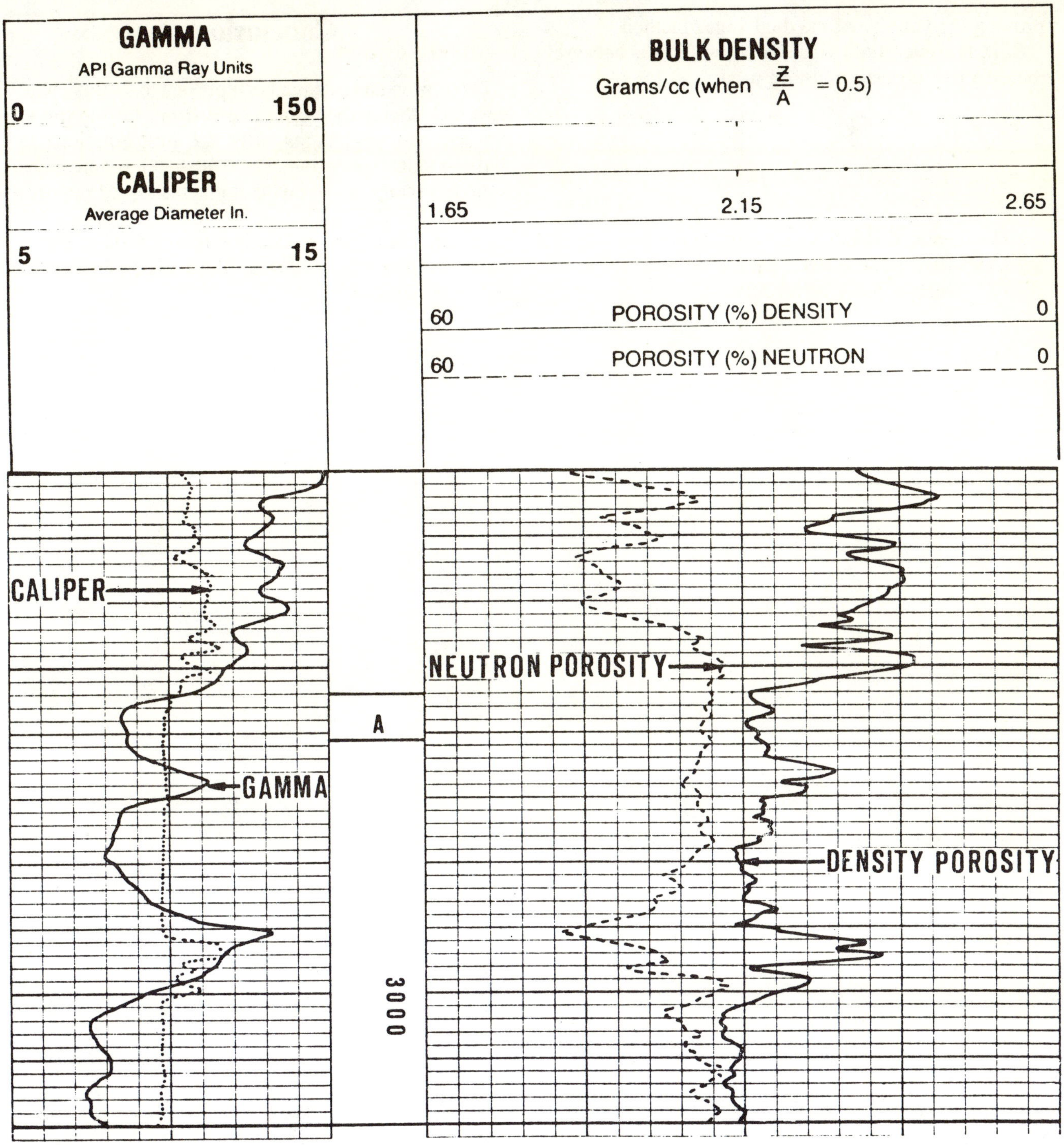

Figure 13-10. Field example of a neutron density log. (Courtesy of Welex.)

proving the validity of the dual water method.

Still, however, both values are low. This is because the effective porosity must be used.

$$\Phi_e = (1 - S_{wb})\ \Phi_t$$

$$\Phi_e = (1 - .22)\ .309 = .24$$

Now we plug this value back into the previous equations and find that the water saturation now becomes 16.6%, which is more believable.

Conclusion

Three procedures have been presented in this chapter which the analyst can choose from. Of the three, the last two procedures offer the best solutions to shaly sand interpretation. The dual water system gains some advantage in the fact that it is faster and requires no value for R_w.

A combination of procedures two and three offers one of the best of two worlds. The analyst is left to his or her preference.

Appendix A

Characteristics of Selected Materials

Material	Resistivity (ohm-m)	Travel (μsec/ft range)	Time (μs/ft avg.)	Density		Neutron (Apparent LM Φ %)
				(range)	(Avg.)	
Aluminum	3×10^{-8}		49		2.7	
Anhydrite	10^{4}–10^{10}		50	2.9–3.0	2.96	
Aragonite			53		2.94	
Barite			77	4.3–4.6	4.46	
Basalt	10^{3}–10^{5}	50–58	53	2.8–3.0	2.9	
Biotite	10^{4}–10^{15}	60–80	60	2.6–3.0	2.85	
Calcite	10^{7}–10^{12}		47.5		2.71	
Carnallite			78		1.57	
Chalk			47.5		2.71	
Chert	10^{6}		55.5	2.6–2.7	2.65	
Chlorite			62		2.77	
Cinnabar	10^{6}–10^{10}			8.0–8.2	8.1	
Coal (anth.)	10^{-4}–10^{-2}	90–120	105	1.4–1.8		50 or more
Coal (bit.)	10–10^{6}	110–140	120	1.3–1.5		50 or more
Copper	10^{-8}		60		8.5	
Diabase	10^{3}–10^{4}	44–46	44.5	2.8–3.2		
Dolomite	10^{8}		43.5	2.8–3.0	2.87	1–3
Dunite			38		3.31	
Feldspar (ortho)	10^{8}	53–82		2.5–2.6	2.57	
Feldspar (plag)	10^{8}	46–53		2.6–2.8	2.69	
Gabbro	10^{3}–10^{5}	42–48	42.4	2.8–3.1	2.98	
Galena	10^{-2}–10^{-5}				7.5	
Granite	10^{3}–10^{6}		51	2.5–3.0		
Gypsum	10^{6}		53	2.3–2.4	2.32	49
Halite	10^{4}–10^{14}		67		2.16	0
Hematite	10^{-3}–10^{6}		43	4.2–5.3	5.25	
Illite	10			2.7–2.9		
Iron	10^{-7}		57		7.87	
Kainite					2.16	
Kaolinite	10	77–210		2.4–2.7	2.64	
Langbeinite			52		2.83	
Lead	2×10^{-8}		140		11.34	
Lignite	50–10^{-6}			0.7–1.1		45 or more
Limestone	10^{7}–10^{12}		47.5		2.71	
Magnetite	10^{-2}–10^{-4}		39	5.0–5.4	5.17	
Mica	10^{8}–10^{12}		64	2.7–3.2		29
Montmorillionite	10			2.0–2.5		
Muscovite	10	38–73		2.8–3.1		
Polyhalite			58		2.78	
Potash					2.04	
Pyrite	0.1–10^{-6}		43	4.9–5.2	5.06	
Quartz	10^{12}–10^{17}	50–56	55.5	2.6–2.7	2.65	
Sandstone	10^{12}–10^{17}	47–56	55.5	2.65–2.7	2.65	−2
Shale	0.3–10^{2}	70–150		2.2–2.8		25–75
Siderite	10–10^{3}				3.88	
Sulfur	10^{9}–10^{16}		122		2.07	0
Trona			65		2.13	40

Appendix B

Equations

Porosity

$$\phi = \frac{\Delta t - \Delta t_{ma}}{\Delta t_f - \Delta t_{ma}}$$

$$\phi = \frac{\rho_{ma} - \rho_b}{\rho_{ma} - \rho_f}$$

$$\Delta t = \phi \, \Delta t_f + (1 - \phi) \, \Delta t_{ma}$$

$$\rho_b = \phi \, \rho_f + (1 - \phi) \, \rho_{ma}$$

$$\phi_e = \sqrt{\frac{\phi_n^2 + \phi_d^2}{2}} \quad \text{for gas}$$

Sulfur

$$\phi_d \,(\text{lime}) = \phi_y + V_s \left(\frac{2.71 - \rho_{sul.}}{2.71 - \rho_f}\right)$$

$$\text{where } \phi_y = \frac{.53 \, \phi_d - .4 \, \phi_{sonic}}{.13}$$

$$V_s = \frac{\phi_d - \phi_n}{.4}$$

$$V_s = \frac{\phi_d - \phi_y}{.4} = \text{sulfur volume}$$

Crossplot Porosity: Algorithm

1. Choose mineral pair (i.e., lime/sand).
2. Calculate two porosities in both lithologies (i.e., $\phi_{sonic} - \phi_{density}$ in lime, $\phi_{sonic} - \phi_{density}$ in sand).

$$\phi_1 a = \frac{\rho_{mal} - \rho_b}{\rho_{mal} - \rho_f}$$

(i.e., $\rho_{mal} = 2.65$ for sand)

$$\phi_2 a = \frac{\rho_{ma}2 - \rho_b}{\rho_{ma}2 - \rho_f}$$

(i.e., $\rho_{ma}2 = 2.71$ for lime)

$$\phi_1 b = \frac{\Delta t - \Delta t_{mal}}{\Delta t - \Delta t_{mal}}$$

(i.e., $\Delta t_{mal} = 55.5$ for sand)

$$\phi_2 b = \frac{\Delta t - \Delta t_{ma}2}{\Delta t - \Delta t_f}$$

(i.e., $\Delta_{ma}2 = 47.5$ for lime)

Note: For the neutron tool, the proper wireline company chart should be used in converting to various lithologies.

3. Calculate the combined matrix porosity for each curve.

Curve A

$$\phi_{a\text{-maxp}} = \frac{\phi_2 a - \phi_1 a}{\phi_2 a - 1}$$

Curve B

$$\phi_{b\text{-maxp}} = \frac{\phi_2 b - \phi_1 b}{\phi_2 b - 1}$$

4. Calculate the crossplot porosity.

$$\phi_{xp} = \frac{\phi a\ \phi_{b\text{-maxp}} - \phi_b\ \phi_{a\text{-maxp}}}{\phi_{b\text{-maxp}} - \phi_{a\text{-maxp}}}$$

where ϕa, ϕb are log porosities

5. If the wrong mineral pair is chosen or a third mineral pair is present, then ϕ_{xp} will be wrong.

Porosity Estimates

$$\phi\left(\frac{aR_w}{S_w{}^n\ R_t}\right)^{1/m}$$

$$\phi = (R_w/R_o)^{1/m}$$

$$\phi = \left(\frac{aR_{mf}}{S_{xo}{}^n\ R_{xoa}}\right)^{1/m}$$

Porosity Corrections

Hydrocarbon

$$\phi_{corr} = \phi_a \times .7 \qquad \text{for gas}$$
$$\phi_{corr} = \phi_a \times .9 \qquad \text{for oil}$$

ϕ_a is density or sonic porosity.

$$\phi_{corr} = \phi_n - \Delta\,\phi_n \qquad \text{(neutron)}$$

$$\Delta\,\phi_n = \frac{\rho_h - .7 + .4P}{1 - .4P} \qquad \text{for oil}$$

$$\Delta\,\phi_n = \frac{2.2\ \rho_h - 1.0 + .4P}{1 - .4P} \qquad \text{for gas}$$

$$\phi_{corr} = \phi_d - \Delta\,\phi_d \qquad \text{(density)}$$

$$\Delta\,\phi_d = \frac{1.07\ \phi_{Shr}\ (1.11 + .65P - 1.24\ \rho_h)}{\rho_{ma} - 1 - .7P}$$

for gas

$$\Delta\,\phi_d = \frac{1.07\ \phi_{Shr}\ (1.11(1 - \rho_h) + .65P - .03)}{\rho_{ma} - X - .7P}$$

for oil

where P = Mud salinity in PPM $\times 10^{-6}$. If the salinity of the mud is less than 50,000 PPM, then P = 0

Shale

$$\phi_e = \phi_a - (V_{sh}\ \phi_{sh}) \qquad \text{(laminated)}$$
$$\phi_e = \phi_a - V_{sh} \qquad \text{(dispersed)}$$
$$\phi_e = \phi_a \qquad \text{(structural)}$$

Formation Factor F

$$F = \frac{a}{\phi^m} \qquad \text{General form}$$

$$F = \frac{1}{\phi^2} \qquad \text{Carbonate}$$

$$F = \frac{.62}{\phi^{2.15}} \qquad \text{Humble}$$

$$F = \frac{.81}{\phi^2} \qquad \text{Modified Humble}$$

$$F = \frac{1.45}{\phi^{1.54}} \qquad \text{Average sands}$$

$$F = \frac{1.65}{\phi^{1.33}} \qquad \text{Shaly sands}$$

$$F = R_o/R_w$$

Resistivity

$R_t = FR_{wa}$

$R_{xo} = FR_{mf}$

$R_o = FR_w$

$R_w = FR_o$

$$R_{wa} = \frac{R_t}{F} = \frac{aR_t}{\phi^m}$$

$$1/R_t = \frac{1 - V_{sh}}{R_{sand}} + \frac{V_{sh}}{R_{shale}}$$

R_w from the SP

R_{mf} to R_{mfe}.

If $R_{mf} > 0.1$ @ 75°F then $R_{mfe} = 0.85R_{mf}$

If $R_{mf} < 0.1$ @ 75°F then $R_{mfe} = (146R_{mf}) - 5$ or $R_{mfe} = (377R_{mf}) + 77$

$$R_{we} = \frac{R_{mfe}}{10^{\left(\frac{-SP}{K}\right)}}$$

Note: −SP is a mathematical quantity and does not imply a negative SP.

$K = 60 + .133T_f$

R_{we} to R_w.

If $R_{we} < 0.12$ @ 75°F, then

$$R_w = \frac{(77\ R_{we}) + 5}{146 - (377\ R_{we})}$$

If $R_{we} > 0.12$ @ 75°F, then

$$R_w = -0.58 + 10^{(.69R_{we} - .24)}$$

yielding R_w @ 75°F

Resistivity Conversion

$$R_2 = R_1 \left(\frac{T_1 + 6.77}{T_2 + 6.77}\right)$$

Temperature

$$T_f = (BHT - T_o)(D_f/T_d) + T_o$$

$$T_f = To + \left(\frac{G_g \times D_f}{100}\right)$$

where BHT = bottom hole temperature
T_o = mean average temperature
D_f = formation depth
T_d = total depth
G_g = geothermal gradient

Permeability

$$K = \sqrt{\frac{250 \times \phi^3}{S_{wirr}}}$$ Medium oils

$$K = \sqrt{\frac{79 \times \phi^3}{S_{wirr}}}$$ Gas

$$K = \sqrt{\frac{8{,}580 \times \phi_e^{4.4}}{S_{wirr}}}$$

$$K = \left[\frac{300 \times \phi^{2w}}{w^4\ (R_w/R_{tcorr})}\right]^2$$

$$w = (3.75 - \phi) + .5(2.2 + \log R_w/R_{tcorr})^2$$

$$R_{tcorr} = R_{tirr}\ (.077 + 1.55\ \rho_h - .637\rho_h)^2$$

$$R_{tirr} = \frac{\left(\frac{R_w}{10^{-6}}\right)}{G\ (\rho_{ma} - 2.6)^3}$$

$$K = \left[\frac{300(.077 + h - .627\, h^2\phi^{2w}}{w^4\,(R_w/R_{tirr})}\right]^2$$

Note: For discussion see Chapter 11.

$$K_{rw} = \left(\frac{S_w - S_{wirr}}{1 - S_{wirr}}\right)^3$$

$$K_{ro} = \left(\frac{0.9 - S_w}{0.9 - S_{wirr}}\right)^2$$

Water Saturation

$$S_w = \sqrt{\frac{R_o}{R_t}} = \sqrt{\frac{FR_w}{R_t}} = \sqrt{\frac{aR_w}{\phi^m R_t}}$$

For shale:

$$S_w = \sqrt{\frac{FR_w}{R_t} + \frac{FR_w V_{sh}{}^2}{2R_{sh}}} - \frac{FR_w V_{sh}}{2R_{sh}}$$

$$S_w = \sqrt{\frac{FR_w}{R_t}} - \frac{V_{sh} R_w}{.4R_{sh}\phi}$$

$$S_{xo} = \sqrt{\frac{FR_{mf}}{R_{xo}}}$$

Appendix C

Estimation of Water Cut

Ideally, the reservoir can be considered to produce 100% water-free oil if the zone is at irreducible water saturation. The test to determine if this is the case is quite simple. Multiply the water saturation by the porosity throughout the interval.

$$C = S_w \Phi$$

This should be done at least every 2 feet by hand, preferably every 6 inches by computer. If the value of C is nearly constant, the reservoir is at irreducible water.

Computers today give you a curve, usually labeled as the "bulk volume of water" (BVW) which is the same product as the constant C. A constant BVW indicates that the reservoir is at irreducible water.

As is the case more often than not, the reservoir will not be at irreducible water and an estimate of water cut is desirable.

Before this can be done, a reasonable estimate of permeability must be known. This value can be obtained through local knowledge of a field, or better yet, through core analysis.

The following average parameters were obtained from a well:

K = 25 millidarcies from sidewall cores
Φ_e = 16% in zone of interest
S_w = 49% in zone of interest
Fluid = medium gravity oil from core analysis

The reservoir is not at irreducible water, therefore the first parameter needed is irreducible water saturation. This value can be estimated using the relationship:

$$S_{wi} = 3.5 \cdot \left(\frac{\Phi^{1.26}}{K^{0.36}}\right) - 1$$

Thus:

$$S_{wi} = 3.5 \left(\frac{0.16^{1.26}}{.025^{0.36}}\right) - 1$$

K is expressed in darcies therefore 0.025 darcies is equivalent to 25 millidarcies.

$$S_{wi} = 3.5 \cdot \left(\frac{0.0994}{0.2650}\right) - 1 = 0.312 \text{ or } 31.2\%$$

The relative permeabilities of water and oil are computed next.

$$K_{rw} = \left(\frac{S_w - S_{wi}}{1 - S_{wi}}\right)^3$$

Thus:

$$K_{rw} = \left(\frac{0.49 - 0.312}{1 - 0.312}\right)^3 = 0.0173$$

And the relative permeability of oil is given by:

$$K_{ro} = \left(\frac{0.9 - S_w}{0.9 - S_{wi}}\right)^2$$

Thus:

$$K_{ro} = \left(\frac{0.9 - 0.49}{0.9 - 0.312}\right)^2 = 0.4862$$

The water-to-oil ratio (WOR) is computed next, taking into account the viscosity of the fluid:

$$WOR = B\,(K_{rw}/K_{ro})$$

where B = 2, light oil, 45 degrees
= 5, light oil, 35 degrees
= 10, medium oil, 27 degrees
= 20, heavy oil, 19 degrees
= 50, heavy oil, 14 degrees

In this case, samples indicate a medium oil so a value of 10 is selected for B.

$$WOR = 10 \cdot \left(\frac{0.0173}{0.4862}\right) \quad \left(\frac{\text{water}}{\text{oil}}\right)$$

$$WOR = 0.356$$

The final computation is the water cut (WC) expressed by:

$$WC = \frac{WOR}{1 - WOR}$$

Thus:

$$WC = \frac{0.356}{1 - 0.356} = 0.553$$

Indicating the water cut to be 55.3%.

It is important to note that the values indicated by this procedure are only estimates based on empirical models.

Appendix D

Picking Curve Values

The hardest part of log analysis to the novice is learning where to select or pick log values.

The philosophy behind choosing a value is simply this:

- If the bed is thin, the curve will peak, or come to a point.
- If the bed is thick and homogeneous, the curve will flat top.

Some curves, such as the nuclear curves, are statistical in nature. A value chosen with these curves should be averaged.

The following examples are submitted for study.

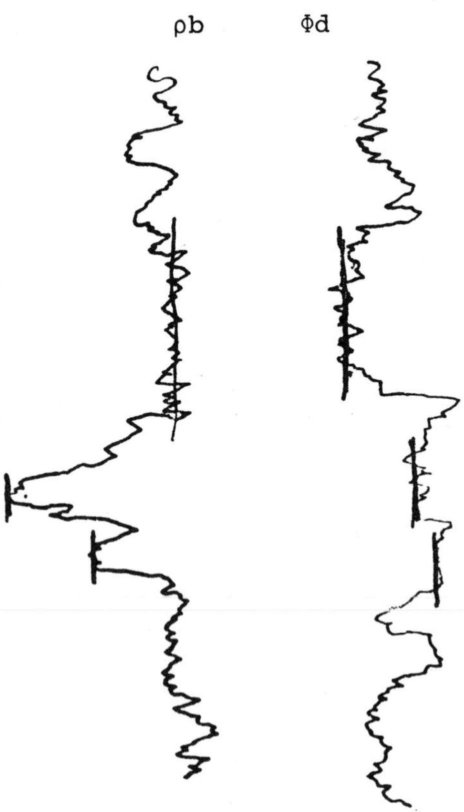

D-1

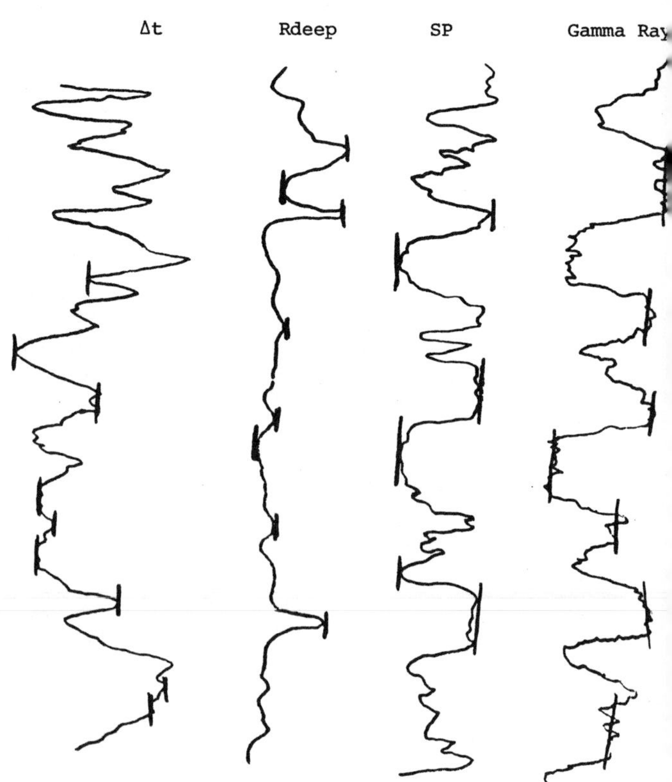

D-2

Appendix E

Supervising a Logging Job

It should be very clear by now that the final product of the analysis is only as good as the information used to derive it. It is my intention here to show the reader what is proper and what is not. I do not intend to imply that any one particular wireline company is at fault, but merely to illuminate what could happen, based upon experience.

The typical wireline crew consists of one field engineer and two equipment operators. The field engineer is the person in charge of the logging operation.

As the crew awaits the rig to trip the pipe out, they begin preparations. All of the tools to be used are laid on the catwalk to be calibrated and checked out. While this is going on, the field engineer is busy readying programs, parameters, and filling out paperwork.

One of the initial measurements to be taken during this time is the mud resistivity. This task should be performed by the engineer, but is all too often delegated to the operator. This is one area which you should supervise closely. The proper procedure is to obtain a sample from the mudpit or flowline.

A portion of the mud will be injected into a mud cell for direct measurement (R_m). The remainder of the sample will be filtered through a press to obtain filtrate and cake. Each of these quantities will be placed in cells and measured. Each sample should be measured for temperature. It is a bad, but extremely common, practice to take current ambient temperature and apply this to R_m, R_{mf}, and R_{mc}. Insist that at least the temperature of the mud be taken.

While the measurements take place, visually examine the mud cells. They must be squeaky clean in order to obtain a proper measurement. If there is contamination from a previous sample, insist that the cell be cleaned thoroughly! It is important to observe this operation because there are times when an engineer or operator may measure only one parameter and estimate the other two. Remember that you may need R_{mf} in order to extract an R_w from the SP. An accurate value is essential. The next step is to supervise the calibrations. Some tools cannot be calibrated in the field. The SP has no calibration other than an electrical check of the circuits for accuracy. The gamma ray tool should be calibrated. A known API source is placed on the tool by a jig constructed to hold the source at a known distance. The counts received by the tool should be close to what the source counts are. Next, the source is taken off and placed far away from the tool. A background radiation check is made. The neutron and density tools are calibrated in the shop by a large water tank for the neutron tools and a large aluminum block for the density. Portable calibrators are taken to the field and both these tools are then calibrated. The values read from the portable calibrators should be very close to the shop calibrations. Insist that these tools be checked for calibration. If you get an answer such as "we calibrated these tools in the shop," it may mean that either this company has no means of field calibration or the two calibrations do not agree. This is the time to be suspicious and ask why or why not. Paying for a log from a tool which exhibits poor response is merely throwing money away.

The sonic tool needs to be calibrated in water. It is not practical to calibrate it in the field. One must take the word of the wireline company for its calibration. There really is not much to calibrate here anyway.

The induction tool is another that cannot be calibrated in the field. This is because the tool requires a conductivity-free environment. The tool is calibrated at the shop by placing it on very high wooden stands away from any metallic objects, such as buildings, fences, pipes and so on. A wheel made of nonconductive material with a calibrating resistance is placed on the tool and calibrated to read that resistance. Some companies calibrate their tools over the entire range of the tool scale. That is 0.2 ohms, 2.0 ohms, 20 ohms, 200 ohms, and 2,000 ohms. Some do a partial scale such as 2 to 200 ohms or worse, 20 to 200 ohms. This is when you need to ask the field engineer how the tool was calibrated. It is preferable to have the tool calibrated over its entire range, otherwise the data obtained may be questionable. Once the induction tool has been calibrated, internal calibrate resistors are set. This in turn is used as a calibration check in the field. The value obtained from these resistors should be close to shop calibration values.

The next phase of supervision is when the tool reaches bottom hole. At this time, the engineer will log 200 feet of hole, drop back down and repeat the operation. This is known as the repeat section. These two logs should overlay each other exactly. The only exception is minor variations due to the statistical nature of radiation curves. Failure to repeat indicates that tool is not really measuring formation parameters.

As the tool comes out of the hole, observe and inquire about any changes that the engineer may make. It has been my unfortunate experience to witness an engineer arbitrarily shift a neutron 6 porosity units simply because the tool was either not properly calibrated or had a poor response. The shift was made according to what the engineer thought it should look like. I've seen undetected scale shifts, matrix shifts, etc., in order to hide one thing or another. It is easy to change almost anything today because of the use of computers, so look closely and ask questions.

When the tool string comes out of the hole, make sure it had a thermometer, and that the bottom hole temperature is read. Sometimes the crew forgets and later estimates it.

When you are sure of log parameters then you can be sure of your analysis. One more thing to consider. Some wireline companies provide a so-called supervisor. Just consider who pays his salary.

Appendix F

Notation

a	constant based on rock texture in F-Φ
A	atomic weight
amf	activity of the mud filtrate
aw	activity of water
B_{cp}	sonic compaction correction
BHT	bottom hole temperature
C	conductivity
C_p	compaction correction, see B_{cp}
D	depth
d_i	diameter of invasion
E_c	potential
EMF	electromotive force
F	formation factor
G_g	geothermal gradient
G	geometrical factor
h	thickness of zone or formation
I	index
I	current (electrical)
K	permeability
K	SP constant proportional to temperature
K_{ro}	relative permeability of oil
K_{rw}	relative permeability of water
L	length
L_e	length of the ion path
LL	laterlog
LL_d	laterlog deep
LL_m	laterlog medium
m	cementation exponent
n	saturation exponent
r	electrical resistance
R	resistivity
R_a	apparent (uncorrected) resistivity
R_m	mud resistivity
R_{mc}	mud cake resistivity
R_{mf}	mud filtrate resistivity
R_{mfe}	equivalent mud filtrate resistivity
R_{mfeq}	same as R_{mfe}
R_o	100% water saturated resistivity
R_t	true resistivity
R_w	water resistivity
R_{wa}	apparent water resistivity
R_{we}	equivalent water resistivity
R_{weq}	same as R_{we}
R_{xo}	flushed zone resistivity
S_i	invaded zone saturation
S_w	water saturation
S_{wi}	irreducible water saturation
S_{wirr}	same as S_{wi}
S_{xo}	flushed zone saturation
T	temperature
T_f	formation temperature
T_o	mean average temperature
t	sonic travel time
t_f	fluid travel time
t_{ma}	matrix travel time
t_{sh}	shale travel time
V	voltage, volume
V_{sh}	shale volume
WC	water cut
WOR	water oil ratio

Greek Symbols

α	proportional to
Δ	delta, difference
Δt	travel time
$\Delta\rho$	density difference, correction curve
μ	micro
μs	microseconds
Φ	porosity
Φ_a	apparent porosity
Φ_e	effective porosity
Φ_d	density porosity
Φ_n	neutron porosity
Φ_s	sonic porosity
ρ	density
ρ_b	bulk density
ρ_f	fluid density
ρ_{ma}	matrix density
ρ_h	hydrocarbon density
ρ_w	water density

Review Questions and Sample Problems

This section contains both questions on the individual chapters and a set of problems requiring an understanding of the material as a whole. The figures referred to by numbers will be found in the chapter indicated by the first number. The figures referred to by letters will be found in this section. The answers to the questions follow them, and the solutions to the problems follow the problems.

Review Questions

Chapter 1

1. Name two types of cementing agents.
2. In a cubic configuration with perfectly round grains, the maximum porosity is ________?
3. By what Greek letter is porosity denoted?
4. Is it possible to produce connate water? Why?
5. What is the unit of permeability?

Chapter 2

1. If you had one cube of water, what parameter would you measure?
2. If you had one cube of formation 100% saturated with water, what parameter would you measure?
3. If the same cube had oil?
4. If you had a water saturation of 30%, what is the hydrocarbon saturation?
5. What is F equal to?
6. Log analysts use the average surface temperatures that are ________ to ________ degrees higher.
7. In the equation $\frac{a}{\Phi^m}$ what does "a" represent?

   What does "m" represent?

Chapter 3

1. Give four uses of SP.
2. What does SSP stand for?

3. What does PSP stand for?

4. Where would you expect to see SSP?

5. Where would you expect to see PSP?

6. An SP deflection to the right indicates what polarity? To the left?

7. How does the formation water compare to the mud filtrate if SP deflection is negative?

8. What important parameter can be derived from the SP?

9. Is the constant K temperature dependent?

10. Using Figures 3-6 and 3-7 and Figures A and B, determine the value of R_w at formation temperature, given:
    Static SP = −100 mv

    R_{mf} = 1.6 @ 75°F

    Bottom hole temp. = 210°

    Total depth = 11,000 ft

    Formation depth = 6,000 ft

    Annual mean temp. = 80°

Chapter 4

1. What type of borehole fluid does a focused-type device need?

2. An induction device?

3. What does a lateral device measure directly?

4. What does an induction device measure directly?

5. How is conductivity related to resistivity and vice versa?

6. Name the three zones that a dual induction focused device measures.

7. Name three influences on the induction signal.

8. Given the following equations:

$$R = \frac{1,000}{C}$$

$$C = \frac{1,000}{R}$$

solve:

a. 30 mmho =

b. 2 mmho =

c. 100 mmho =

d. 40 ohms =

e. 20 ohms =

f. 10 ohms =

g. Given R_t = 18 ohm/m, porosity = 16%, a = 1, m = 2, solve for R_{wa}.

h. Given F = 14, and R_w = 0.05, solve for R_o.

i. Given a porosity of 15%, calculate F in a sandstone. In a carbonate.

Chapter 5

Given the equation:

$$\Phi = \frac{\Delta t - \Delta t_{ma}}{\Delta t_f - \Delta t_{ma}}$$

1. What is Δt?

2. What is Δt_f?

3. What is its value?

4. What is Δt_{ma}?

5. What would happen to Δt if the formation had gas?

6. How can you compensate porosity for gas?

7. For oil?

8. What value is used to decide if the formation is compacted?

9. State the compaction correction.

10. Using Figure C, calculate the porosity, given:

    a. Sandstone matrix

    b. Δt shale of 110 μs

(text continued on page 219)

ESTIMATION OF FORMATION TEMPERATURE
(Linear Gradient Assumed)

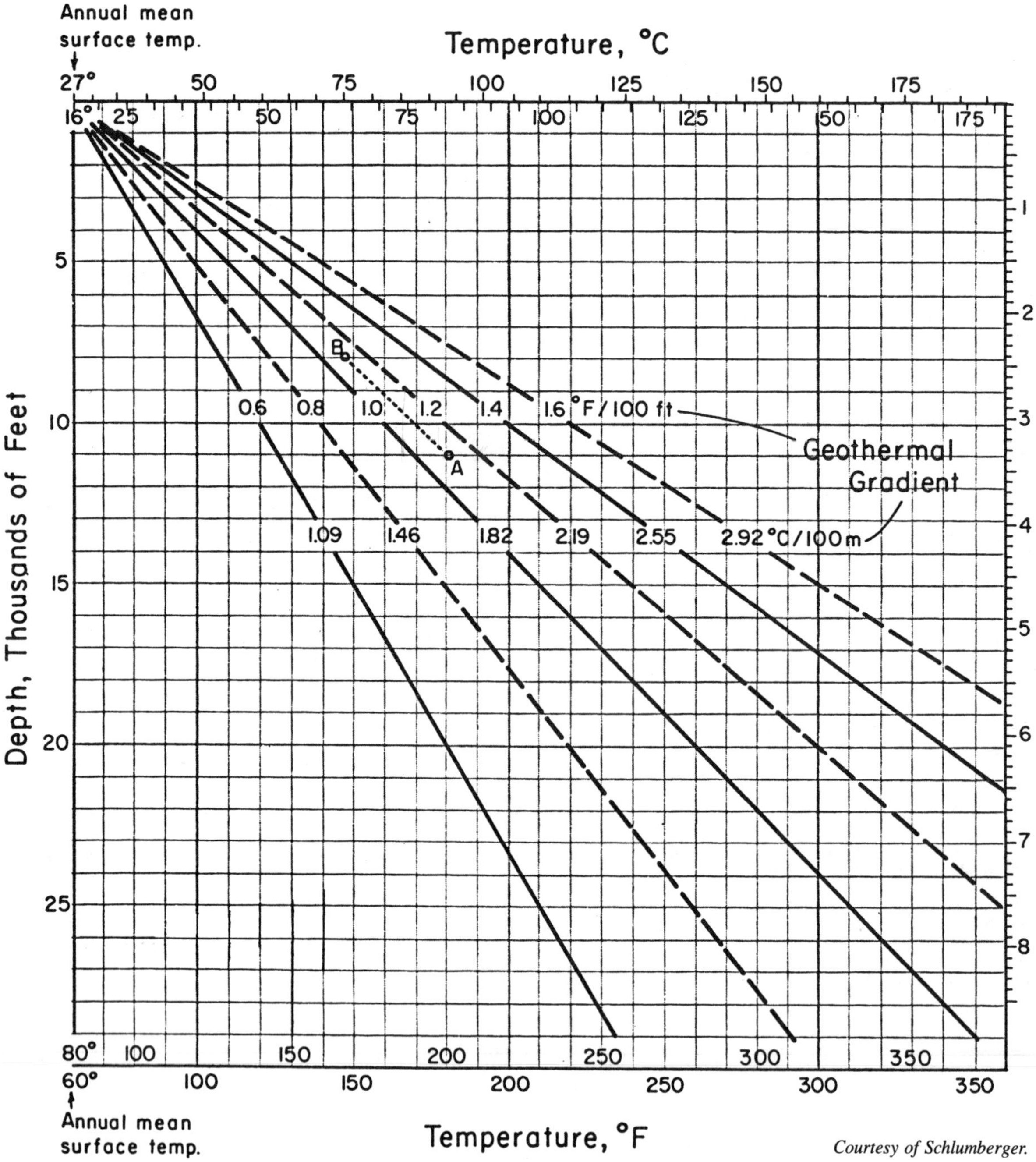

Courtesy of Schlumberger.

Figure A

RESISTIVITY NOMOGRAPH FOR NaCl SOLUTIONS

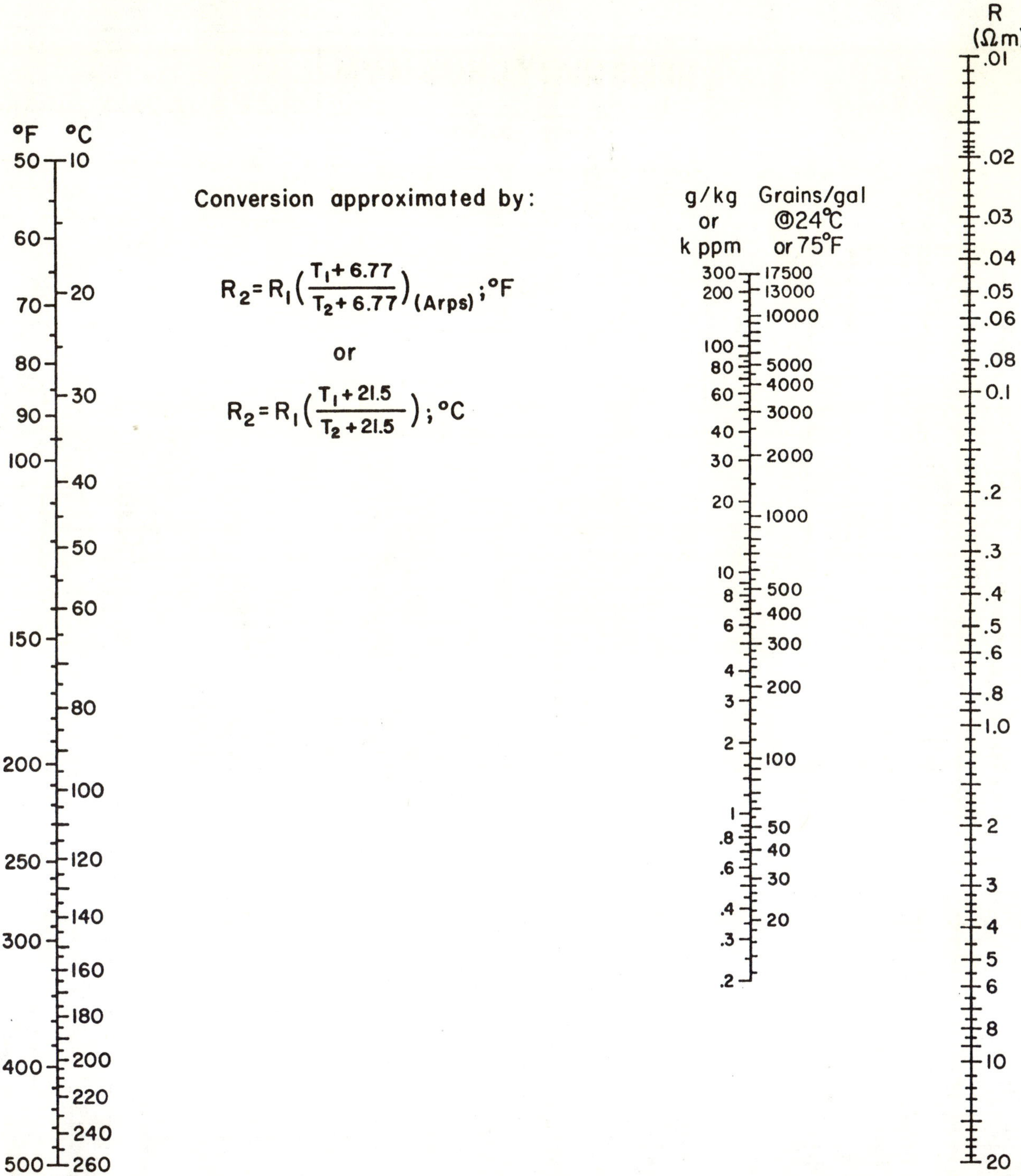

Courtesy of Schlumberger.

Figure B

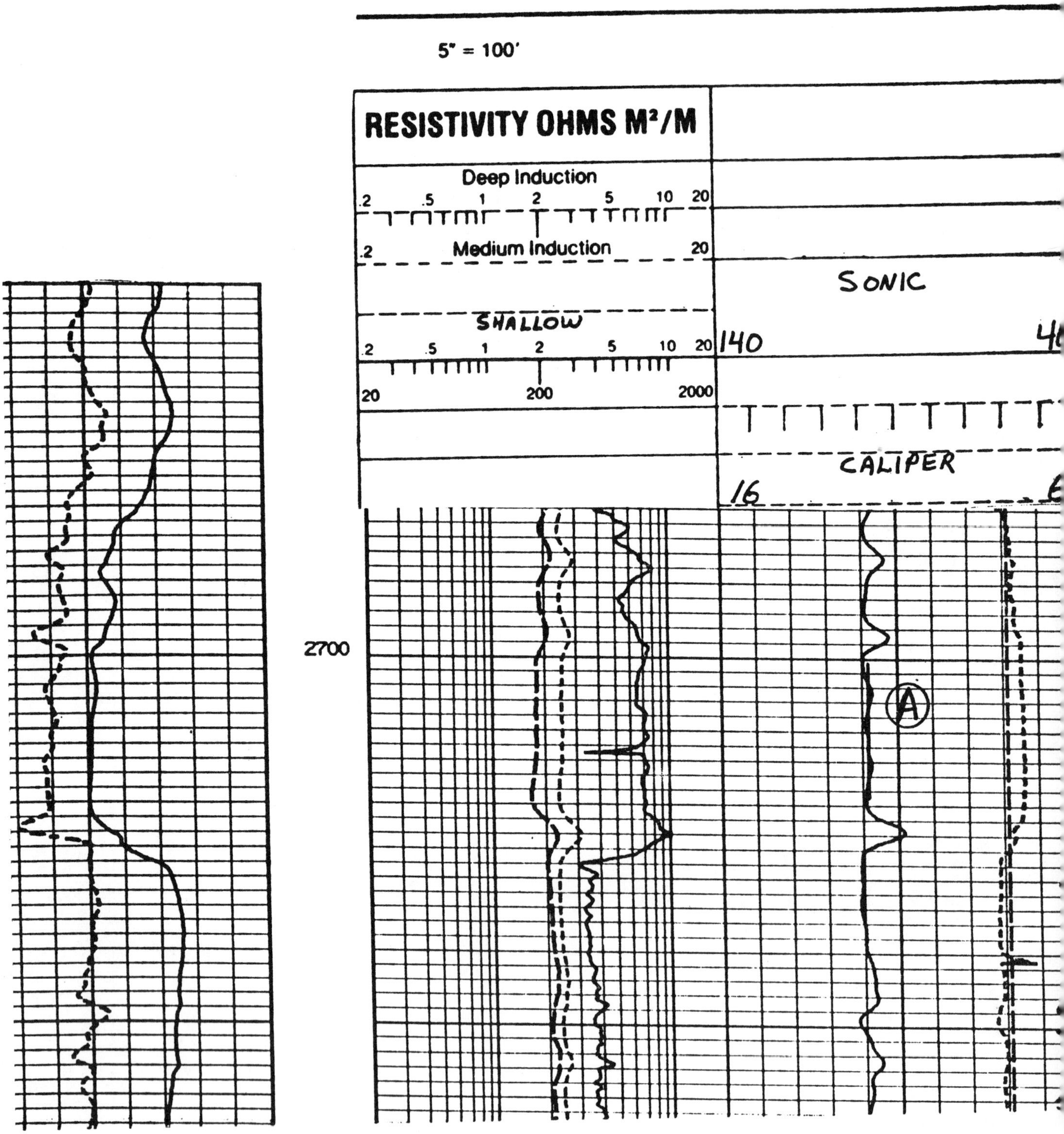
5" = 100'
RESISTIVITY OHMS M²/M
Deep Induction
.2 .5 1 2 5 10 20
.2 Medium Induction 20
SHALLOW
.2 .5 1 2 5 10 20
20 200 2000
SONIC
140
CALIPER
16
2700
A

Figure C

Chapter 6

1. The bed boundary is chosen from the SP at what point?

2. Name three sources for R_w.

3. What are the two equations used in the quick look technique?

4. R_t comes from what curve on the log?

5. How is R_t obtained?

6. What is F equal to?

7. What is R_o equal to?

8. What is S_w equal to?

9. What types of resistivity logs are needed to determine invasion?

10. What type of chart is used for correcting R_t for invasion?

11. Given the following:

    R_{mf} = 0.2 ohms @ 75°F

    R_m = 0.29 ohms @ 75°F

    TD = 10,000 ft

    BHT = 200°F

    Matrix = sandstone

    Mean avg. temp. = 80°

    Bit size is 7⁷/₈ in.

    Find S_w for zone B 7,036 ft. (See Figure D.) List your parameters (borehole correction chart follows, Figure E).

    a. Formation temperature =

    b. R_{mf} @ T_f =

    c. SP =

    d. R_{mfe} =

    e. R_{we} =

    f. R_w @ T_f =

    g. R_d =

    h. R_m @ T_f =

    i. R_t =

    j. Δt =

    k. Φ_a =

    l. Φ_e =

    m. F =

    n. R_o =

    o. S_w =

12. Using the procedure in Appendix C, estimate the water cut from the previous exercise. Given from *core data*:

    Permeability (K) = 180 md

    Oil gravity = 35

    calculate:

    a. S_{wi}

    b. K_{rw}

    c. K_{ro}

    d. WOR

    e. WC

Chapter 7

Given the following equation:

$$\Phi = \frac{\rho_{ma} - \rho_b}{\rho_{ma} - \rho_f}$$

1. What is ρ_{ma}?

2. What is ρ_b?

3. If the water is fresh, what is the value of ρ_f? If it is salty?

4. How do you correct density porosity for gas?

5. If a neutron collided with a lead atom, would it lose much energy?

6. If it collided with a hydrogen atom?

7. V_{sh} can be obtained from which single curve?

8. What better way can we determine V_{sh}?

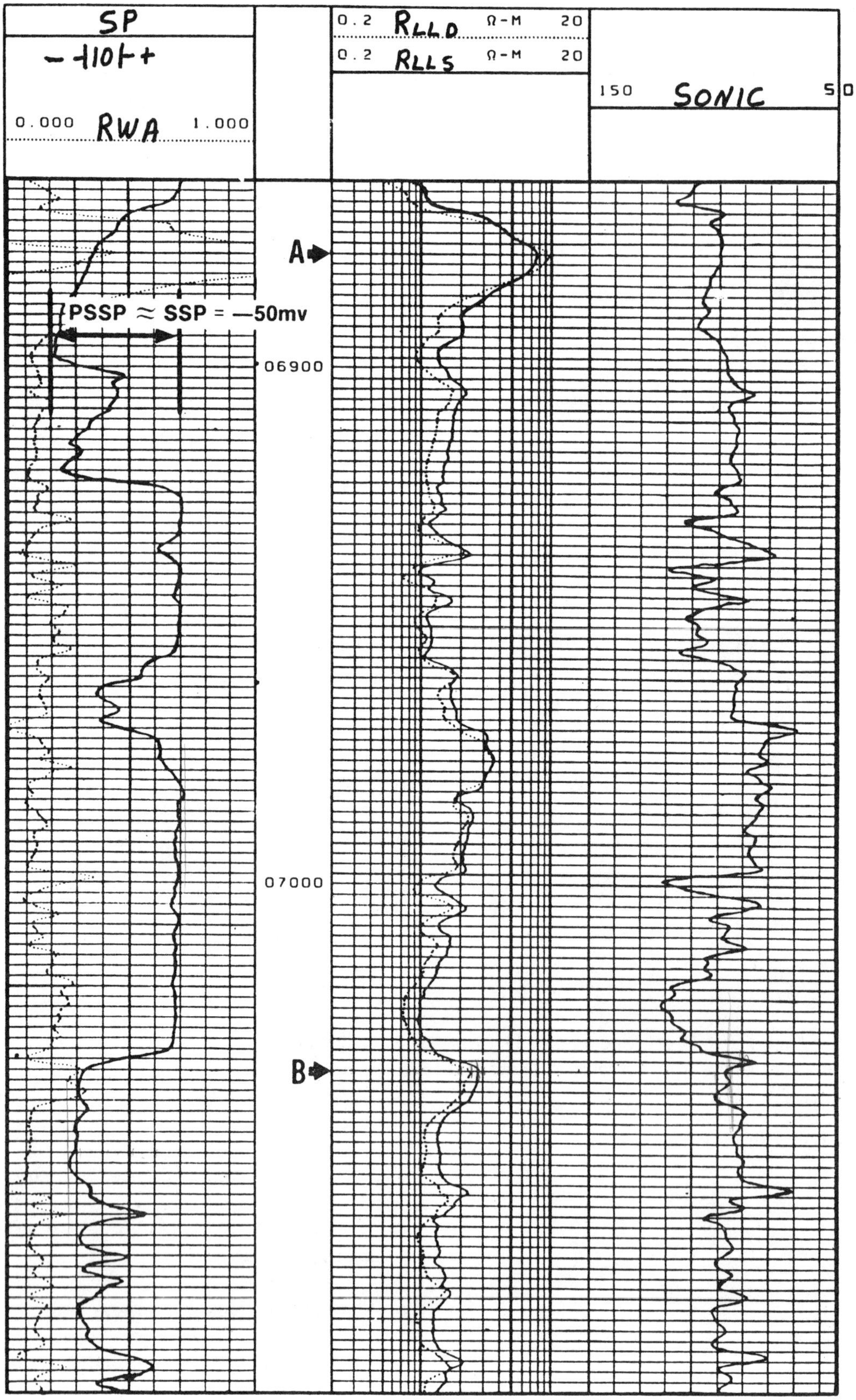

Courtesy of Gearhart.

Figure D

DUAL LATEROLOG—BOREHOLE CORRECTION

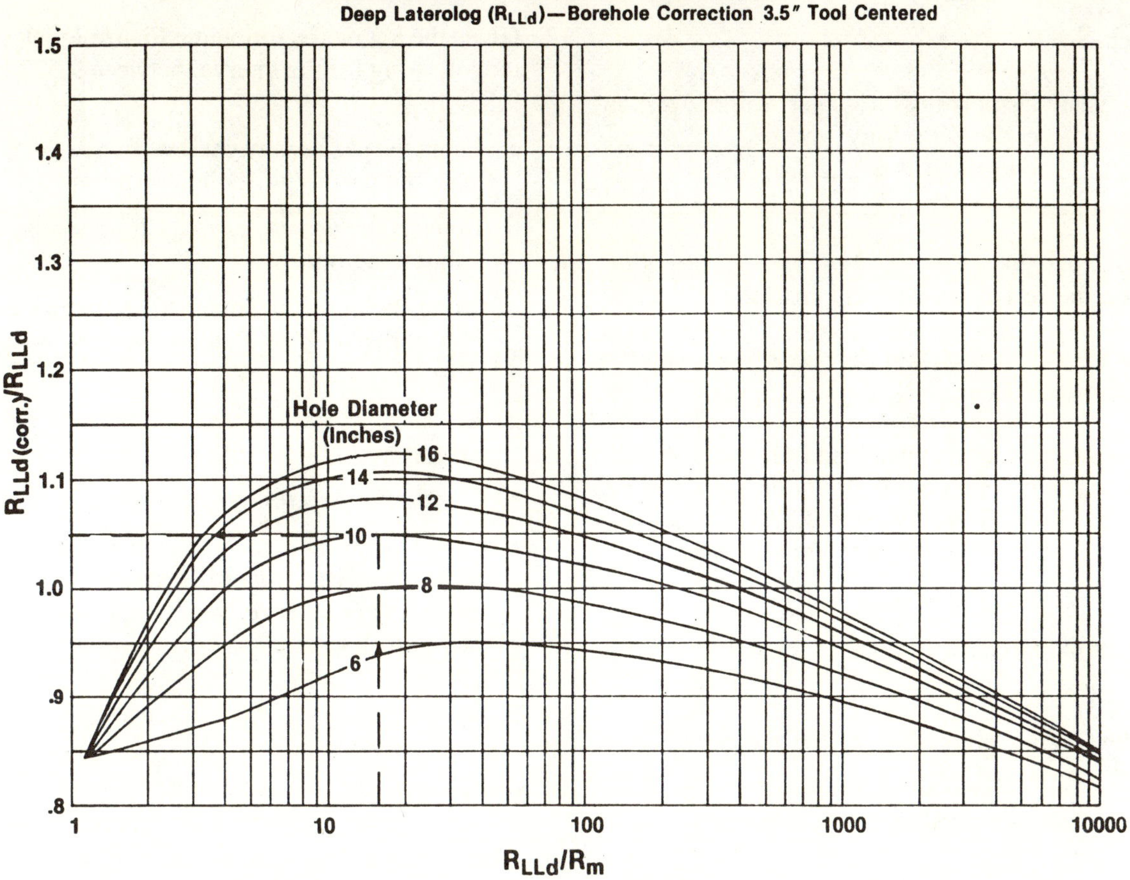

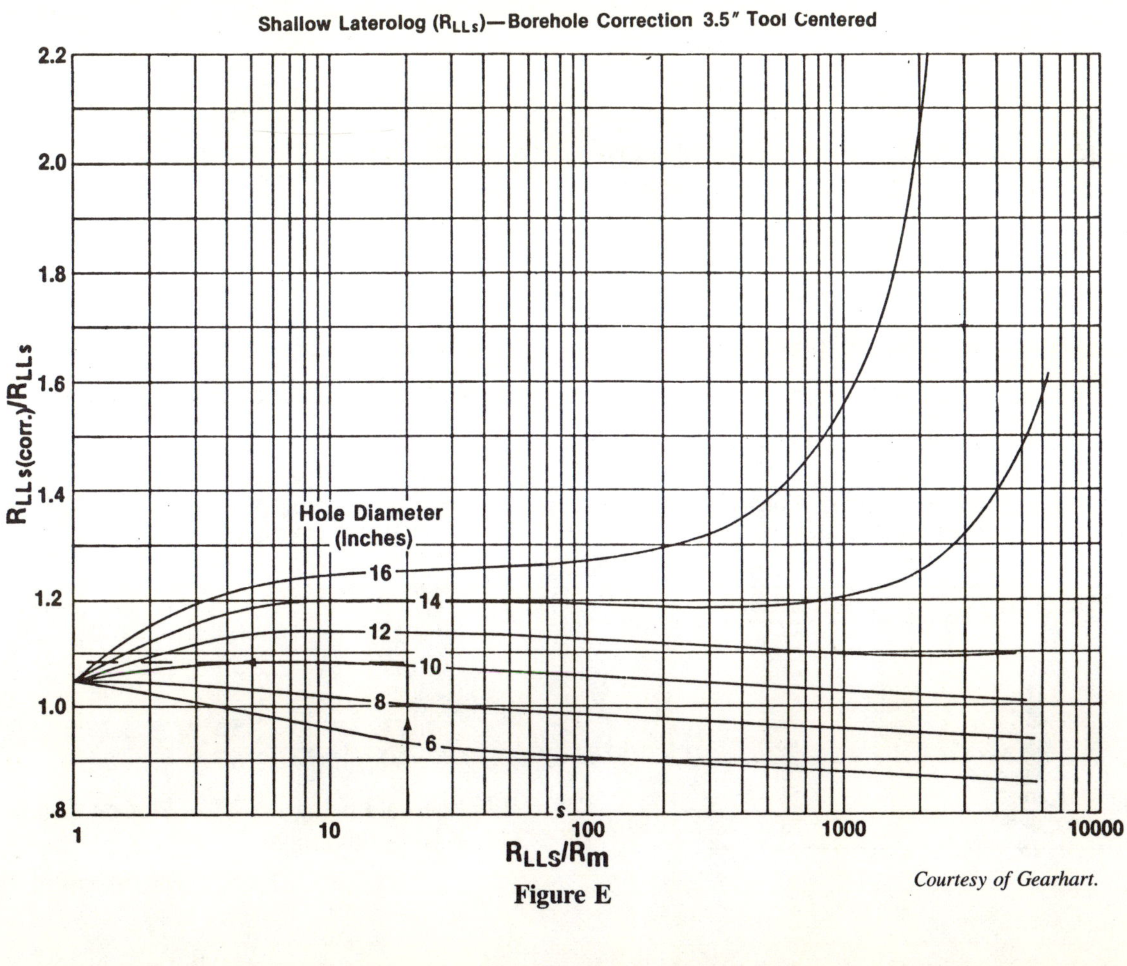

Courtesy of Gearhart.

Figure E

Chapter 8

1. The resistivity-porosity crossplot works well in cases where the lithology and R_w ________ ________, and there is a ________ ________ ________ zone.

2. At what percentage do we draw a water saturation line on the resistivity-porosity crossplot?

3. What are three errors that can occur when a tool is miscalibrated?

4. Using the logs in Figures F and G, plot the bulk density as a function of resistivity on Figure H for the depth interval of 2,190 ft to 2,230 ft. After you have plotted the data, answer the following:

   a. What is the value of R_o?

   b. What is the matrix value?

   c. What is the value of R_w?

   d. How many points fall below the 50% water saturation line?

   e. Will this zone produce hydrocarbons?

Chapter 9

1. When using a two-porosity crossplot, one assumes that the lithology of the formation is what?

2. The porosity curve must be corrected for what two problems before entering the porosity crossplot?

3. What depth of investigation do neutron and density tools have?

4. What is the average value of a lime matrix and a dolomite matrix?

5. Residual hydrocarbon saturation is equal to what?
   1 - Rxo

6. Crossplot the data below on the following Figure I and determine which type of shale is present:

Bulk Density	Neutron Porosity
2.4	27
2.38	25
2.41	28
2.45	30
2.35	24
2.39	26
2.40	26
2.38	26

7. Given the log on the following Figure J, determine the following for the interval between 9,522 ft and 9,530 ft.

   a. Using the gas density chart of Figure 9-15, estimate the density of the hydrocarbon.

   b. Convert the density porosity from sandstone to a limestone porosity.

   c. Using the chart in Figure K, convert the neutron porosity from a sandstone to a limestone porosity.

   d. Given S_{hr} = 96%, correct the density porosity for gas.

   e. Correct the neutron porosity for gas.

   f. Crossplot the two porosities on Figure I. What is the effective porosity? If you had used a root-mean square average, would you have an accurate porosity value? Why?

Chapter 10

1. Where is sulfur usually found?

2. The Schlumberger Mid-Plot is a good indicator of what type of porosity?

3. Given the following data:

   ρ_{ma} = 2.65

   ρ_b = 20.5

   R_w = 0.04

   R_t = 0.5

   Gamma = 7 API units

   a. Will this zone produce hydrocarbons? Explain why.

   b. What mineral is present in a formation with a sonic travel time of 118 and a bulk density of 2.02? What is its volume?

   c. Given:

Δt	ρ_b	Φ_n
65	2.55	24.2
41	2.70	24.0
73	2.50	24.0
60	2.65	22.0

Plot the data using the M-N Plot on Figure 10-5. What is its lithology?

(text continued on page 228)

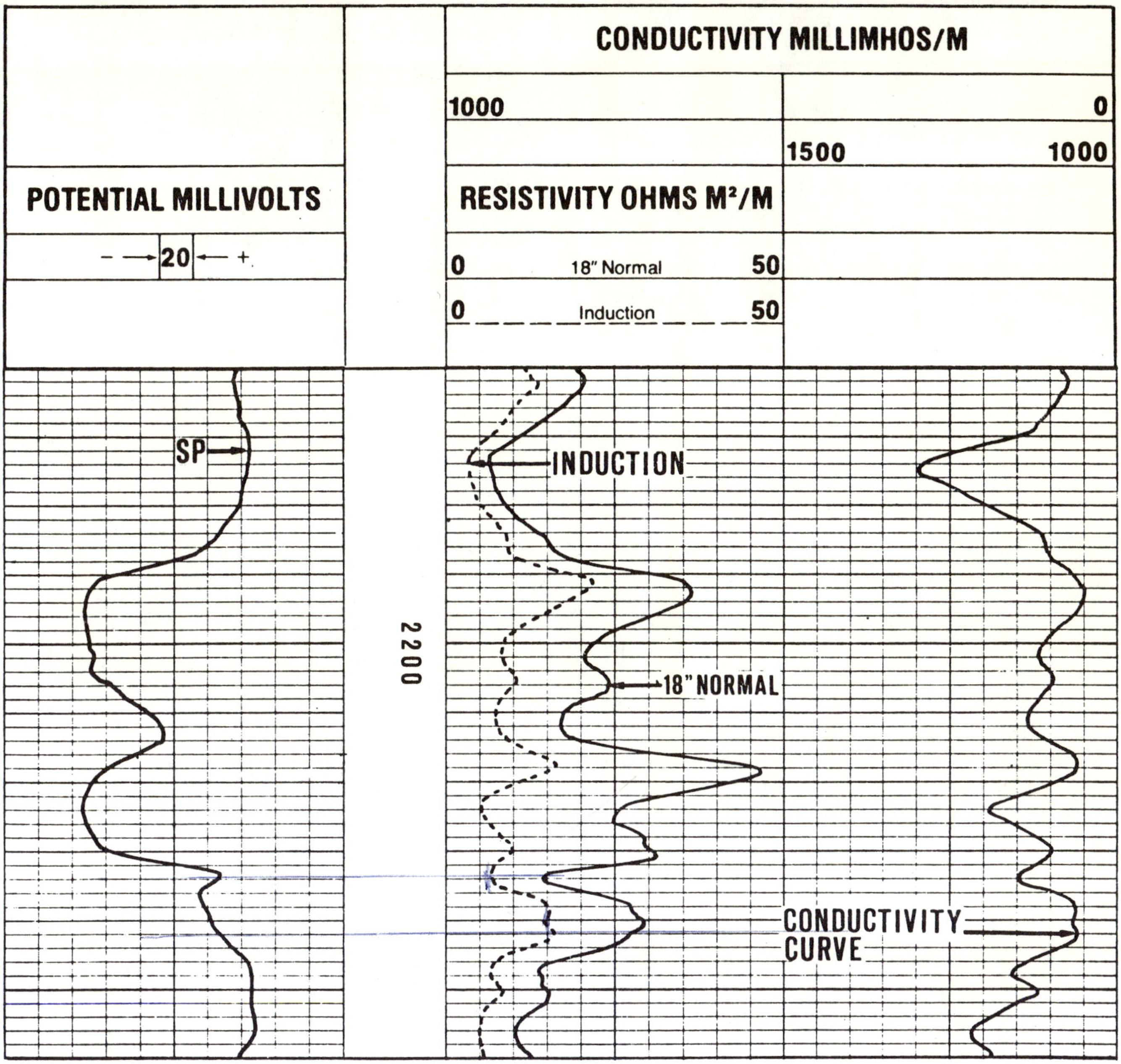

Courtesy of Welex.

Location: Okmulgee Co., Oklahoma

Formation: Ducher Sand (Calcarious)

Figure F

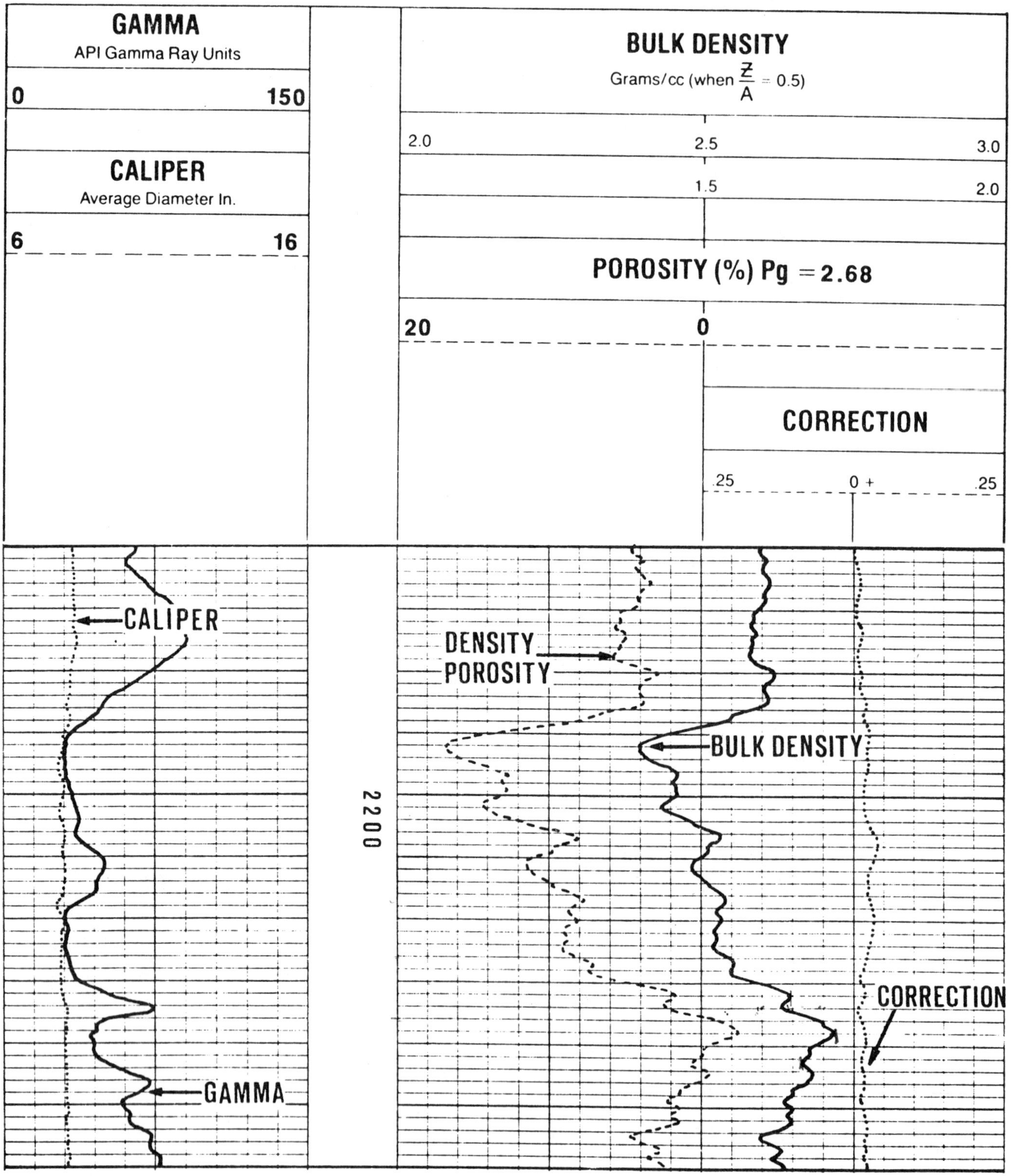

Courtesy of Welex.

Figure G

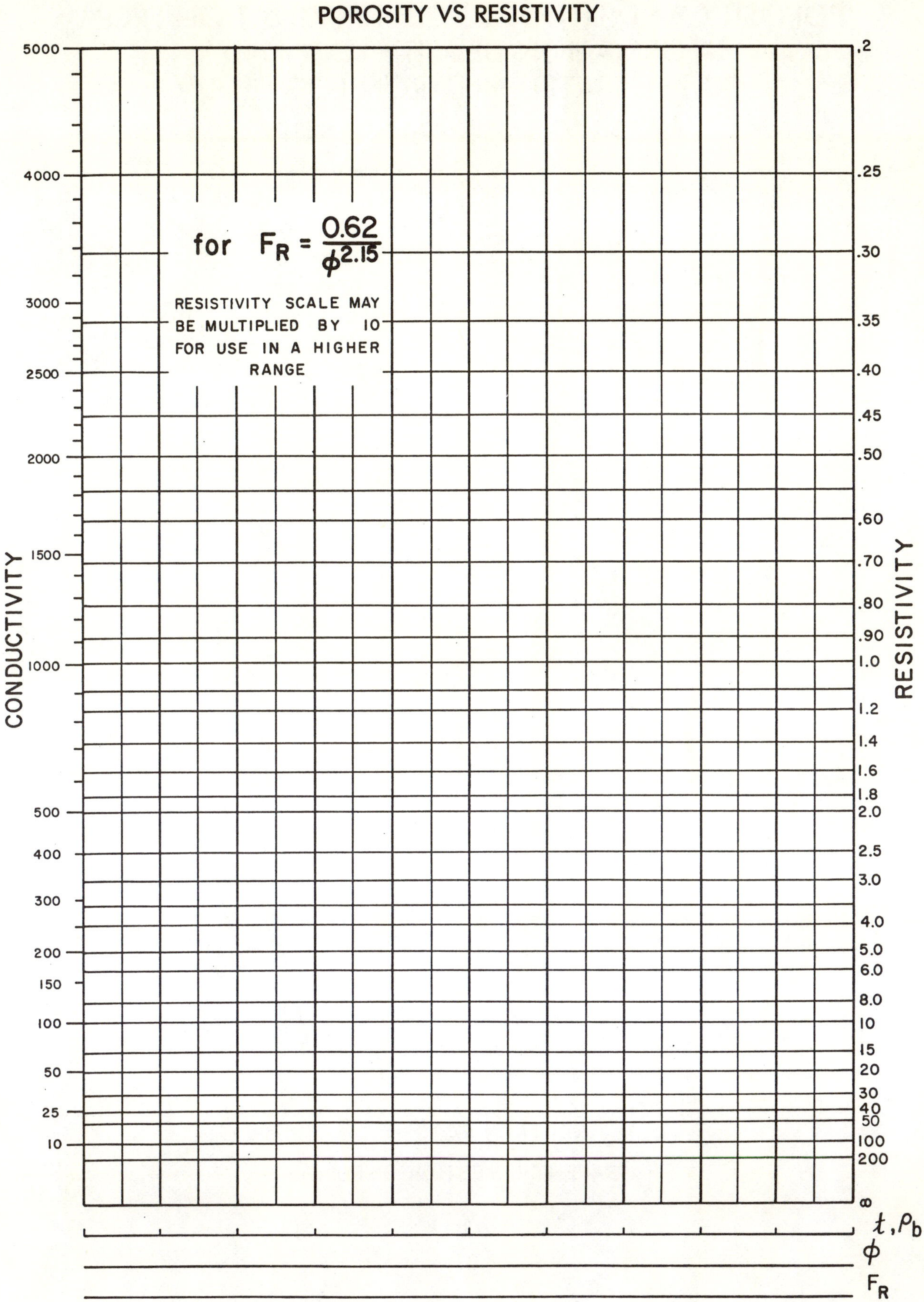

Figure H

Courtesy of Schlumberger.

POROSITY AND LITHOLOGY DETERMINATION FROM FORMATION DENSITY LOG AND COMPENSATED NEUTRON LOG (CNL*)

FRESH WATER, LIQUID-FILLED HOLES

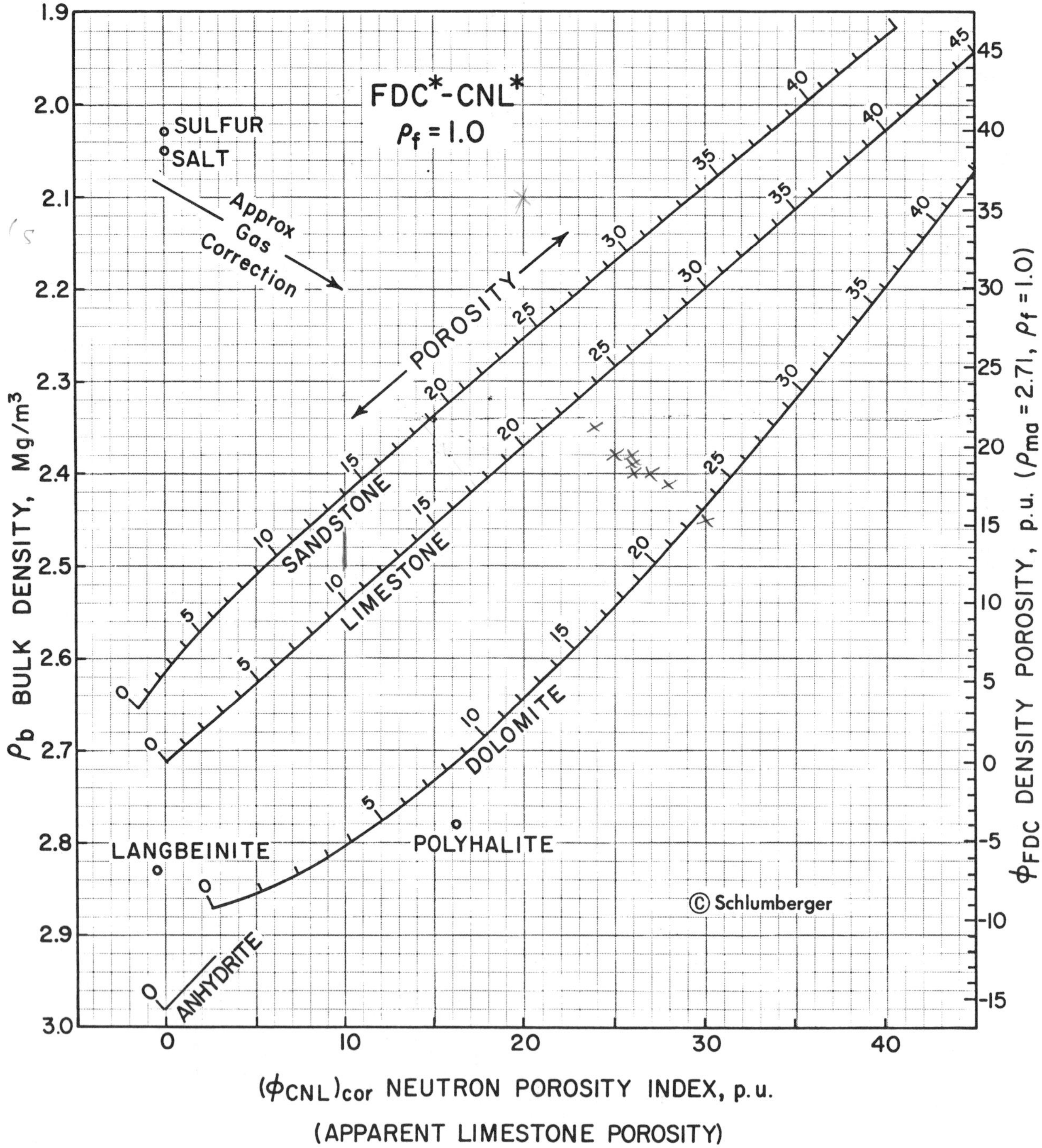

Courtesy of Schlumberger.

Figure I

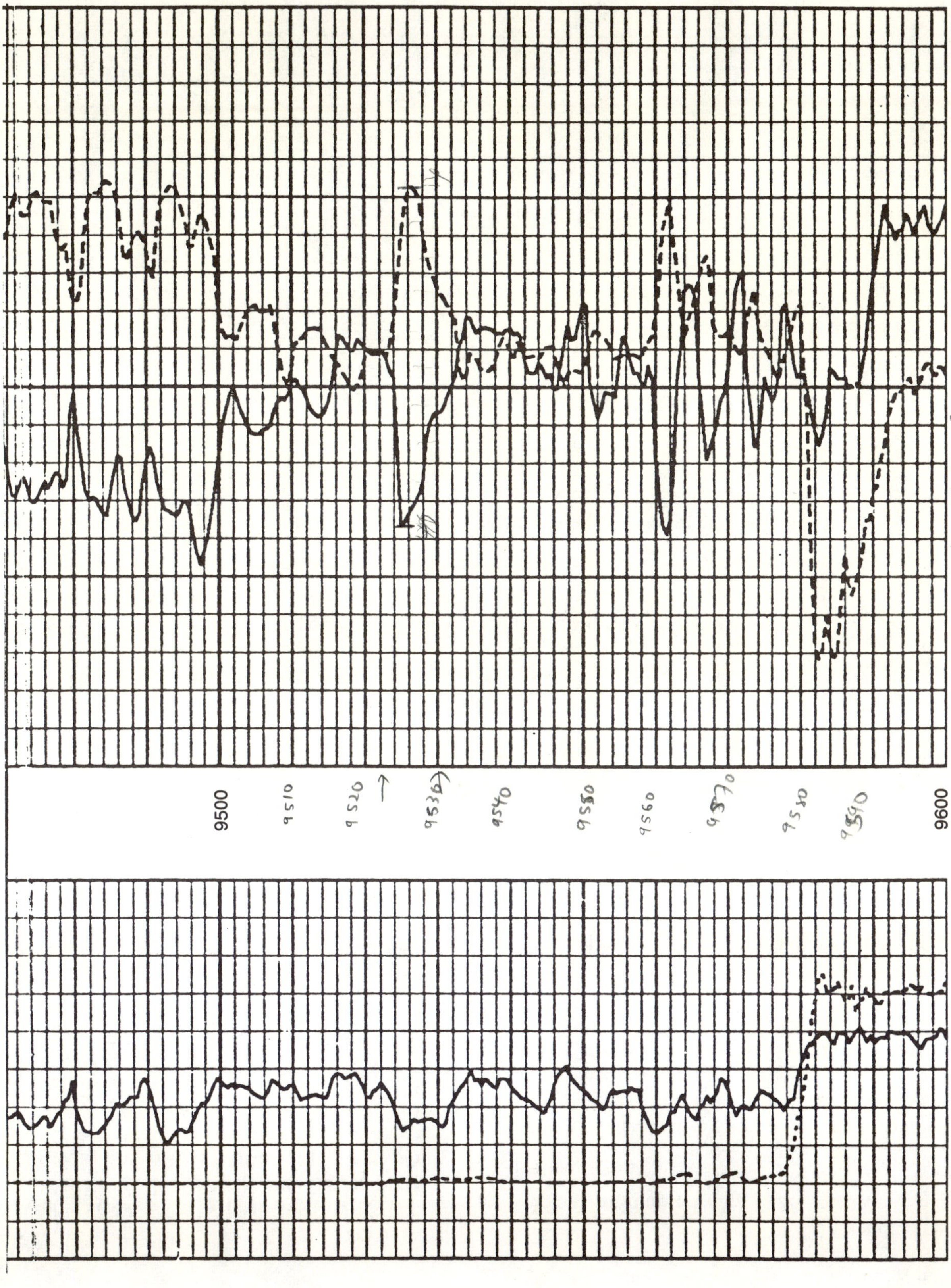

Courtesy of Welex.

Figure J

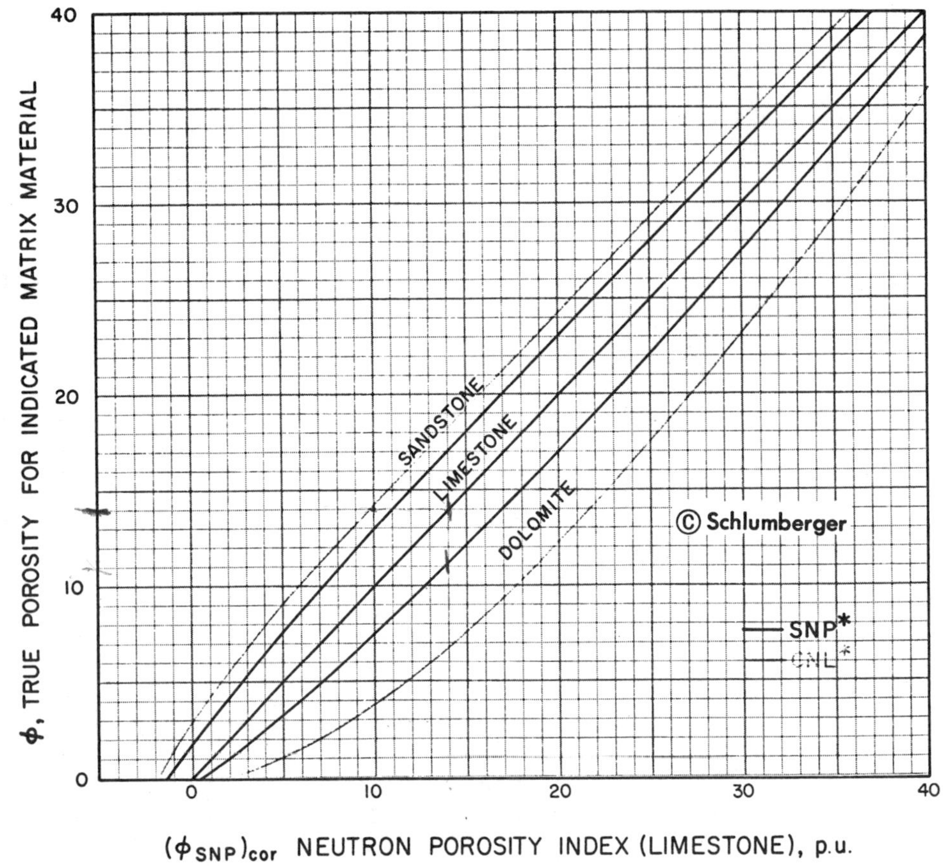

Courtesy of Schlumberger.

Figure K

Chapter 11

1. The microlog can indicate ________ but is a poor indicator of ________.

2. If the product of porosity and water saturation is constant over an interval, what does this indicate?

3. A shallow gas sand has 17% porosity and 21% irreducible water. What is its estimated permeability?

4. Using the contact log in Figure L indicate the permeable zones.

5. Given R_w = 0.05, is the log in Figure M at irreducible water?

Chapter 12

1. When reading a gamma ray log, what is sometimes mistaken for shale?

2. The bulk density measurement uses the ________ ________ ________ because it is not sensitive to the ________ ________ _____ _____ ________.

3. Given a bulk density of 2.46 in a lime matrix, a fluid density of 1.0, a CNL porosity of 11%, and an EPT porosity of 11%, calculate the bulk volume of water.

4. Name the three components of the GRS curve.

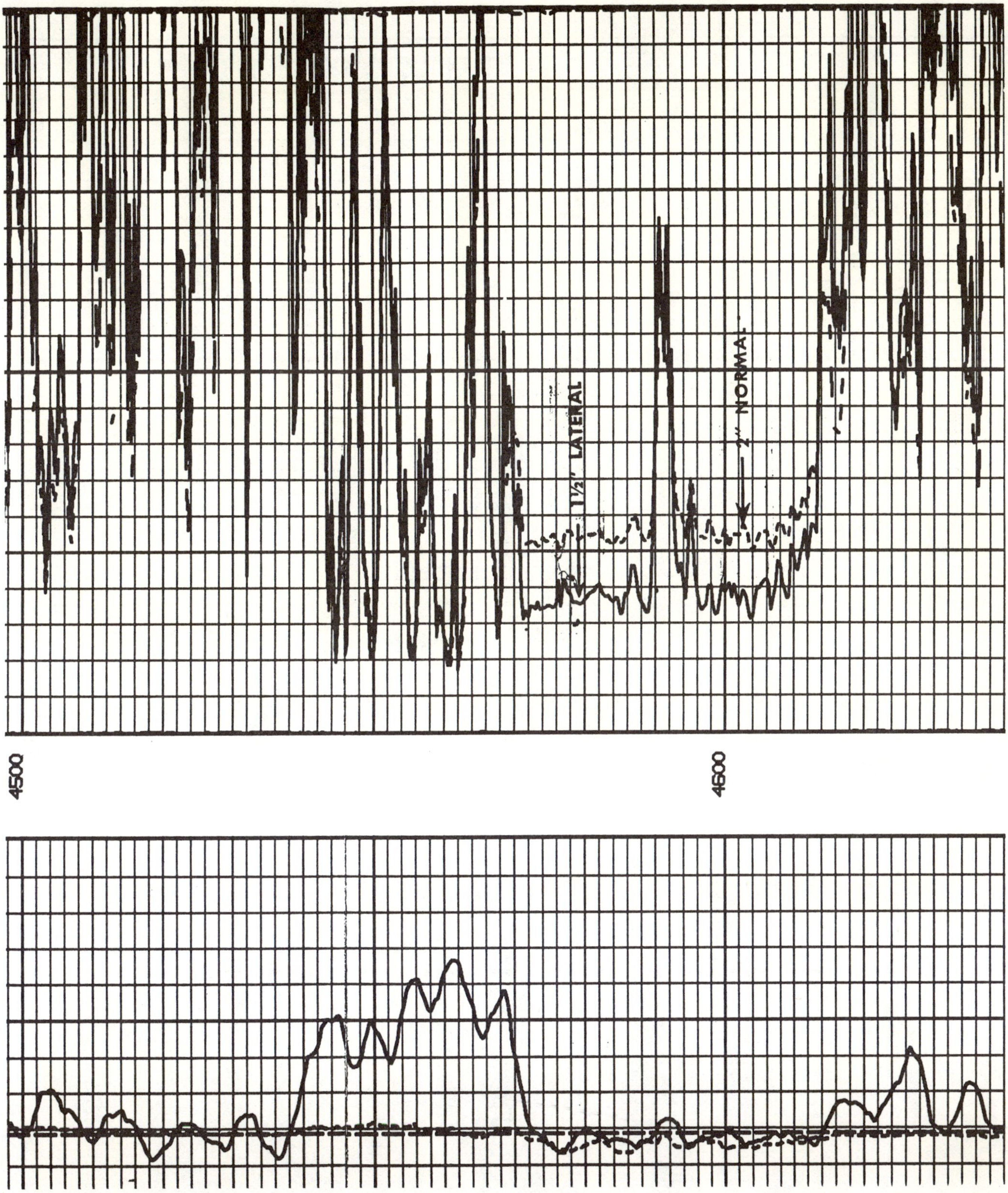

Courtesy of Welex.

Figure L

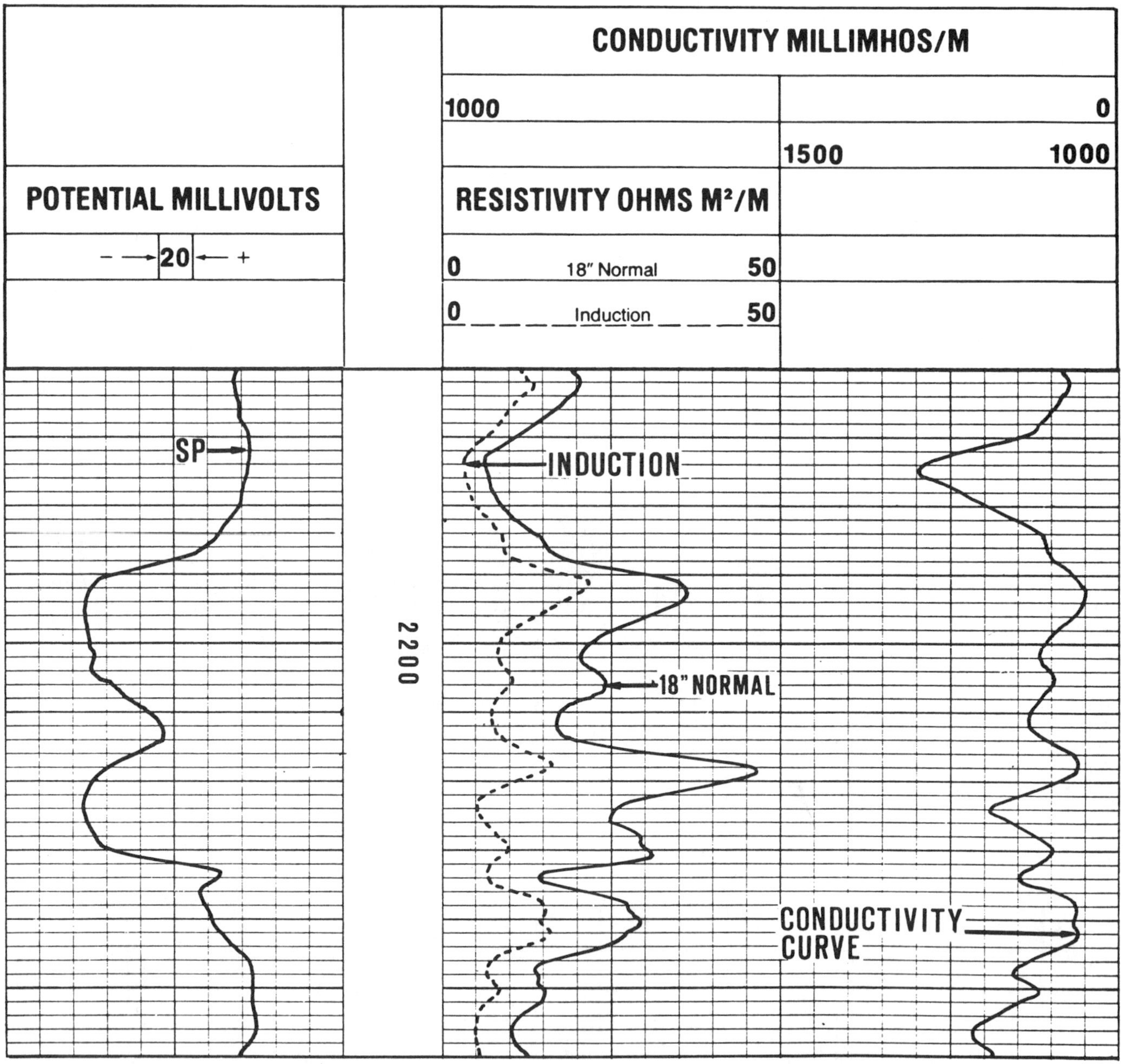

Location: Okmulgee Co., Oklahoma Formation: Ducher Sand (Calcarious)

Figure M

5. Given the following P_e values, identify the matrix.
   a. 1.81
   b. 3.14
   c. 5.08
   d. 5.05

6. In a pure lithology, what happens to the P_e curve as porosity increases? On which matrix is this effect most pronounced?

7. When would you recommend the use of the EPT tool?

Chapter 13

1. Name the three ways shale is distributed in sand.

2. In a shaly sand, the SP does not reach what condition?

3. The two-porosity shale crossplot combines which two curves?

4. What two types of water does the dual water model identify?

5. The dual water model does not need which value?

6. What is meant by bound water?

7. Is S_w-S_{wirr} equivalent to free water?

8. Given a neutron porosity of 45% and a density porosity of 9%, construct a shale crossplot and fill in the missing answers.

Φ_n	Φ_d	Φ_{corr}	V_{sh}
30%	22%	____	____
32%	14%	____	____
42%	24%	____	____
29%	16%	____	____

9. Given a sandstone lithology and

   Φ_n = 38%
   Φ_d = 20%
   R_w = 0.05 ohms
   R_{sh} = 2.0 ohms
   R_t = 14.0 ohms

   Using the crossplot constructed in the above exercise, find the corrected porosity and shale volume. Then compute the water saturation.

Answers to Review Questions

Chapter 1

1. a. Quartz.
   b. Calcite.
2. 47.6%.
3. Φ or phi.
4. No. The water adheres by surface tension.
5. Darcy.

Chapter 2

1. R_w.
2. R_o.
3. R_t.
4. 1 − .3 = .7 or 70%.
5. $R_o/R_w\ \frac{a}{\Phi^m}$.
6. 5 *to* 10.
7. a. Rock texture.
   b. Cementation exponent.

Chapter 3

1. a. To detect permeable beds.
   b. To locate bed boundaries.
   c. To determine R_w.
   d. As a shale indicator.
2. Static spontaneous potential.
3. Pseudo static potential.

4. In a thick, clean permeable bed.
5. In a thick, shaly permeable bed.
6. a. Positive.
   b. Negative.
7. The formation water is more saline than the mud filtrate.
8. R_w.
9. Yes. $K = (60 + 0.133T)$ where T is formation temperature.
10. a. To determine T_f, enter 11,000 ft and 250°F on Figure N, the formation temperature chart (points a and b). Using the gradient line of 1.2 as indicated, enter 6,000 ft and read a formation temperature of 150 (points c and d).
    b. Convert R_{mf} to formation temperature. Using Figure O, the resistivity nomograph, plot 75° and 1.6 crossing the salinity line at 3.5 K ppm (points a and b). Plot 150° (point c) and 3.5 K ppm extending the line to point d. Thus R_{mf} @ T_f is 0.85 ohm/m.
    c. Using Figure P, the R_w vs. R_{weq} chart, convert R_{mf} to R_{mfeq} as illustrated by points a, b, and c. The 150° line is used, and point c shows R_{mfeq} to be 0.42.
    d. On Figure Q, the R_{weq} determination chart, plot the −100 mv of SP past the formation temperature of 150° and continue until point b. Plot point b to R_{mfeq} (point c) of 0.42, extending the line to point d. Read R_{weq} at point d. It is 0.025.
    e. Go back to Figure P, the R_w vs. R_{weq} chart, and plot points d, e, and f using the 150° line. Read an R_w of 0.045 at point f.
    f. R_w @ T_f = 0.39 @ 150°.

Chapter 4

1. Any conductive fluid.
2. A conductive or nonconductive fluid.
3. Resistivity.
4. Conductivity.
5. a. $R = \frac{1}{C}$.
   b. $C = \frac{1}{R}$.
6. a. Invaded.
   b. Transition.
   c. Virgin.
7. a. Bed thickness.
   b. Hole diameter.
   c. Invasion.
   d. Highly conductive mud.
8. a. 30 mmho = 33.3 ohms
   b. 2 mmho = 500 ohms
   c. 100 mmho = 10 ohms
   d. 40 ohms = 25 mmho
   e. 20 ohms = 50 mmho
   f. 10 ohms = 100 mmho
   g. $R_{wa} = \frac{18 \times 0.16^2}{1} = 0.46$
   h. $R_o = 14 \times 0.05 = 0.7$
   i. (1) $\frac{0.62}{0.15^{2.15}} = 36.6$

      (2) $\frac{1}{0.15^2} = 44.4$

Chapter 5

1. Travel time through the formation recorded on the log.
2. Travel time through water.
3. 189 microseconds per foot.
4. Travel time through the rock matrix.
5. It would be slowed down.
6. Multiply by 0.7.
7. Multiply by 0.9.
8. 100 μsec in nearby shale.
9. 100/Δtsh.
10. a. For line A, Δt reads 98 μsec/ft.
    b. Δt in a nearby shale is 100 μsec/ft, indicating a compaction correction. Thus:

$$\Phi = \frac{\Delta t - \Delta t_{ma}}{\Delta t_f - \Delta t_{ma}} \times \frac{100}{\Delta t_{sh}}$$

where Δt_{ma} is 57 for a sandstone.

$$\Phi = \frac{98 - 57}{189 - 57} \times \frac{100}{110} = 0.282 \text{ or } 28.2\%$$

(text continued on page 237)

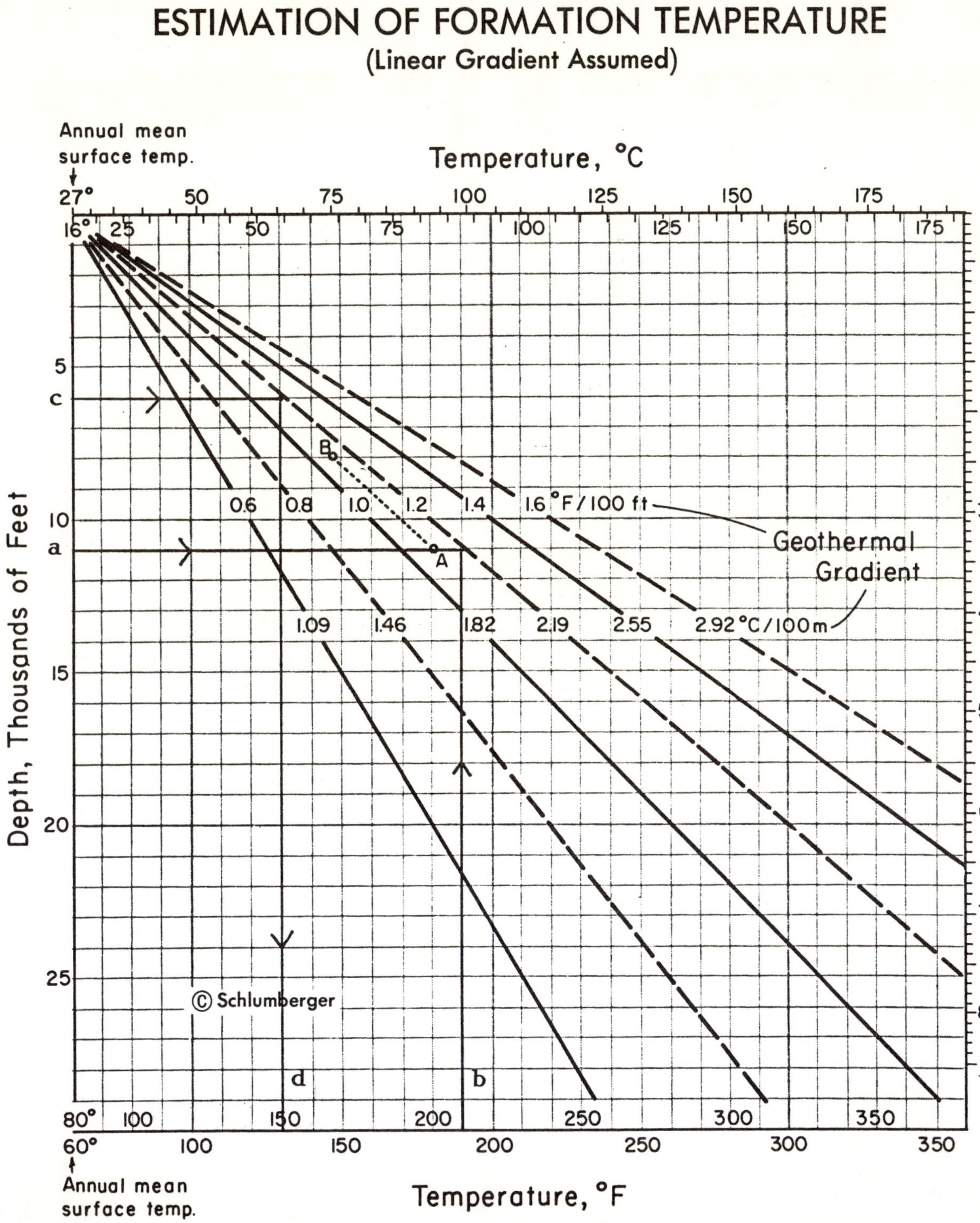

Courtesy of Schlumberger.

Figure N

RESISTIVITY NOMOGRAPH FOR NaCl SOLUTIONS

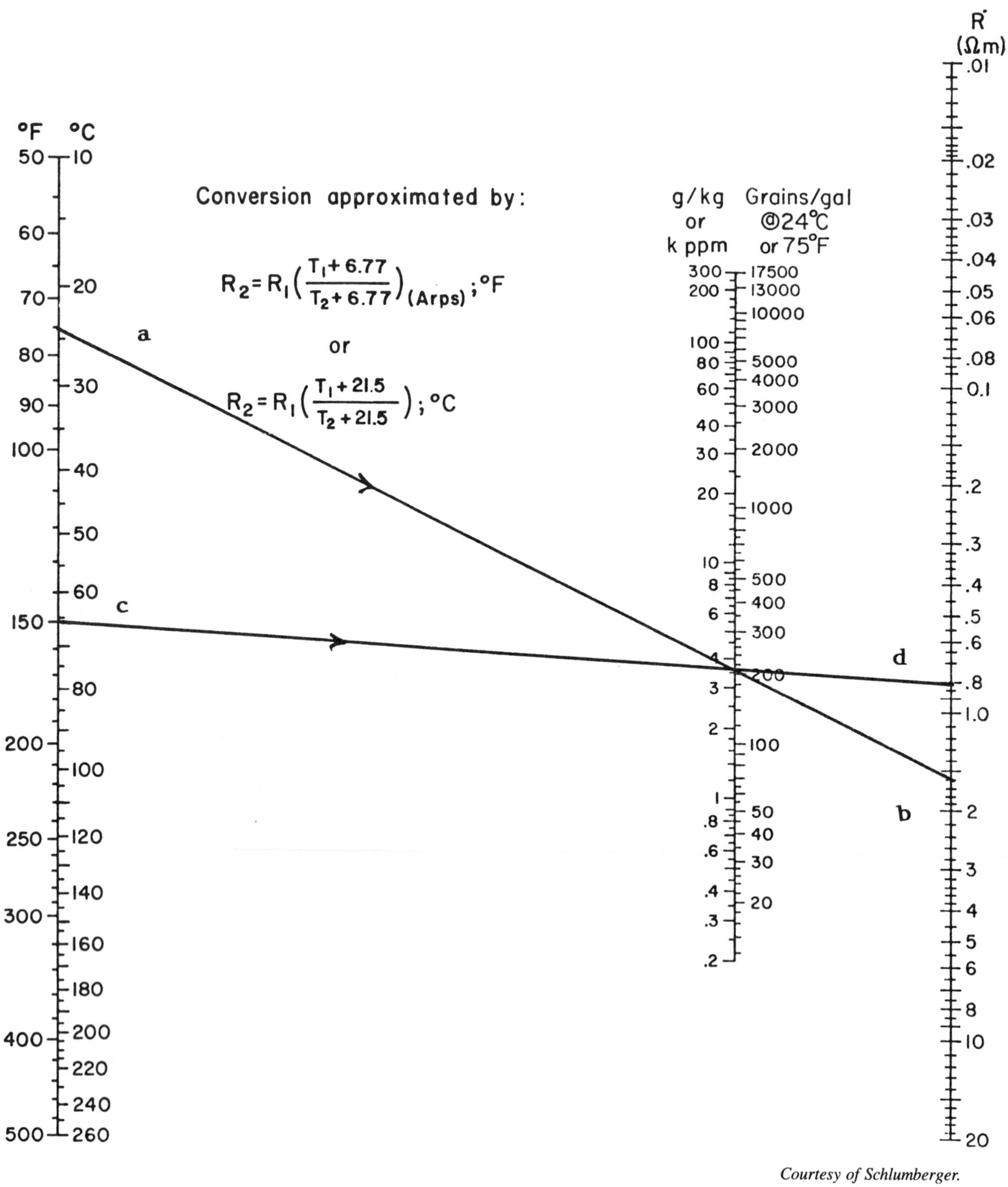

Courtesy of Schlumberger.

Figure O

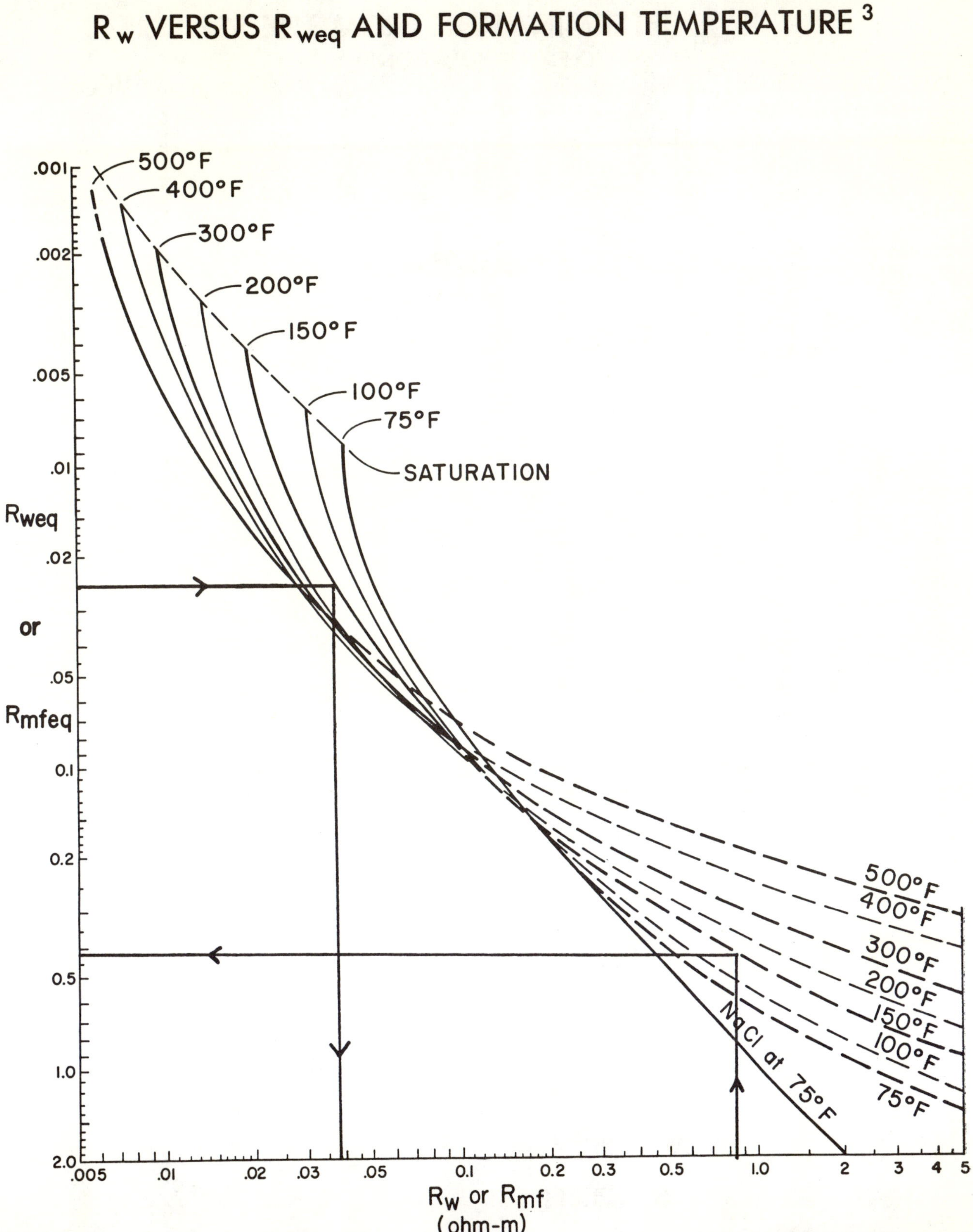

Courtesy of Schlumberger.

Figure P

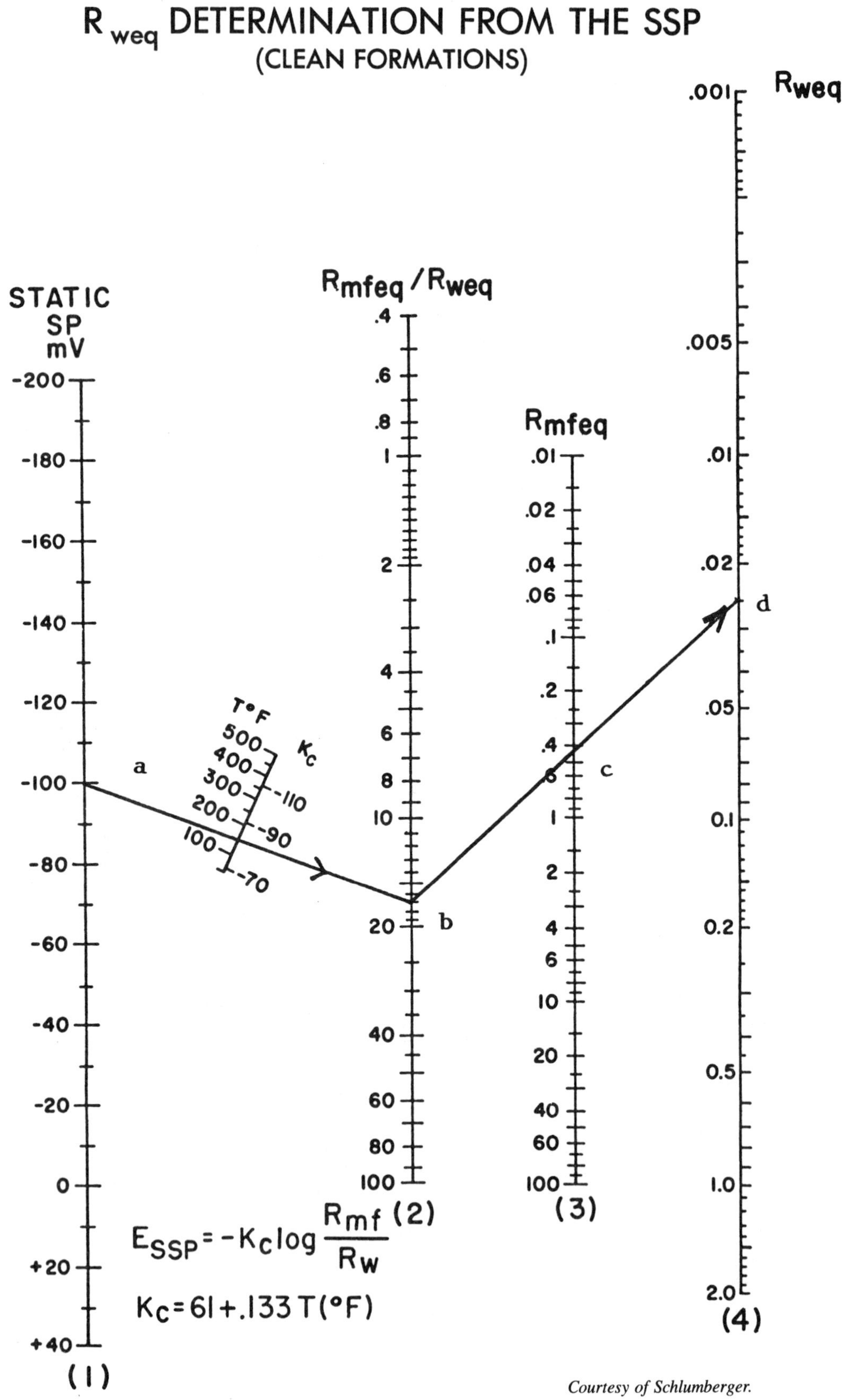

Courtesy of Schlumberger.

Figure Q

Chapter 6

1. Point of inflection.
2. a. Local knowledge.
   b. SP.
   c. Produced samples.
3. a. $S_w = \sqrt{\frac{R_o}{R_t}}$.

   b. $S_w = \sqrt{\frac{R_w}{R_{wa}}}$.
4. R_{deep}.
5. R_{deep} is corrected for borehole effects, bed thickness, and invasion, yielding R_t.
6. $F = \frac{a}{\Phi^m}$.
7. $R_o = F\ R_w$.
8. $S_w = \sqrt{\frac{R_o}{R_t}} = \sqrt{\frac{FR_w}{R_t}}$
9. Dual induction.
10. Tornado chart.
11. a. Formation temperature = 164

    b. R_{mf} @ $T_f = 0.096$

    $$R_2 = 0.2\left(\frac{75 + 6.77}{164 + 6.77}\right) = 0.096$$

    c. SP = −43

    d. $R_{mfe} = 0.8$ @ 164° or 0.17 @ 75°
    $R_{mfe} = 0.2 \times .85 = .17$ @ 75°

    e. $R_{we} = 0.03$ @ 164°, or .05 @ 75°

    $$R_{we} = \frac{R_{mfe}}{10^{\left(\frac{-SP}{K}\right)}}$$

    where $K = 60 + .133T$
    $K = 60 + .133 \times 164 = 81.812$

    $$R_{we} = \frac{\frac{0.17}{47}}{10^{81.812}} = 0.05$$

    f. $R_w = 0.034$ @ T_f

    $$R_w = \frac{(77 \times .05) + 5}{146 - (377 \times .05)} = 0.07 \text{ @ } 75°$$

    $$R_2 = \left(\frac{75 + 6.77}{164 + 6.77}\right) = 0.034 \text{ @ } T_f$$

    g. $R_d = 2.4$ ohms from log

    h. $R_m = 0.139$ @ T_f

    i. $R_t = 2.4$

    Taking the ratio of $R_d/R_m = 17$, bit size is 7⁷/₈ entering the borehole; correction chart shows that R_d is R_t.

    j. $\Delta t = 96\ \mu s/ft$ from log

    k. $\Phi_a = \frac{96 - 55.6}{189 - 55.6}$ matrix is sandstone, thus $\Delta t_{ma} = 55.6$

    $\Phi_a = 30.3\%$

    l. Hydrocarbon appears to be oil. This statement is made by judging the sonic curve in the water wet portion of the reservoir. The change is small. If it were gas, the travel time would be higher. Correcting for oil:

    $\Phi_e = 0.9 \times .303 = .273$ or 27.3%

    m. Using the Humble Relationship:

    $$F = \frac{0.62}{.273^{2.15}} = 10.11$$

    n. $R_o = FR_w = 10.11 \times 0.034 = 0.344$

    o. $S_w = \sqrt{\frac{R_o}{R_t}} = \sqrt{\frac{0.344}{2.4}} = 0.379$ or 37.9%
12. a. $S_{wi} = 26.4\%$

    $$S_{wi} = 3.5 \times \left(\frac{0.273^{1.26}}{0.18^{0.36}}\right) - 1 = .264 \text{ or } 26.4\%$$

    b. $K_{rw} = 0.0038$

    $$K_{rw} = \left(\frac{0.379 - 0.264}{1 - 0.264}\right)^3 = 0.0038$$

    c. $K_{ro} = 0.6711$

$$K_{ro} = \left(\frac{0.9 - 0.379}{0.9 - 0.264}\right)^2 = 0.6711$$

d. WOR = 0.283

WOR = B (K_{rw}/K_{ro})

where B = 2, light oil, 45 degrees
= 5, light oil, 35 degrees
= 10, medium oil, 27 degrees
= 20, heavy oil, 19 degrees
= 50, heavy oil, 14 degrees

therefore

B = 5

WOR = 5 (0.0038/0.6711) = 0.0283

e. WC = 2.9%

$$WC = \left(\frac{0.0283}{1 - 0.0283}\right) = 0.029$$

i.e., nearly water-free production.

Chapter 7

1. Density of the rock matrix.
2. Bulk density.
3. a. 1.0.
   b. 1.1.
4. Multiply porosity by 0.7.
5. No. FE loss is about 2%.
6. Yes. FE loss can be as high as 100%.
7. Gamma ray.
8. Crossplot neutron density for V_{sh} and effective porosity.

Chapter 8

1. *Lithology and* R_w remain constant, *and there is a* 100% water saturated *zone*.
2. 50%.
3. a. The tool is not properly zeroed.
   b. The tool sensitivity is in error.
   c. A combination of these two problems.
4. See Figure R.
   a. 0.35
   b. 2.82
   c. $\Phi = \frac{2.82 - 2.7}{2.82 - 1.0} = 6.6\%$

   $$F = \frac{.62}{.066^{2.15}} = 214$$

   $$R_w = \frac{R_o}{F} = \frac{10}{214} = 0.047$$

   Discussion: The data point was chosen strictly for convenience. A bulk density of 2.7 corresponds to an R_o of 10 ohms. Any point on the R_o line could have been chosen with the same results. It is evident that a dolomite is present. It would be wise to re-plot the data on a carbonate chart such as the one in Figure 8-6.

   d. 11 points. Recall that $R_t = 4R_o = 1.4$. Draw a line from 2.82 to 1.4, which defines the 50% water saturation line.
   e. Yes, there are a significant number of points below 50%.

Chapter 9

1. Either pure or a binary mixture.
2. Gas and shale.
3. Shallow.
4. Lime is 47.6 μs, and dolomite is 43.5 μs.
5. One minus the flushed zone saturation, or, $1 - S_{xo}$.
6. Laminated. See illustration in Figure 9-8.
7. Density porosity = 39% and neutron porosity = 14%. For the zone of interest:
   a. 0.18 g/cc
   b. $\Phi_d = 39\%$ sandstone
   Sandstone matrix value = 2.65
   Limestone matrix value = 2.71
   Fluid density = 1.0

   $$\rho_b = \Phi \times \rho_f + (1 - \Phi)\ \rho_{ma}$$
   $$\rho_b = .39 + (1 - .39) \times 2.65 = 2.0$$

   $$\Phi = \frac{\rho_{ma} - \rho_b}{\rho_{ma} - 1}$$

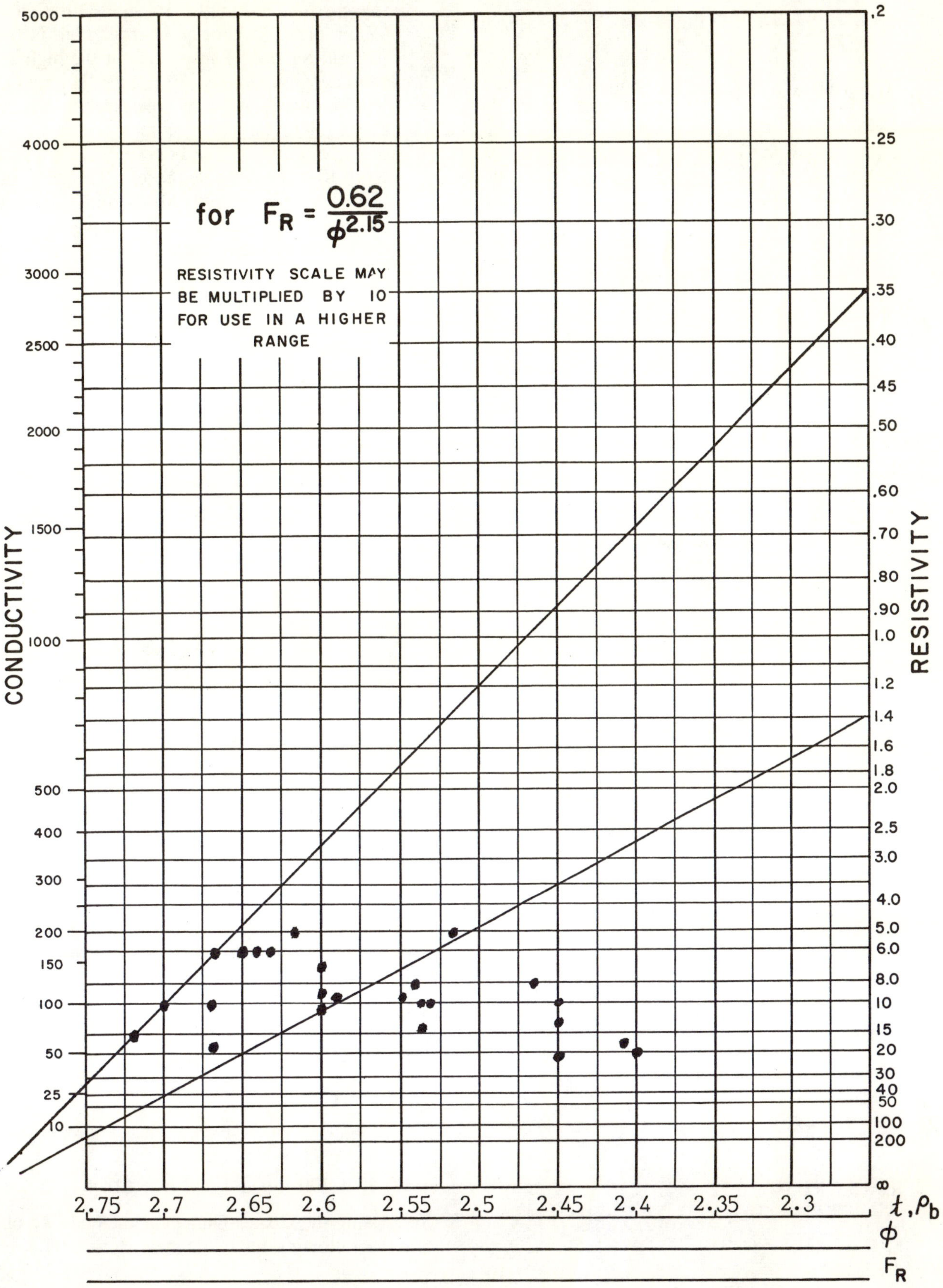

Figure R *Courtesy of Schlumberger.*

$$\Phi = \frac{2.71 - 2}{1.71} = 41.5\% \text{ limestone}$$

c. 10% neutron porosity in limestone.

d. $\Phi_{dcorr} = \Phi_d - \Delta\Phi_d$

$$\Delta\Phi_d = \frac{1.07 \times \Phi \times S_{hr}\ (1.11\ (1 - \rho_g) - 0.03)}{\rho_{ma} - 1}$$

$$\Delta\Phi_d = 0.219$$

$$\Phi_{dcorr} = 41.5\% - 21.9\% = 19.6\%$$

e. $\Phi_{ncorr} = \Phi_n - \Delta\Phi_n$

$$\Delta\ \Phi_n = \Phi_n \times S_{hr}\ (2.2 \times \rho_g - 1)$$

$$\Delta\ \Phi_n = -5.8\%$$

$$\Phi_{ncorr} = 10\% - (-5.8\%) = 15.8\%$$

f. (1) Crossplot porosity = 18.9%.
(2) Root mean square average

$$\Phi = \sqrt{\frac{\Phi_n^2 + \Phi_d^2}{2}}$$

$$\Phi = 30\%$$

(3) You would not have an accurate porosity using the average because it does not take into consideration the matrix value. Clearly this case is a binary lime-sand mixture. Only a crossplot porosity should be used.

Chapter 10

1. In the pore spaces of limestone, and in the caprock of some salt domes.
2. Secondary porosity.
3. a. Density porosity is 36.4%

$$F = \frac{.62}{36.4^2} = 5.5$$

$$R_o = 5.5 \times 0.04 = 0.22$$

$$S_w = \sqrt{0.22/.5} = 66.3\%$$

Discussion: Normally, a water saturation of 66% produces little or no hydrocarbon. In this case, one should be suspicious of the high apparent density porosity. Since there is only one porosity curve, we may crossplot it with the gamma ray to see if there is some reason for the excessive porosity. The bottom chart found in Figure 10-1 shows halite present. This accounts for the low bulk density. Also, halite can cause low resistivity readings, thus making the zone look like it will not produce.

b. (1) Sulfur.

(2) Volume of sulfur $= \dfrac{\Phi_s - \Phi_l}{0.53}$

where $\Phi_l = \dfrac{0.53\ \Phi_d - 0.4\ \Phi_s}{0.13}$

Using a lime matrix:

$\Phi_d = 0.4$
$\Phi_s = 0.5$
$\Phi_l = 0.092$
Volume = 0.77

c. (1)

M	N
.8	.48
.88	.44
.78	.5
.79	.47

(2) Gypsum.

Chapter 11

1. *Indicate* permeability *but is a poor indicator of* porosity.
2. $S_{wirr} = S_w$.
3. 1.36 darcies.

$$K = \sqrt{\frac{79\ \Phi^3}{S_{wirr}}}$$

4. 4,561 ft–4,590 ft and 4,952 ft–4,614 ft.
5. Yes. Using the Coates-Dumanoir method, we obtain the following for two zones. $R_w = 0.05$.

Zone A			Zone B		
depth	R_t	R_w/R_t	depth	R_t	R_w/R_t
2190	22	0.0023	2216	13	0.0038
2192	22	0.0023	2218	16	0.0031
2194	18	0.0028	2220	10	0.005
2196	15	0.0033	2222	5	0.01
2198	10	0.005	2224	5	0.01
2200	8	0.0063	2226	6	0.0083
2202	8	0.0063	2228	8	0.0063
2204	10	0.005	2230	10	0.005
2206	10	0.005			
2208	8	0.0063			
2210	7	0.0071			

Zone A

$(R_w/R_t)_{min} = 0.0023$
$(R_w/R_t)_{average} = 0.0047$
From Table 11-1 you find the multiplier = 3.5.

$(R_w/R_t)_{max} = (R_w/R_t)_{min} \times \text{multiplier} = 0.0081$

Zone B

$(R_w/R_t)_{min} = 0.0031$
$(R_w/R_t)_{average} = 0.0064$
multiplier = 3.0
$(R_w/R_t)_{max} = 0.0093$

Both zones are at irreducible water, since each average is less than $(R_w/R_t)_{max}$.

Chapter 12

1. Uranium streaks.
2. Compton scattering principle *because it is not sensitive to the* specific material of the formation.
3. Using a crossplot chart such as the one found on Figure 10-11, the crossplot porosity is found to be 13%.

$$S_{xo} = \frac{.11}{.13} = .846$$

$$BVW = .846 \times .13 = .11$$

4. a. Thorium
   b. Uranium
   c. Potassium
5. a. 1.81 quartz
   b. 3.14 dolomite
   c. 5.08 calcite
   d. 5.05 anhydrite
6. a. P_e decreases as porosity increases.
   b. Limestone (see Figure 12-3).
7. a. When you have a zone containing very fresh water where the hydrocarbon/water contrast is extremely low or not apparent.
   b. When you have a zone where the oil is very dense and thus affects its producibility (relative permeability).

Chapter 13

1. a. Laminated.
   b. Dispersed.
   c. Structural.
2. Static condition.
3. Neutron and density.
4. Bound and free.
5. R_w
6. Water bound by shale.
7. No. Free water is water that is not bound by a shale.
8.

Φ_{corr}	V_{sh}
20%	23%
10%	50%
20%	50%
11.5%	37%

9. $\Phi_{corr} = 16\%$
   $V_{sh} = 50\%$

   Using the Simandoux equation:

$$S_w = \sqrt{\frac{.62 \times .05}{.0194 \times 14} + \left(\frac{.62 \times .05 \times .5}{2 \times 2 \times .0194}\right)^2} - \frac{.62 \times .05 \times .5}{2 \times 2 \times .0194}$$

$$S_w = \sqrt{\frac{.031}{.2723} + \left(\frac{.0155}{.0776}\right)^2} - \frac{.0155}{.0776}$$

$$S_w = \sqrt{.1138 + .0399} - .1997$$

$$S_w = .392 - .1997 = .1923 \text{ or } 19.2\%$$

Sample Problems

The problems presented here reinforce what has been learned in all the chapters. In order to benefit the most from these problems, a few suggestions follow:

- A good scientific calculator capable of raising a digit to any power is highly recommended.
- Keep digits to two or three decimal places rounding off where necessary.
- A Welex and a Schlumberger chartbook will greatly facilitate the problem-solving procedure. All necessary charts can be found here or in previous chapters if the chartbooks are not available.
- A good engineering ruler scaled in tenths of inches or a metric ruler is recommended.
- Linear and logarithmic chart paper is required for some of the problems.

Except where noted, all log examples are courtesy of Welex.

Problem 1

Given:

- Depth is 2,224 ft
- $R_{mf} = .78$ @ 93°F
- BHT = 93°F
- TD = 2,300 ft
- See Figure S.

Calculate:

R_w from SP.

$$SP = -(60 + 0.133T) \log \frac{R_{mfe}}{R_{we}}$$

SP =

T =

R_{mf}/R_{we} =

R_{mfe} =

f =

R_w =

Problem 2

Given:

- $\Phi_{shale} = 0.2$
- $T_o = 70°$
- $R_{mf} = 2.21$ @ 90°
- BHT = 123°
- TD = 3,300 ft
- m = 2
- a = 1
- $\gamma_{clean} = 28$
- $\gamma_{shale} = 150$
- See Figures T and U.

Find:

SP =

T_f =

R_{mfe} @ 75° =

R_{we} @ 75° =

R_w @ T_f =

V_{sh} (using the gamma ray)

Using the laminated shale model, correct the density porosity for shale.

Φ_e =

Calculate the water saturation:

F =

R_o =

R_t =

S_w =

Problem 3

Given:

- The following three curves:

  R_{sg} = resistivity shallow guard (Same as an SFL)
  R_{med} = resistivity medium induction
  R_d = resistivity deep induction

- The tornado chart (Figure V) and Figure W

(text continued on page 248)

INDUCTION ELECTRIC LOG

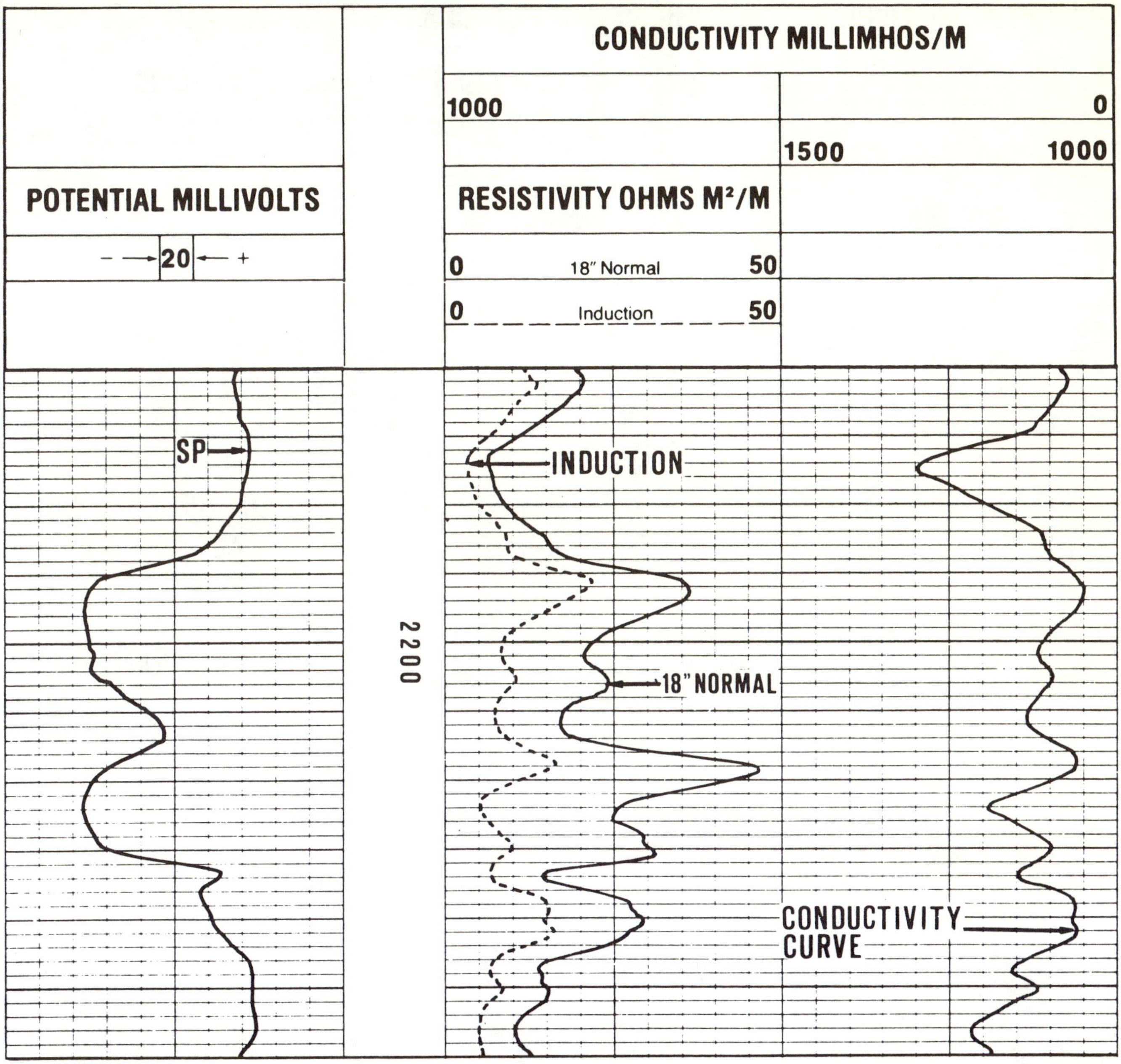

Courtesy of Welex.

Figure S

INDUCTION ELECTRIC LOG

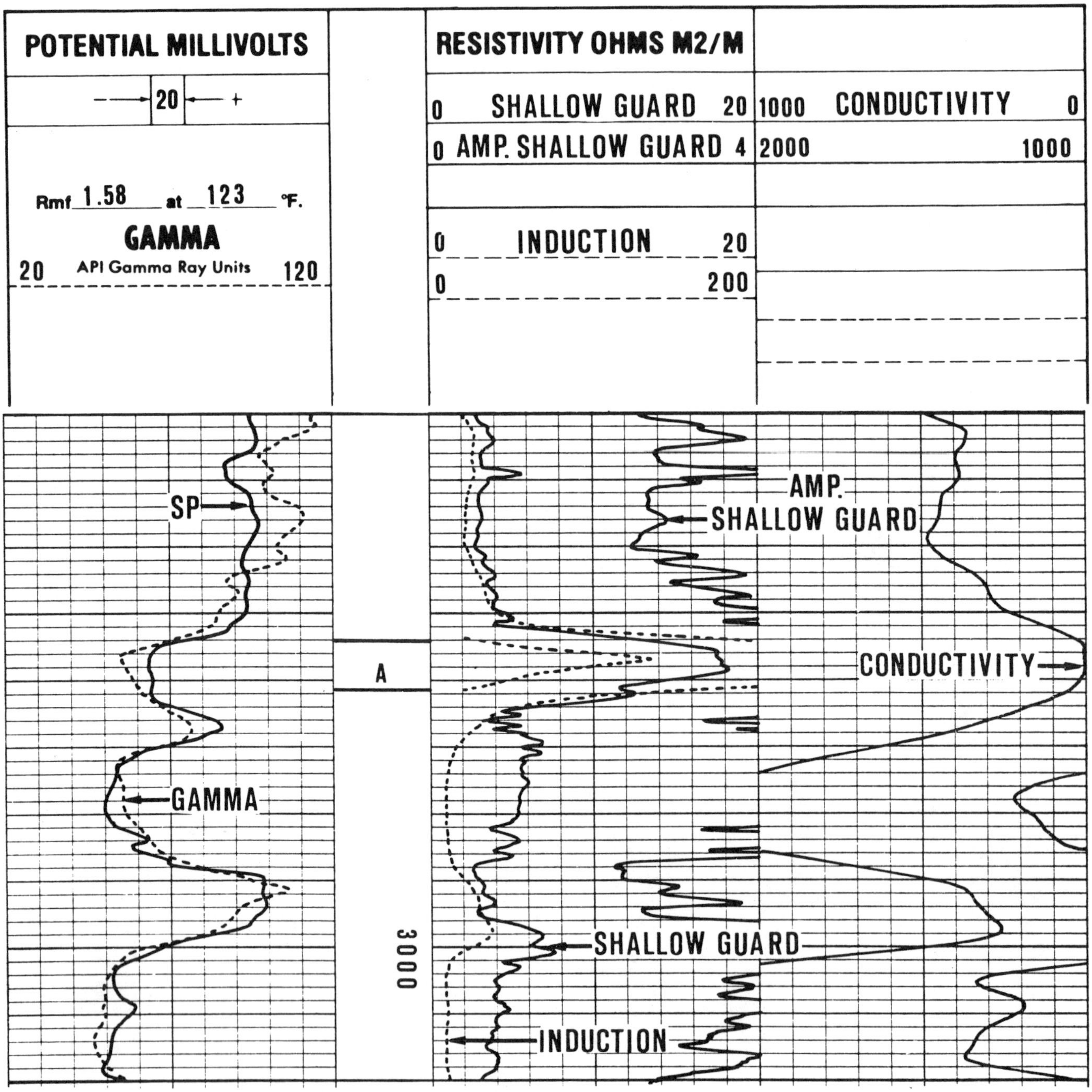

Courtesy of Welex.

Figure T

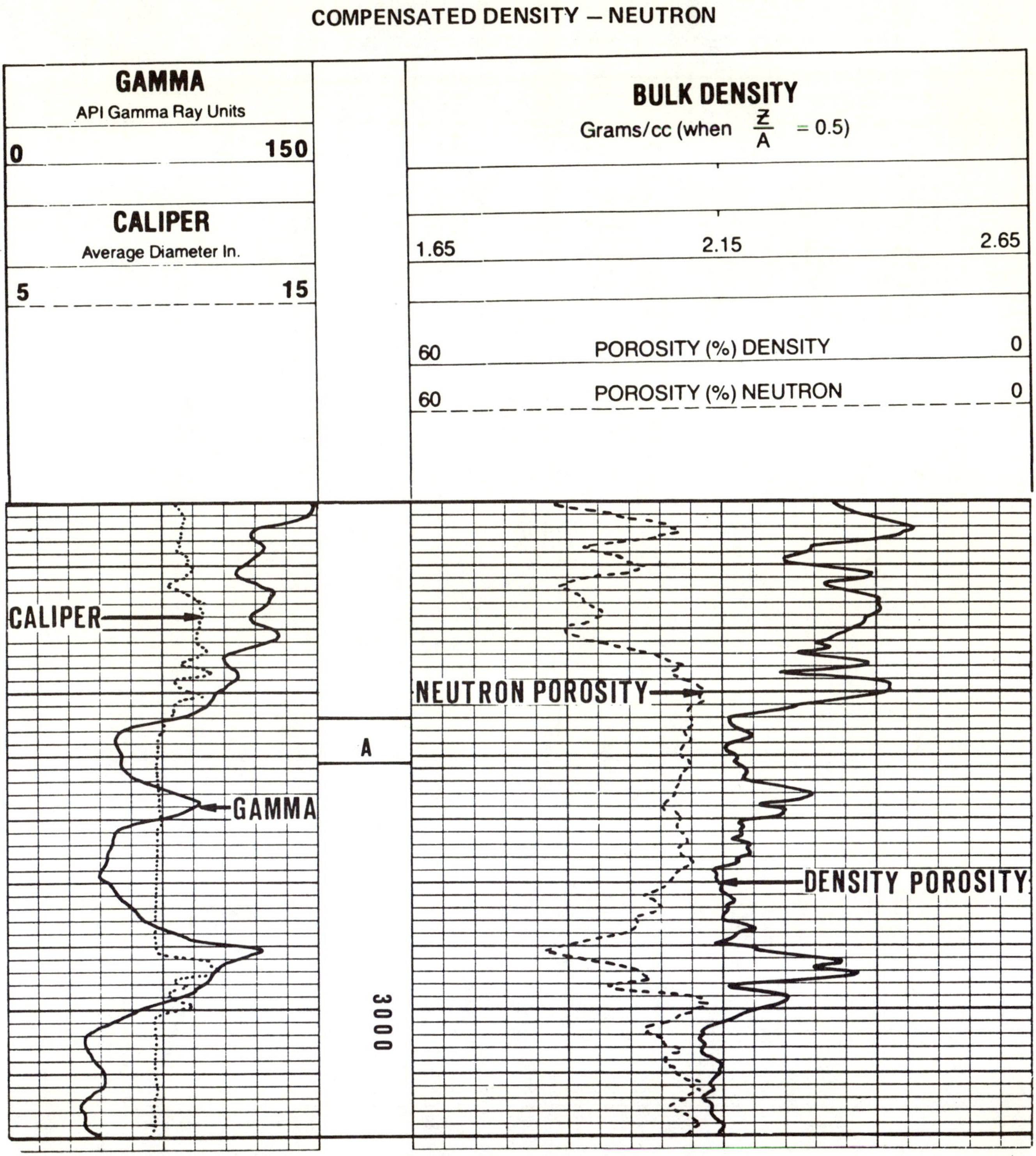
COMPENSATED DENSITY – NEUTRON
GAMMA
API Gamma Ray Units
0
150
CALIPER
Average Diameter In.
5
15
BULK DENSITY
Grams/cc (when $\frac{Z}{A}$ = 0.5)
1.65
2.15
2.65
60
POROSITY (%) DENSITY
0
60
POROSITY (%) NEUTRON
0
CALIPER
GAMMA
A
3000
NEUTRON POROSITY
DENSITY POROSITY

Figure U

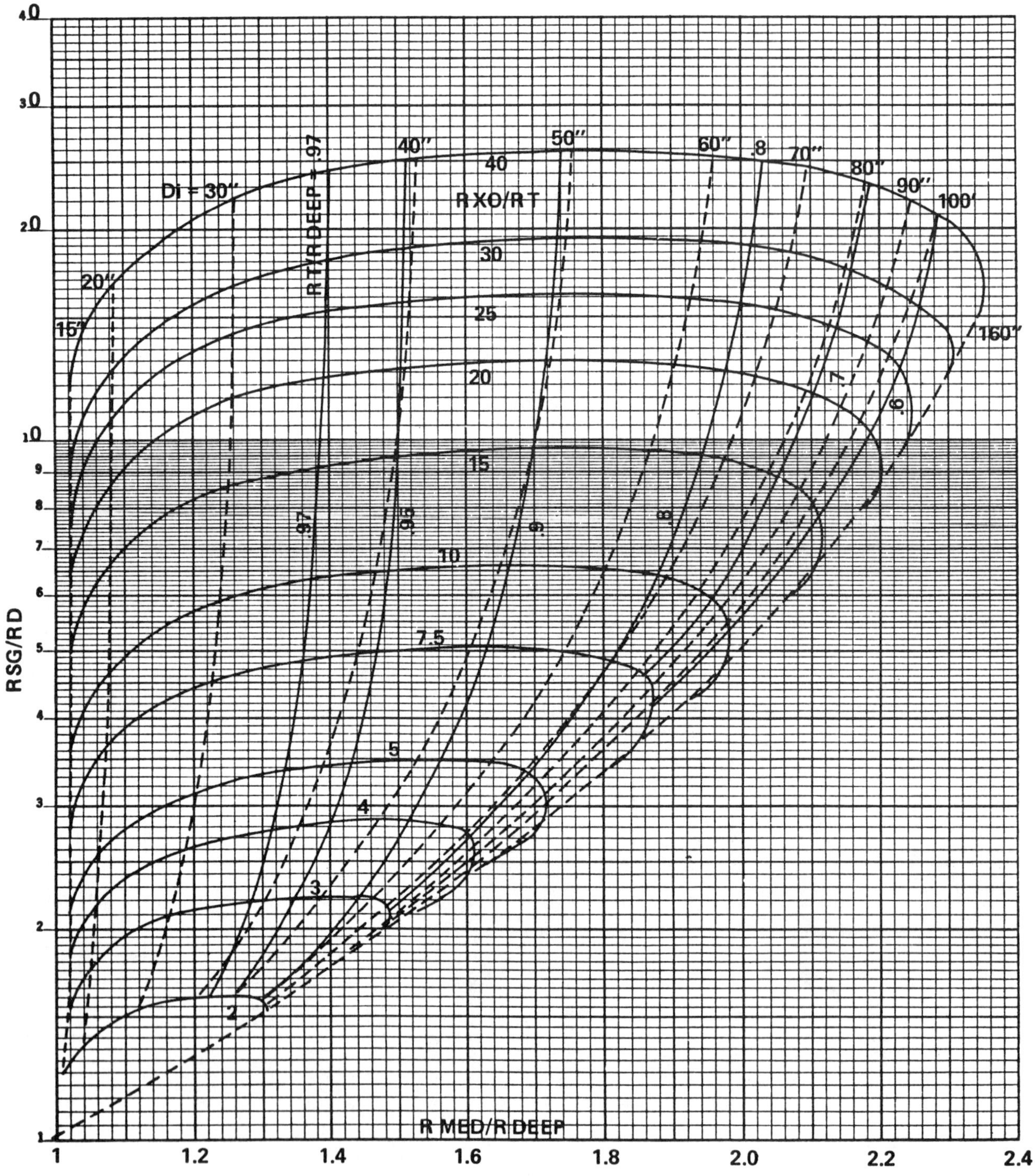
RSG/RD
R MED/R DEEP
RT/RDEEP = .97
RXO/RT
Di = 30"
15"
20"
40"
50"
60"
70"
80"
90"
100"
160"
40
30
25
20
15
10
7.5
5
4
3
2
.97
.95
.9
.8
.7
.6
1
1.2
1.4
1.6
1.8
2.0
2.2
2.4

Figure V

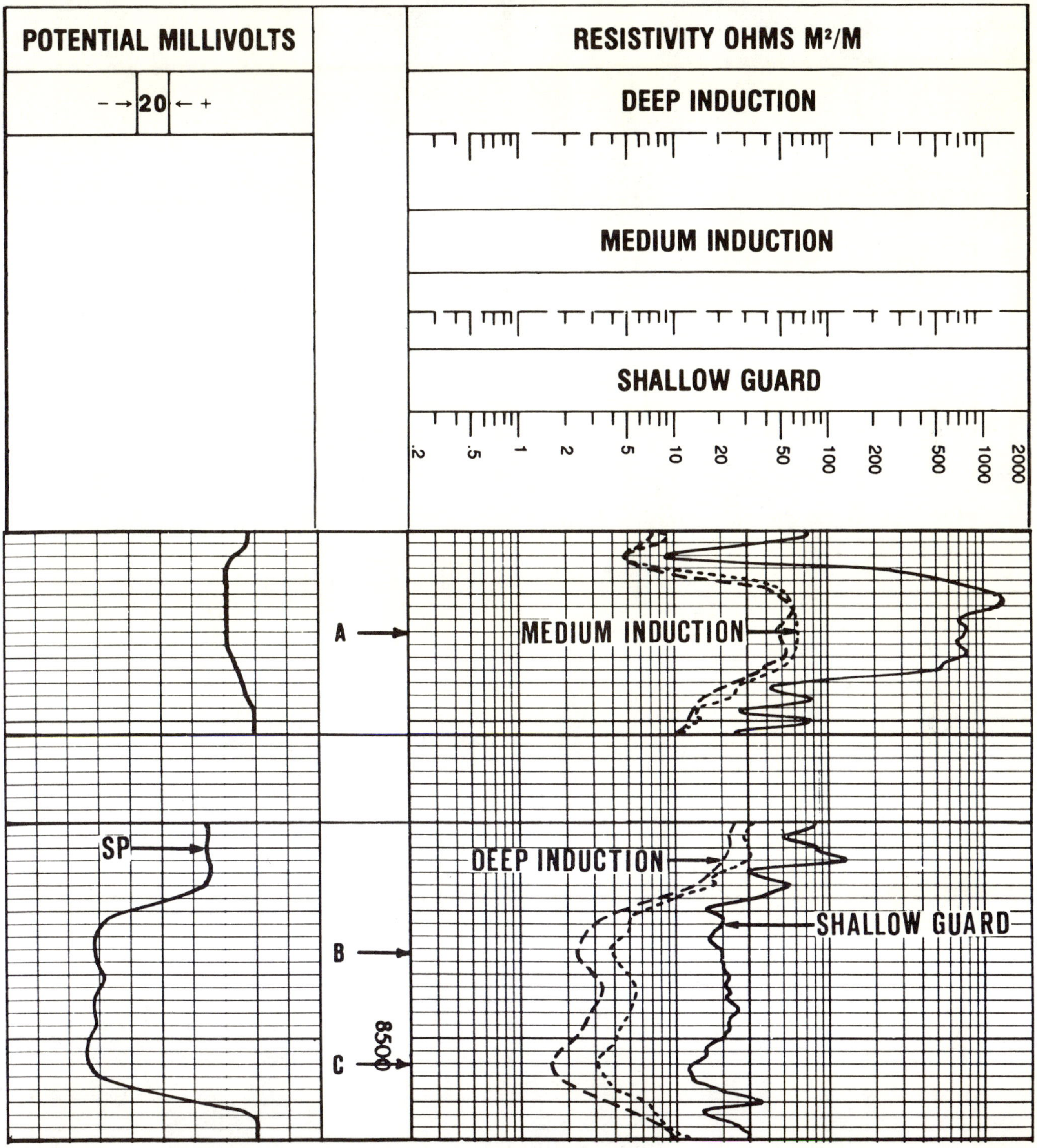

POTENTIAL MILLIVOLTS
– → 20 ← +
RESISTIVITY OHMS M²/M
DEEP INDUCTION
MEDIUM INDUCTION
SHALLOW GUARD
.2
.5
1
2
5
10
20
50
100
200
500
1000
2000
A
MEDIUM INDUCTION
SP
DEEP INDUCTION
SHALLOW GUARD
B
8500
C

Figure W

Find:

	Zone A	Zone B	Zone C
R_{sg}			
R_{med}			
R_d			
R_{sg}/R_d			
R_{med}/R_d			
Di			
R_t/R_d			
R_t			
R_{xo}/R_t			
R_{xo}			

Discuss the significance of this exercise.

Problem 4

Given:

- Dual induction acoustic velocity log (Figure X) and the tornado chart, Figure V.
- R_{mf} =
- T_d =
- Δt_{ma} = 57.14 μs
- Δt_{ma} = 100 μs
- Δt_f = 189 μs
- T_f = 171
- Humble Equation

Find:

Using SP at 7,328 ft

SP =
R_{mf} @ 75° =
R_{mfe} =
$R_{mfe}R_{we}$ =
R_{we} =
R_w @ T_f =

	Zone A	Zone B	Zone C
R_{sg}			
R_{med}			
R_d			
R_{sg}/R_d			
R_{med}/R_d			
Di			
R_t/R_d			
R_t			
R_{xo}/R_t			
R_{xo}			
Δt log			
Φ_v			
F			
R_o			
S_w%			

Problem 5

Given:

- $R_t = R_d$
- $R_w = 0.16$
- Neutron-density log (Figure Y)
- Dual induction log (Figure Z)
- Dual-spaced neutron-density crossplot chart (Figure AA)
- a = 1
- m = 2

Find:

	Zone A			
Depth	4,347	4,350	4,352	4,354
Φ_d				
Φ_n				
Φ_{xp}(crossplot)				
F				
R_o				
R_t				
S_w%				

What is the binary lithology mixture?

(text continued on page 253)

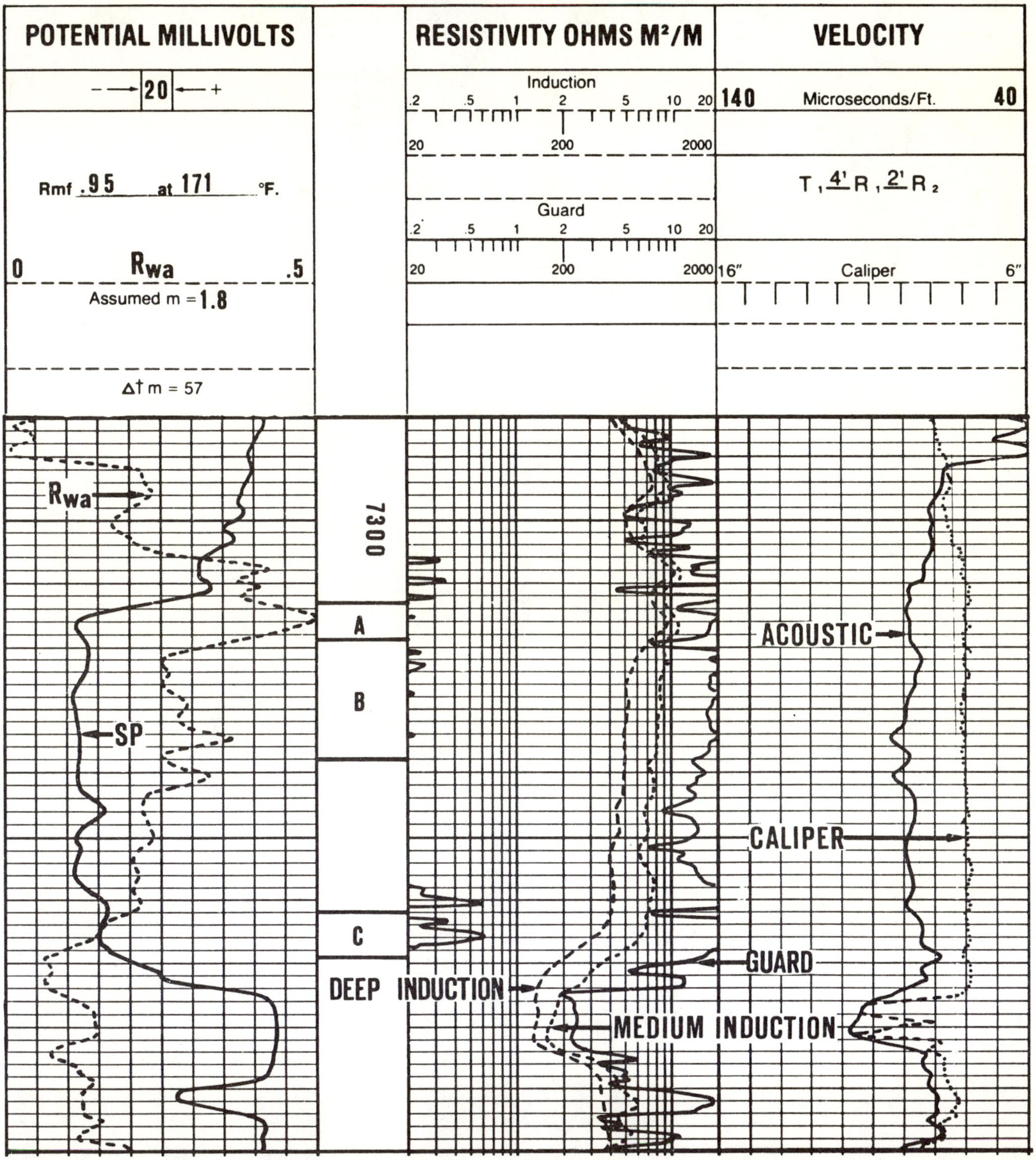

POTENTIAL MILLIVOLTS
20
Rmf .95 at 171 °F.
0 Rwa .5
Assumed m = 1.8
Δt m = 57
RESISTIVITY OHMS M²/M
Induction
.2 .5 1 2 5 10 20
20 200 2000
Guard
.2 .5 1 2 5 10 20
20 200 2000
VELOCITY
140 Microseconds/Ft. 40
T 1 4' R 1 2' R 2
16" Caliper 6"
7300
A
B
C
Rwa
SP
ACOUSTIC
CALIPER
GUARD
DEEP INDUCTION
MEDIUM INDUCTION

Figure X

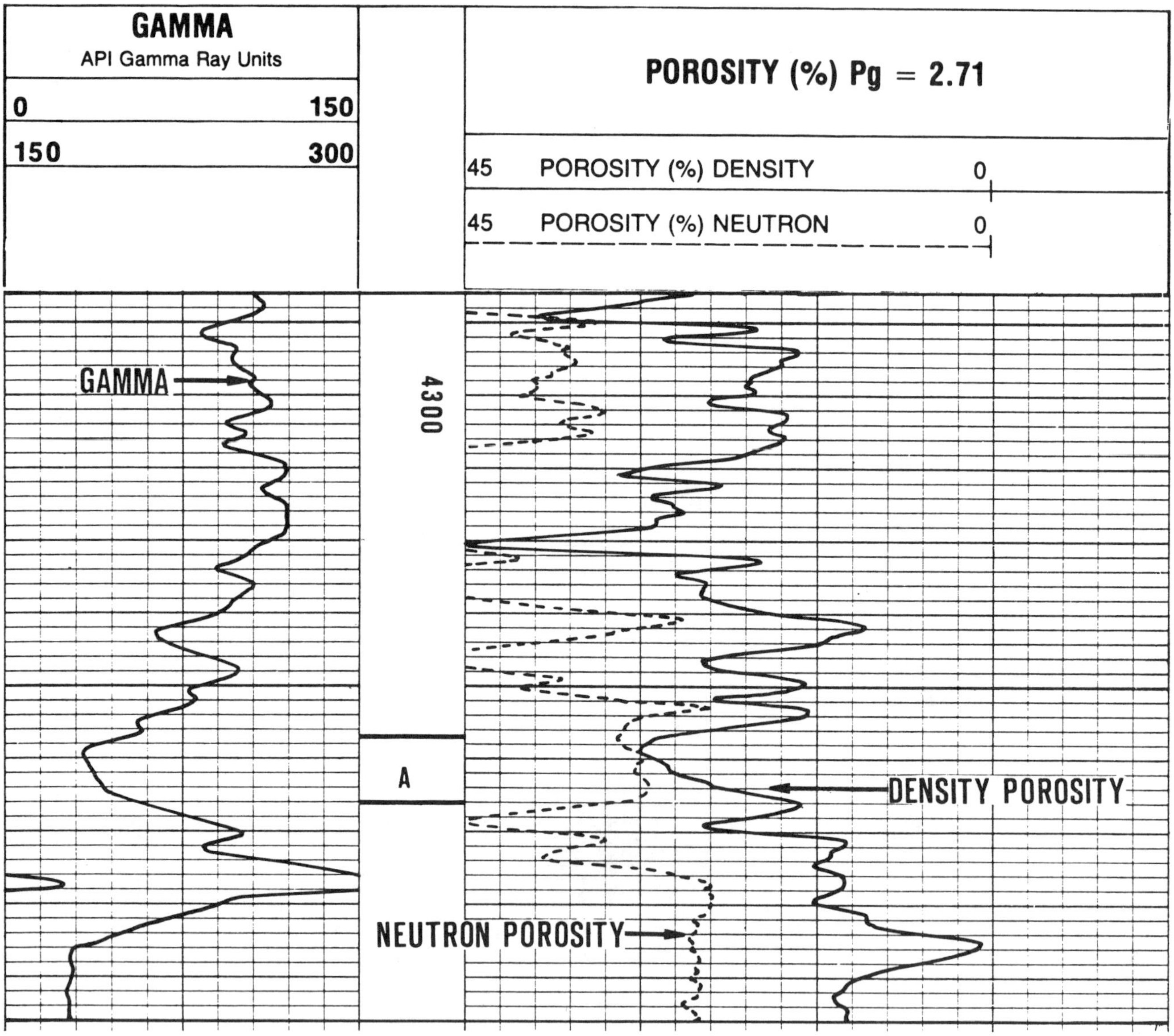
GAMMA
API Gamma Ray Units
0
150
150
300
POROSITY (%) Pg = 2.71
45 POROSITY (%) DENSITY 0
45 POROSITY (%) NEUTRON 0
GAMMA
4300
A
DENSITY POROSITY
NEUTRON POROSITY

Figure Y

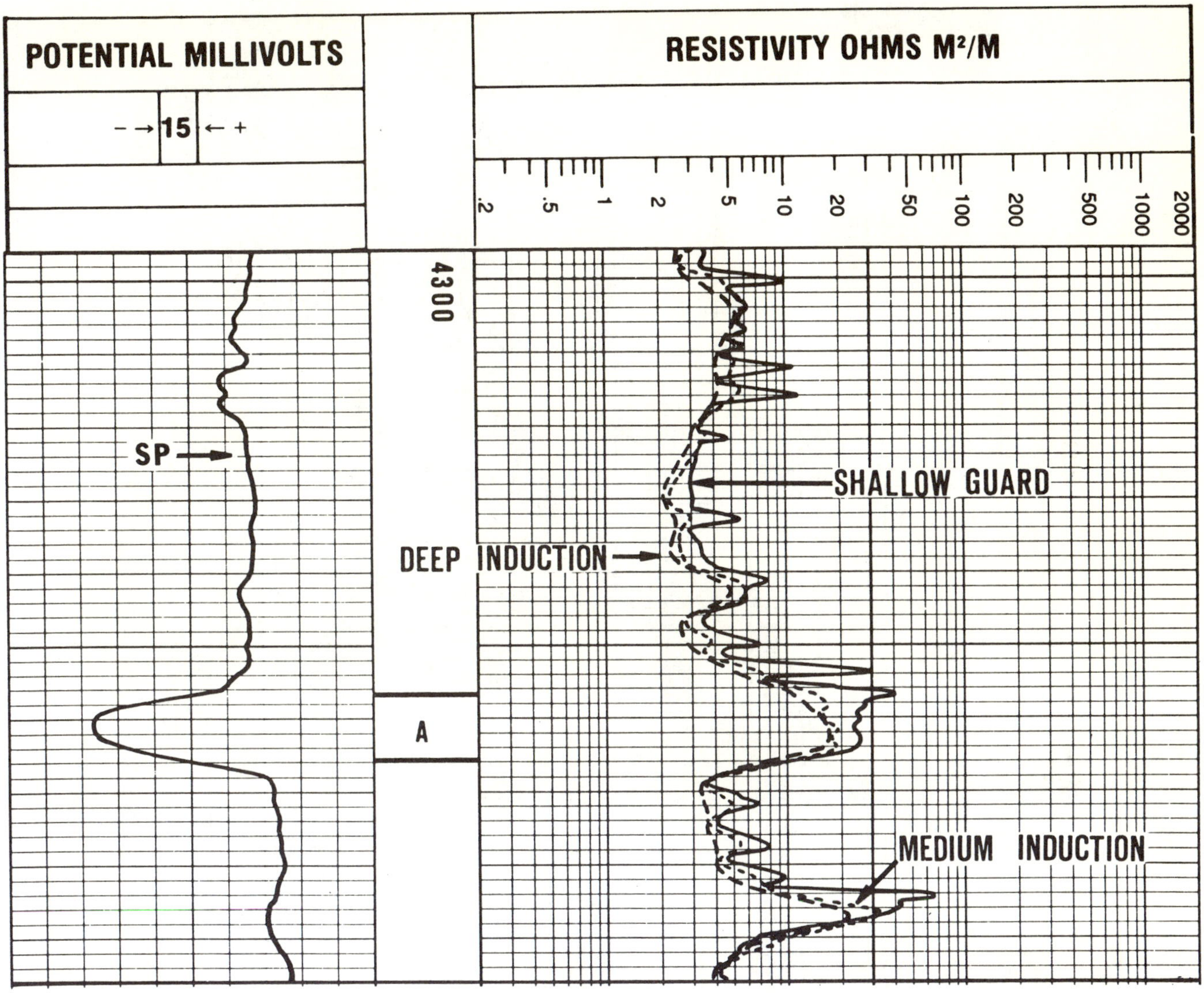

POTENTIAL MILLIVOLTS
− → 15 ← +
RESISTIVITY OHMS M²/M
.2
.5
1
2
5
10
20
50
100
200
500
1000
2000
4300
A
SP
SHALLOW GUARD
DEEP INDUCTION
MEDIUM INDUCTION

Figure Z

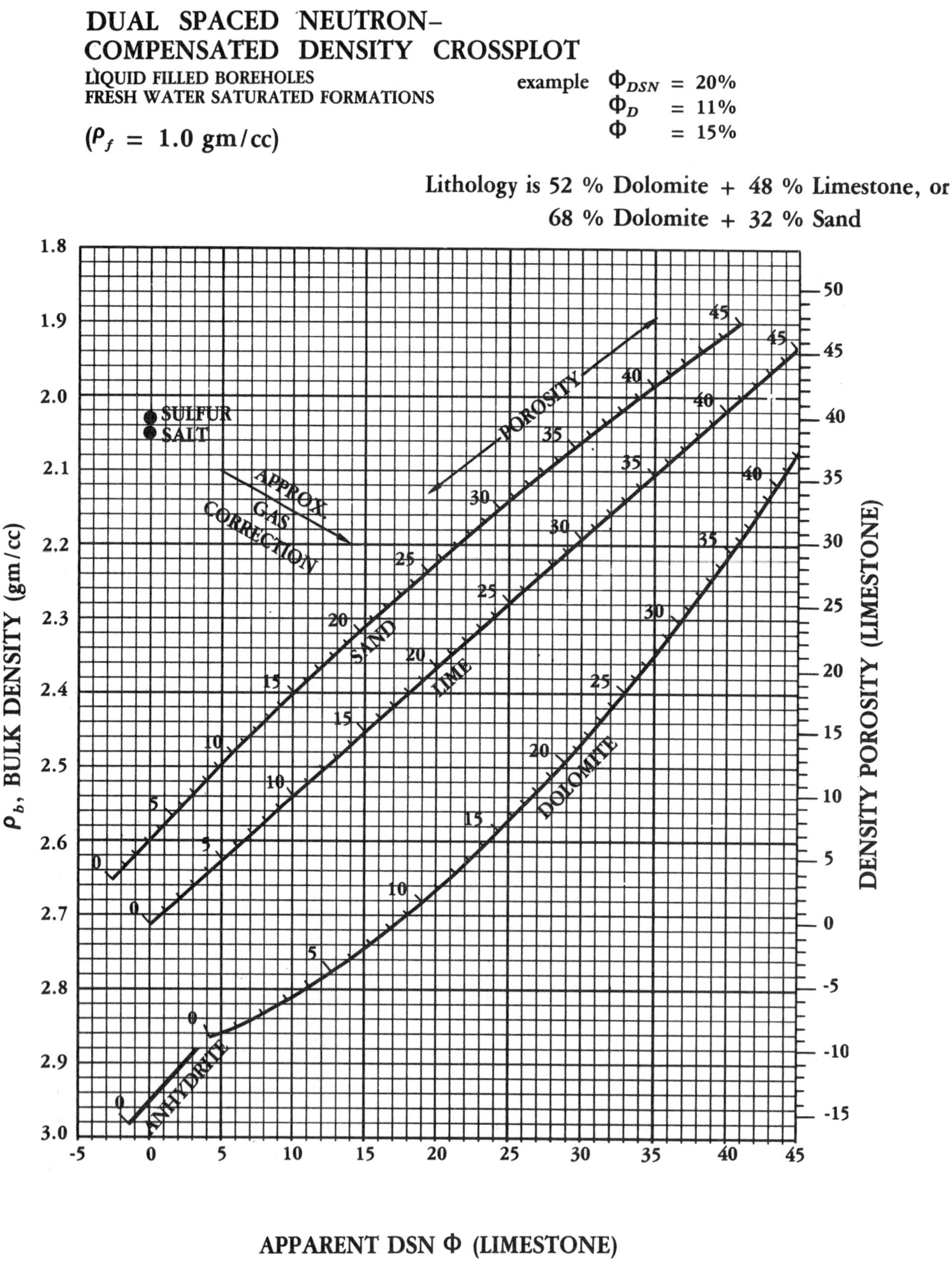
DUAL SPACED NEUTRON–
COMPENSATED DENSITY CROSSPLOT
LIQUID FILLED BOREHOLES
FRESH WATER SATURATED FORMATIONS
(ρ_f = 1.0 gm/cc)
example Φ_{DSN} = 20%
Φ_D = 11%
Φ = 15%
Lithology is 52 % Dolomite + 48 % Limestone, or
68 % Dolomite + 32 % Sand
SULFUR
SALT
APPROX. GAS CORRECTION
POROSITY
SAND
LIME
DOLOMITE
ANHYDRITE
ρ_b, BULK DENSITY (gm/cc)
DENSITY POROSITY (LIMESTONE)
APPARENT DSN Φ (LIMESTONE)

Figure AA

Problem 6

Given:

- $R_t = R_d$
- R_w @ $T_f = 0.15$
- Sandstone matrix
- $\rho_f = 1.0$
- Density log (Figure BB)
- Induction log (Figure CC)
- Hydrocarbon is GAS

Find:

Zone A

$\rho_b =$
$\Phi_d =$
$\Phi_{dcorr} =$
$F =$
$R_o =$
$R_t =$
$S_w\% =$

Problem 7

Given:

- $R_w = 0.045$ @ T_f
- Limestone matrix
- $a = 1$
- $m = 1.3$
- $R_t = R_d$
- Dual induction log (Figure DD)
- Density log (Figure EE)

Find:

Depth	Φ_d	F	R_o	R_t	S_w
5,031–36					
5,036–38					
5,038–42					
5,042–47					
5,047–55					
5,055–59					
5,059–62					
5,062–64					
5,064–70					
5,070–78					
5,078–83					
5,083–87					
5,087–90					
5,090–95					
5,095–01					

Problem 8

Given:

- Induction-electric log in Figure S and the corresponding density log in Figure FF.
- Porosity-resistivity crossplot chart for sandstones (Figure GG).
- Depth interval 2,210–2,248 ft
- $R_t = R_{deep}$

Find:

R_w using the porosity-resistivity crossplot.

Discussion:

1. Compare the value of R_w found in Problem 1 with the value obtained here.
2. How did you choose the R_o line?
3. Was this step necessary?

Problem 9

Given:

- 3-cycle log-log graph paper
- $R_w = 0.026$
- The following data:

$\Phi\%$	R_t
6.0	1.0
10.0	2.0
5.6	0.6
3.0	1.3
6.0	2.1
3.2	2.7
4.0	1.3
15.0	7.0
7.0	6.5
12.8	7.0
12.0	6.5
4.0	9.0
2.9	1.5
4.3	7.0
4.4	2.1
15.0	9.0
13.0	7.5
10.0	7.2
12.8	7.0
15.0	7.0

(text continued on page 260)

COMPENSATED DENSITY — NEUTRON LOG

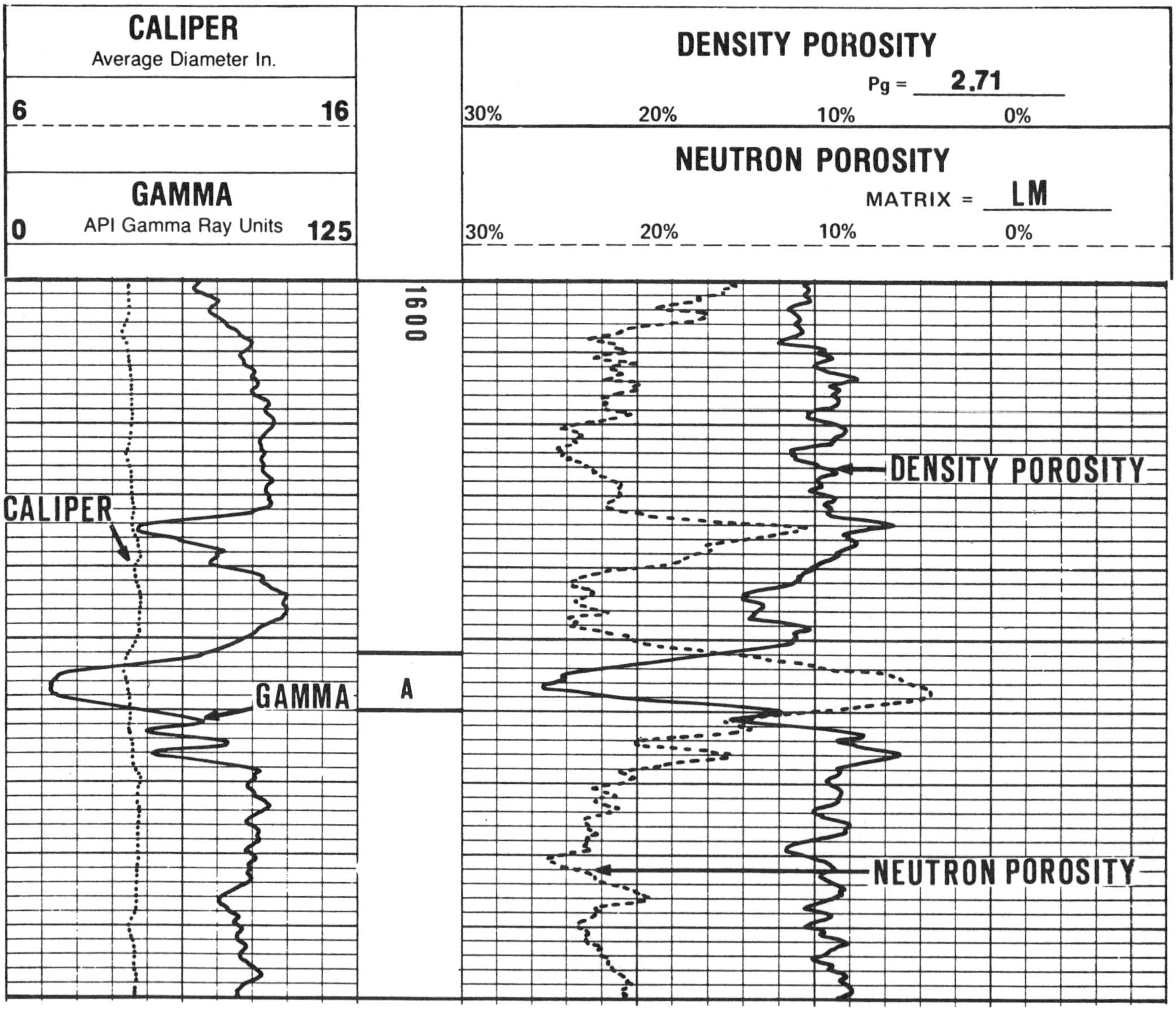

Figure BB

INDUCTION LOG

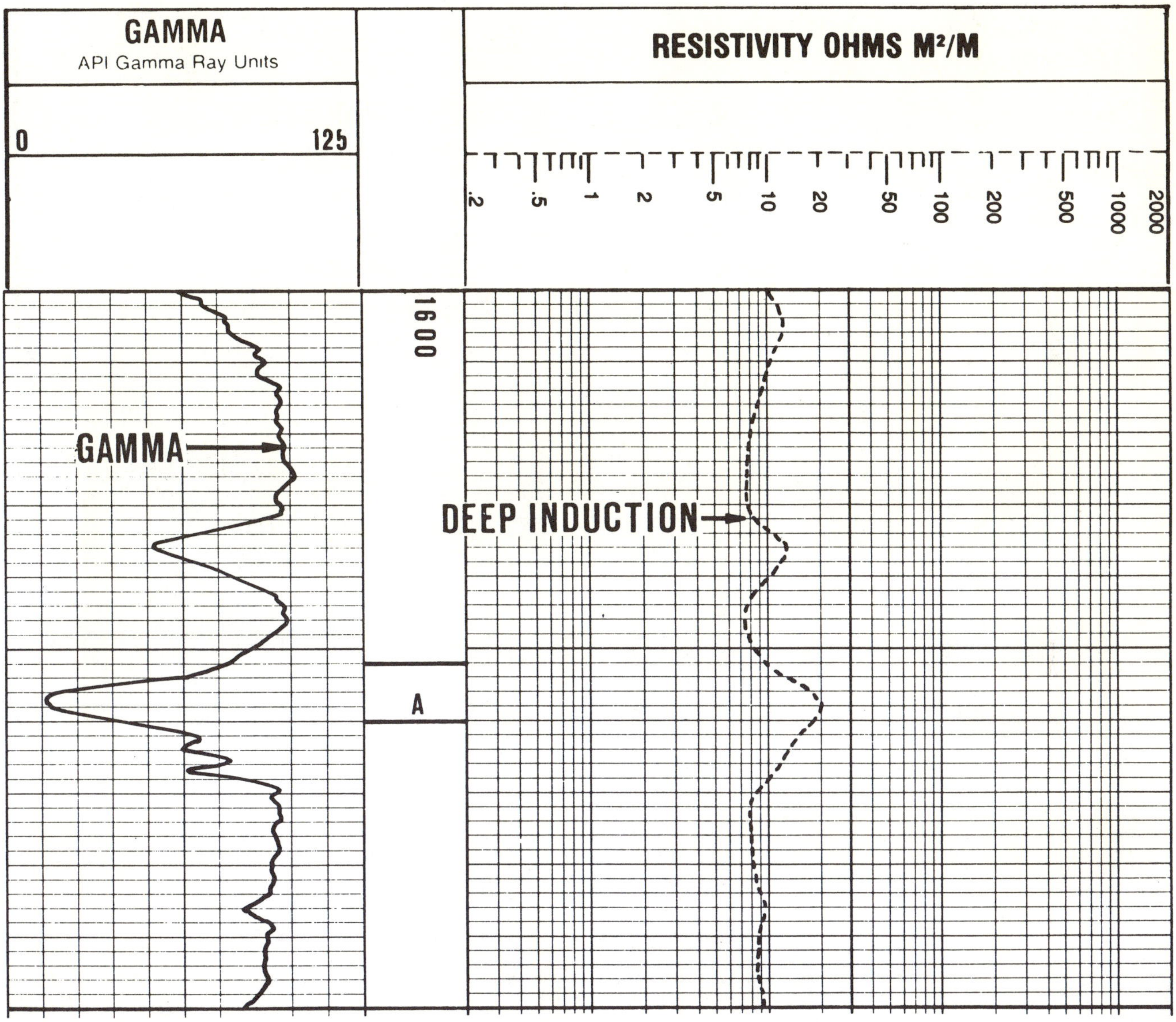

Figure CC

DUAL INDUCTION GUARD LOG

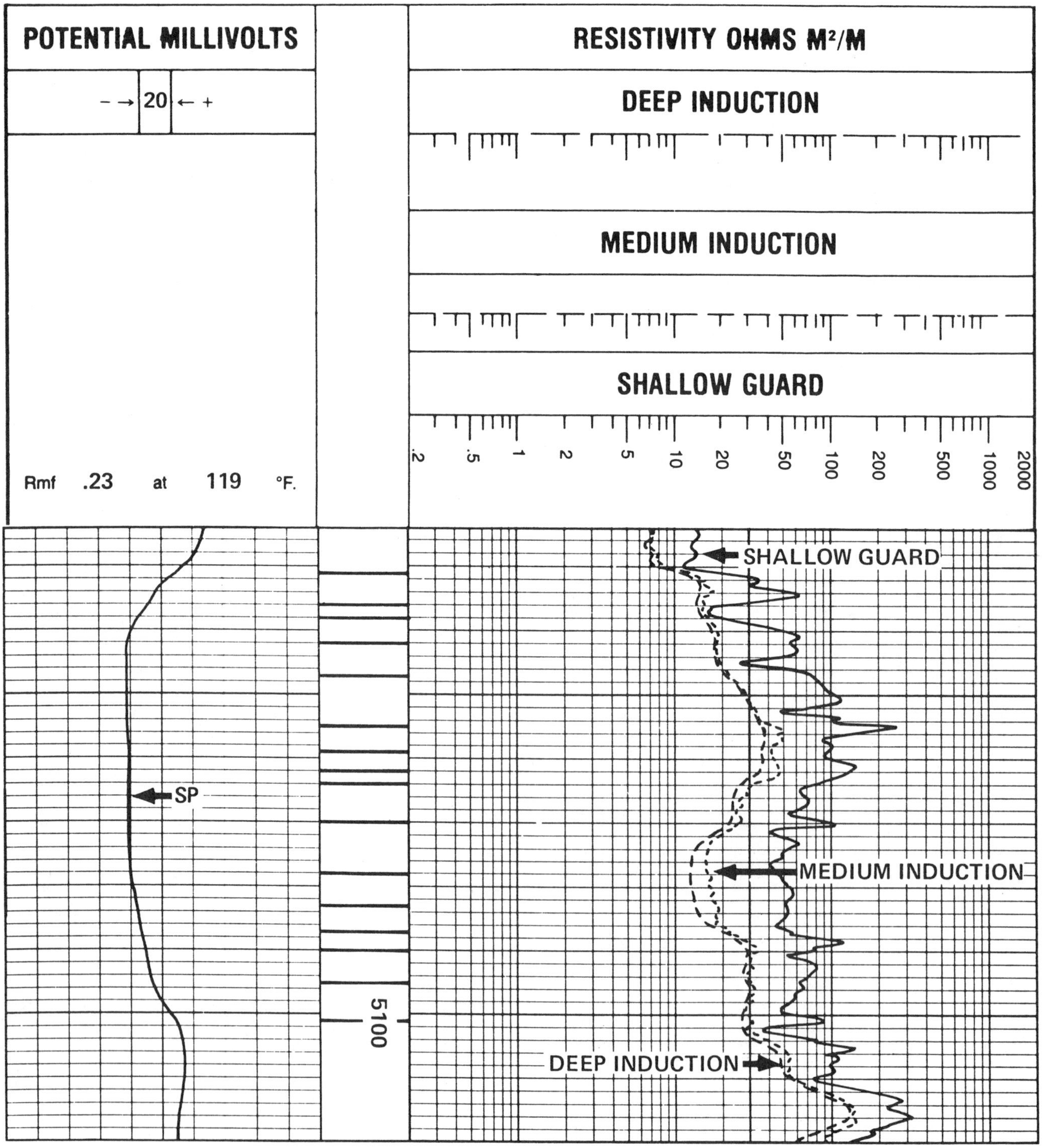

Figure DD

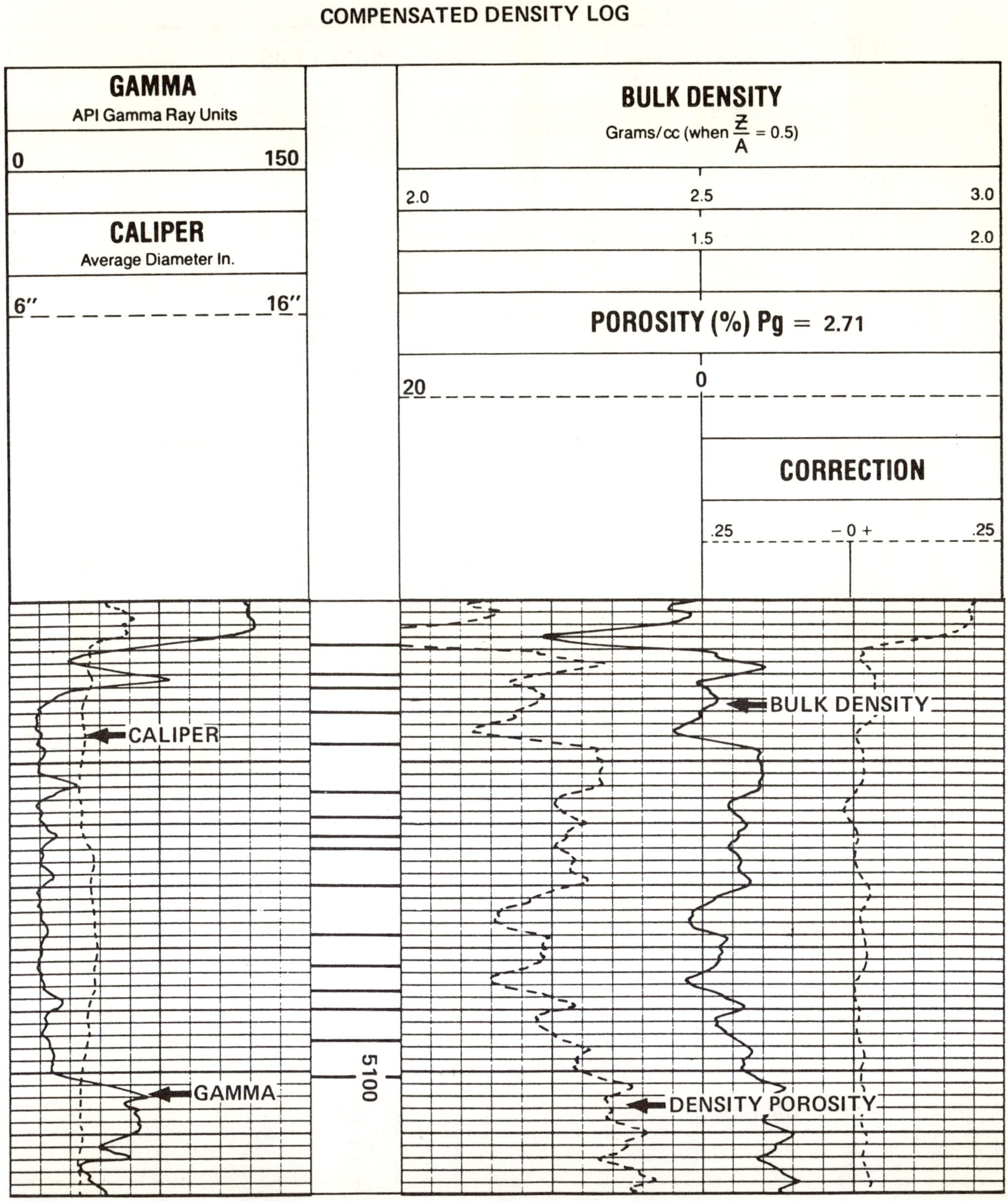
COMPENSATED DENSITY LOG
GAMMA
API Gamma Ray Units
0
150
CALIPER
Average Diameter In.
6"
16"
BULK DENSITY
Grams/cc (when Z/A = 0.5)
2.0
2.5
3.0
1.5
2.0
POROSITY (%) Pg = 2.71
20
0
CORRECTION
.25
− 0 +
.25
CALIPER
GAMMA
BULK DENSITY
DENSITY POROSITY
5100

Figure EE

COMPENSATED DENSITY LOG

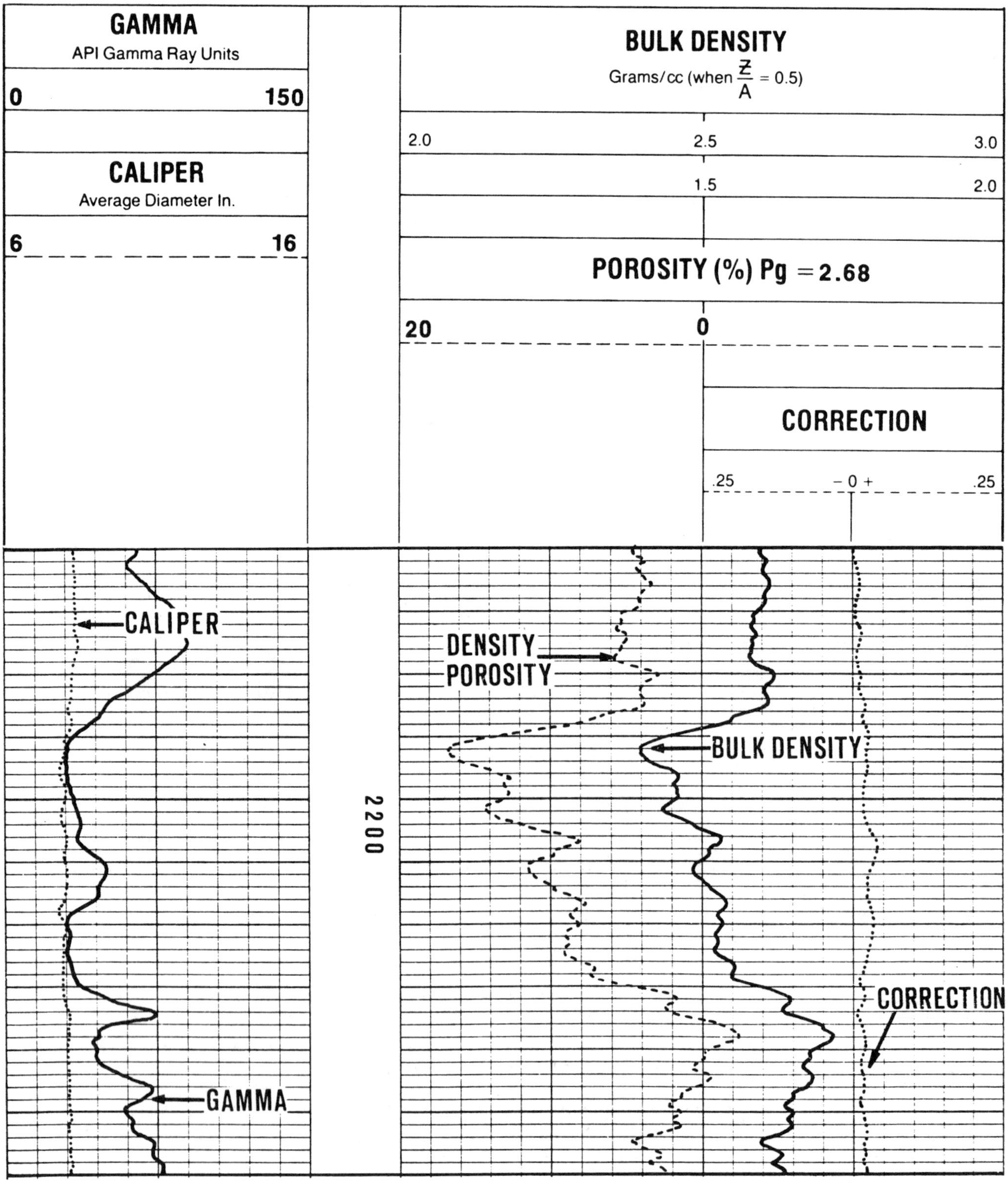

Figure FF

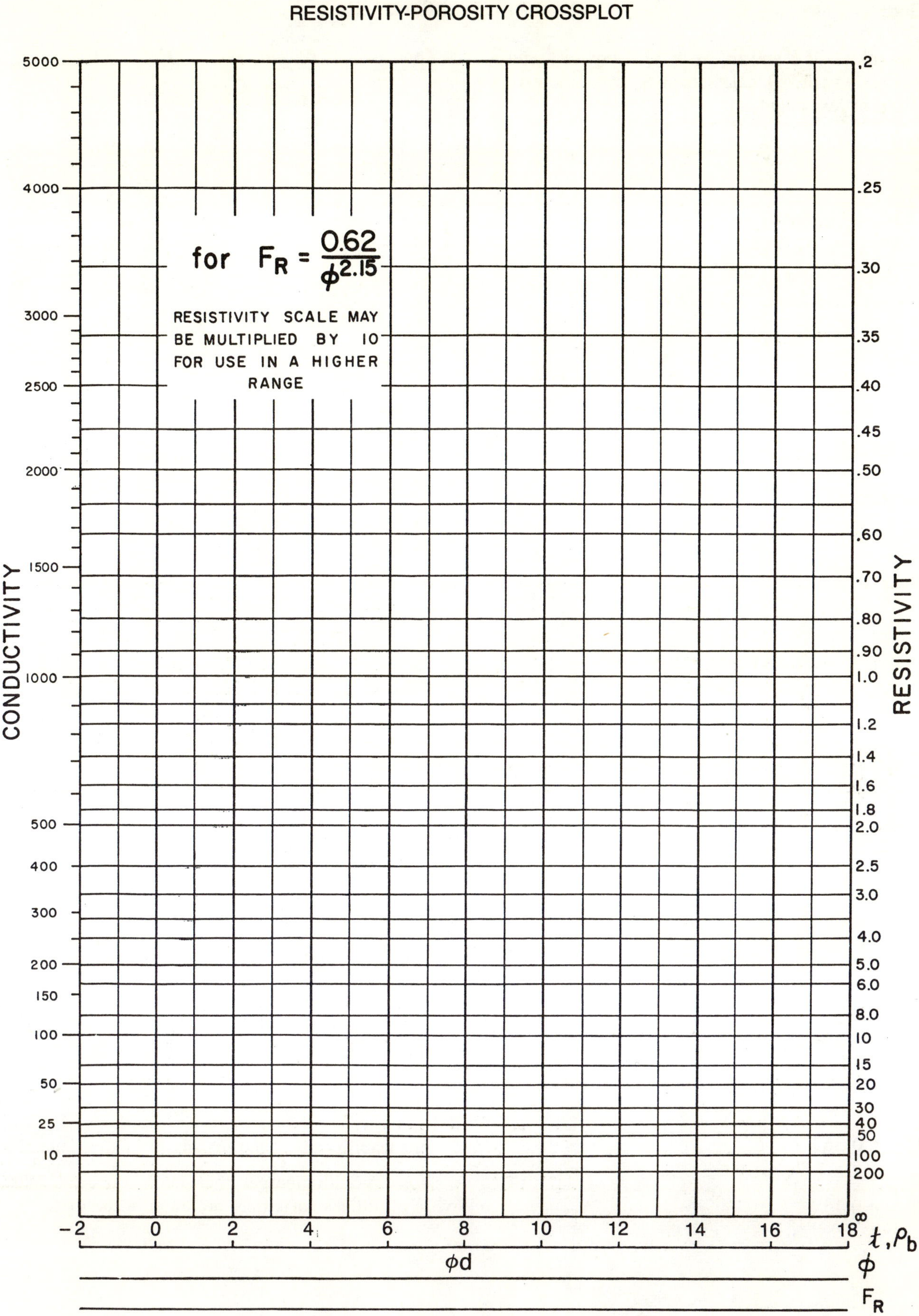
RESISTIVITY-POROSITY CROSSPLOT
for $F_R = \frac{0.62}{\phi^{2.15}}$
RESISTIVITY SCALE MAY BE MULTIPLIED BY 10 FOR USE IN A HIGHER RANGE
CONDUCTIVITY
5000
4000
3000
2500
2000
1500
1000
500
400
300
200
150
100
50
25
10
RESISTIVITY
.2
.25
.30
.35
.40
.45
.50
.60
.70
.80
.90
1.0
1.2
1.4
1.6
1.8
2.0
2.5
3.0
4.0
5.0
6.0
8.0
10
15
20
30
40
50
100
200
∞
-2
0
2
4
6
8
10
12
14
16
18
t, ρ_b
ϕd
ϕ
F_R

Figure GG

Find:

The values of a and m. With 12.8 porosity and a true resistivity of 7.0, calculate an S_w using the values of a and m just found.

a =
m =
$S_w\%$ =

Now calculate S_w using the following:

If a = 1 and m = 2, $S_w\%$ =
If a = 0.62 and m = 2.15, $S_w\%$ =

Discuss your results.

Problem 10

Given:

- $R_w = 0.061$ @ T_f
- $m = 1.9$
- $a = 1$
- $R_t = R_d$
- Gamma guard log (Figure HH)
- Density-neutron log (Figure II)

- $V_{sh} = \dfrac{I_{GR}}{3 - (2 \times I_{GR})}$

- $I_{gr} = \dfrac{GR - GR_{clean}}{GR_{shale} - GR_{clean}}$

- $GR_{clean} = 35$

- $GR_{shale} = 170$

- $\Phi_{xp} = \dfrac{\Phi_{ncorr} + \Phi_{dcorr}}{2}$

- $\Phi_{corr} = \Phi_a - (\Phi_{sh} \times V_{sh})$ laminated shale model

- $\Phi_{n\ sh} = 0.36$

- $\Phi_{d\ sh} = 0.15$

Find:

Zone	A	B	C	D	E
GR					
I_{GR}					
V_{sh}					
Φ_n					
$\Phi_{n\ corr}$					
Φ_d					
$\Phi_{d\ corr}$					
Φ_{xp}					
F					
R_o					
R_t					
$S_w\%$					

Problem 11

Given:

- Density-neutron log (Figure JJ)
- $\Phi_n = 35\%$
- $\Phi_d = 4\%$ (shale point or wet clay point)

Find:

The shale model types for the depth interval of 4,870 ft–4,904 ft, using a sheet of linear graph paper.

Problem 12

Given:

- $\Phi_n = 4.8\%$ (CNL)
- $\Phi_d = 12.0\%$
- $R_{mf} = 0.95$ @ T_f
- $R_w = 0.05$ @ T_f
- $R_d = 67.4$ @ depth 6,792
- $R_m = 89.7$
- $R_{SFL} = 156$ (shallow)
- $R_d = 47$ @ depth 6,800
- Dual-induction SFL tornado chart (Figure KK)

Find:

Permeability using the resistivity gradient method

Note: A Schlumberger chart book is helpful, otherwise all other charts may be found in the text.

(text continued on page 265)

GAMMA - GUARD LOG

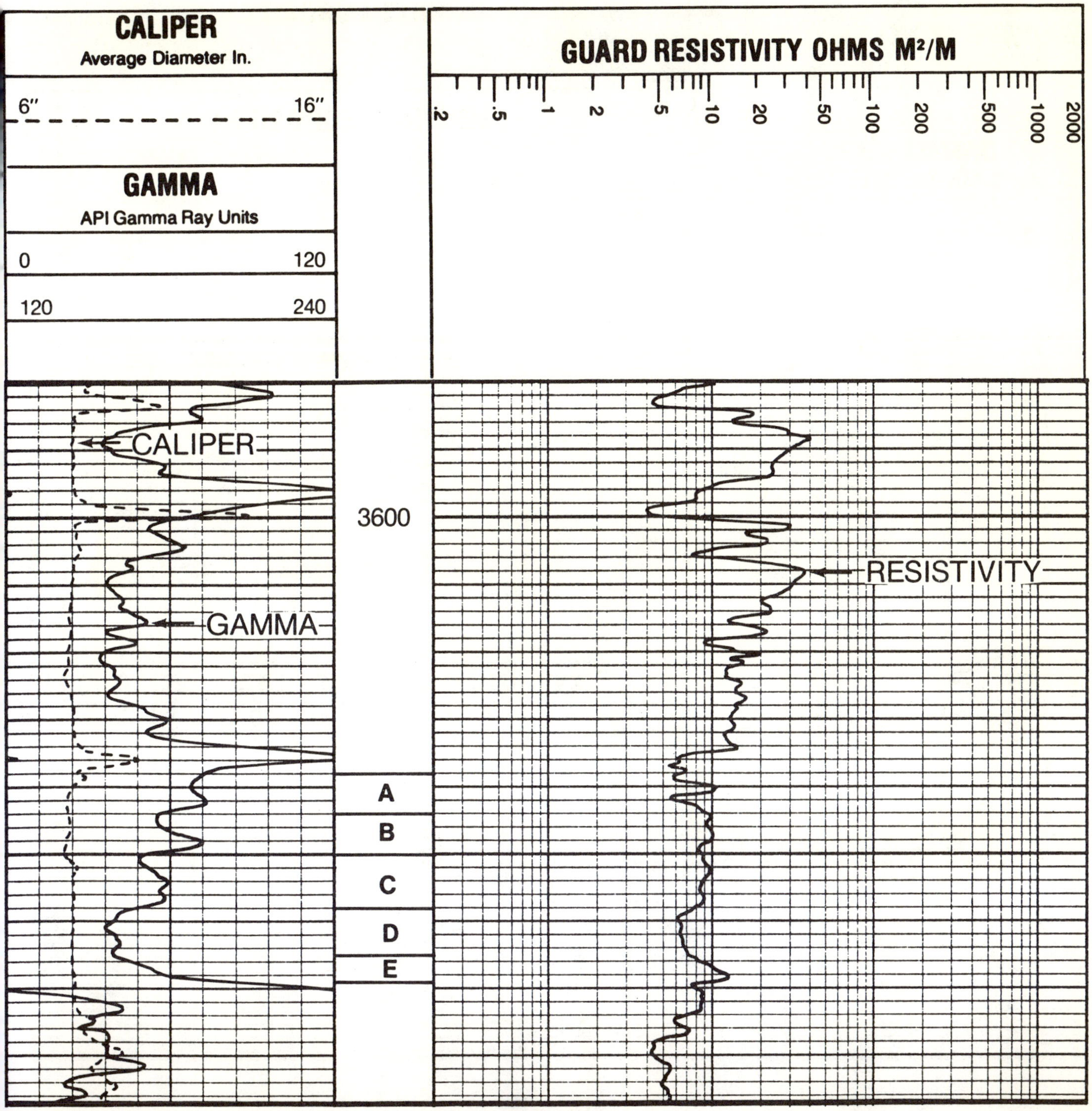

Figure HH

DENSITY - NEUTRON LOG

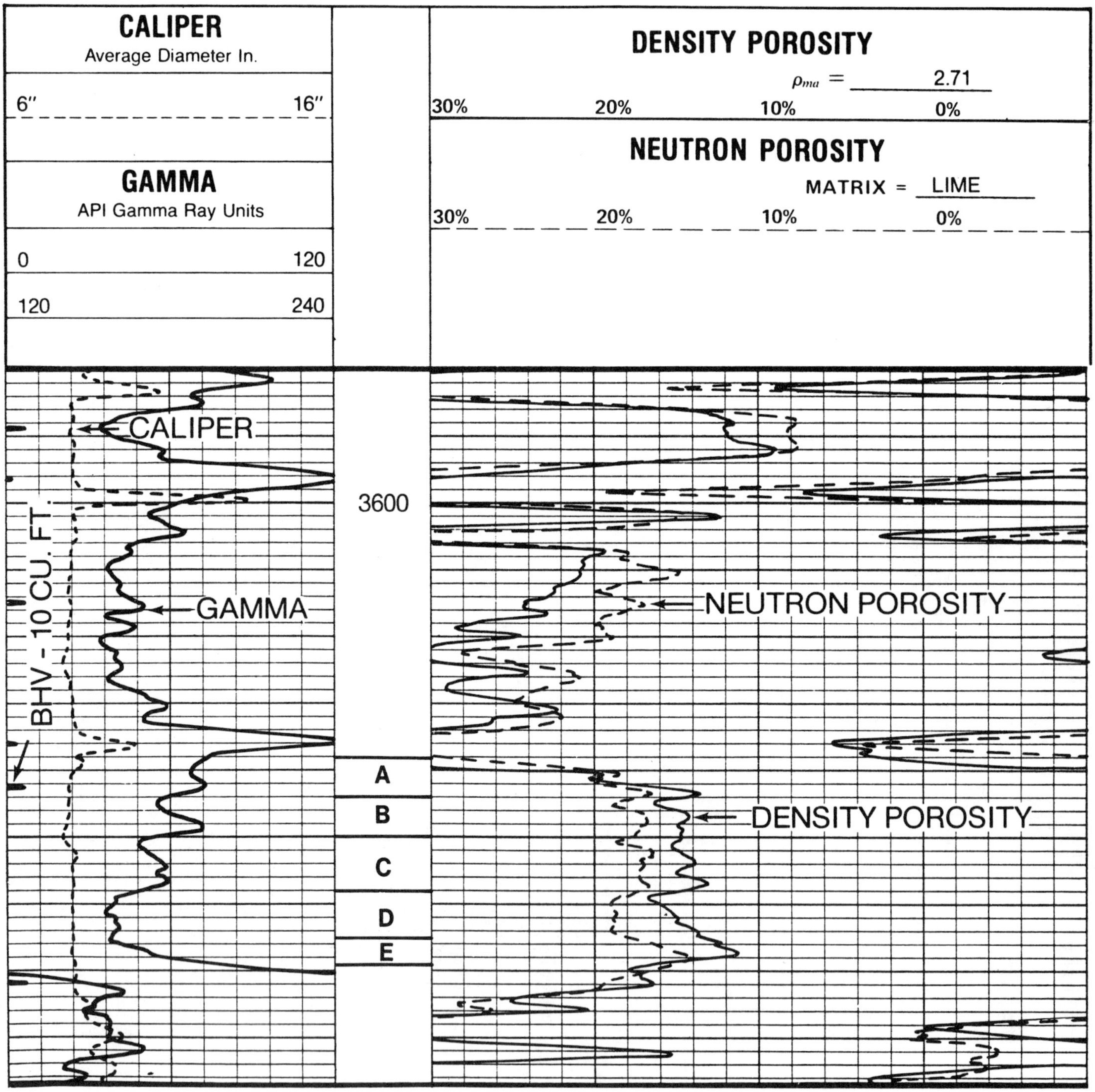

Figure II

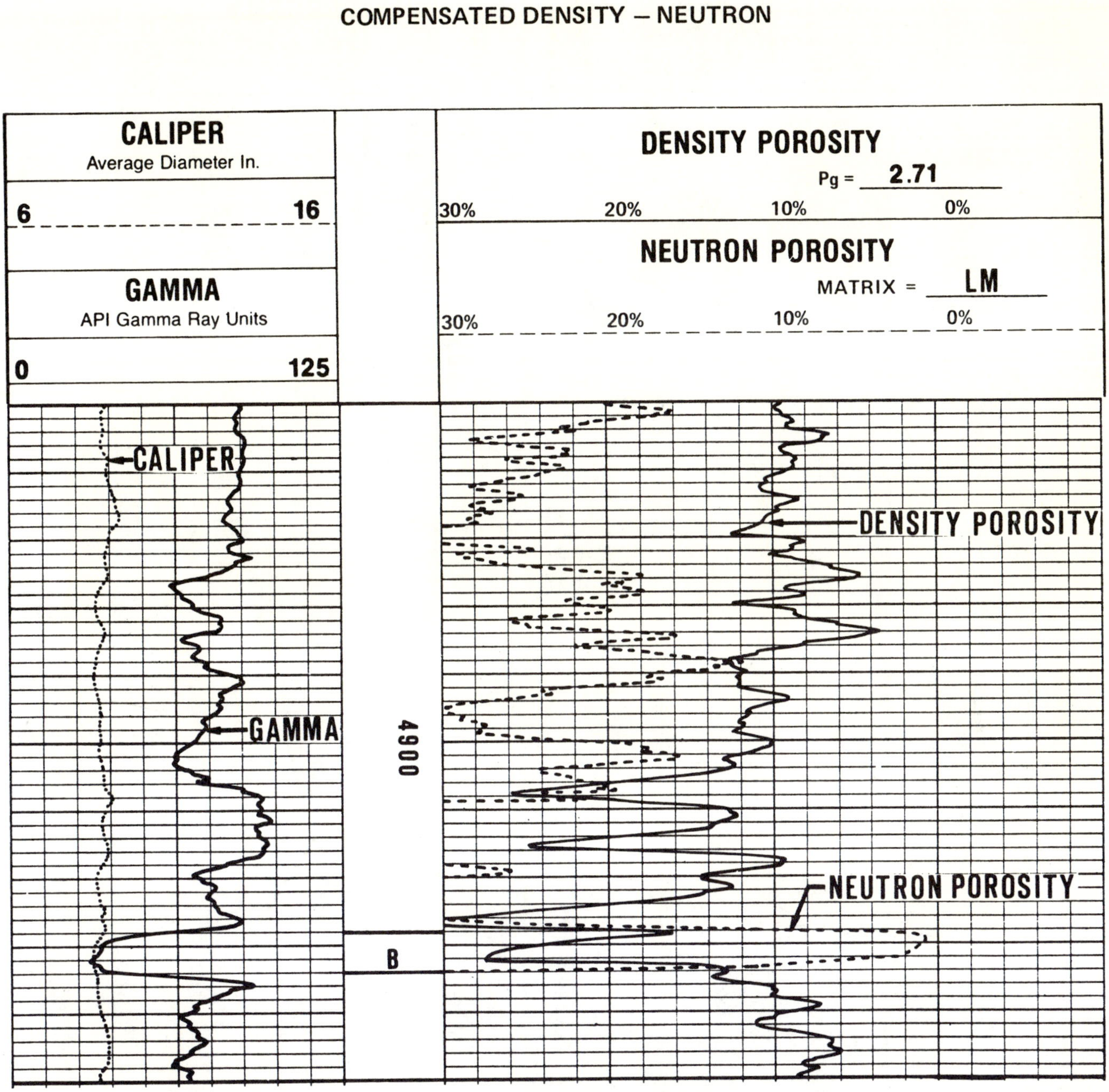
COMPENSATED DENSITY – NEUTRON
CALIPER
Average Diameter In.
6
16
GAMMA
API Gamma Ray Units
0
125
DENSITY POROSITY
Pg = 2.71
30%
20%
10%
0%
NEUTRON POROSITY
MATRIX = LM
30%
20%
10%
0%
CALIPER
GAMMA
4900
B
DENSITY POROSITY
NEUTRON POROSITY

Figure JJ

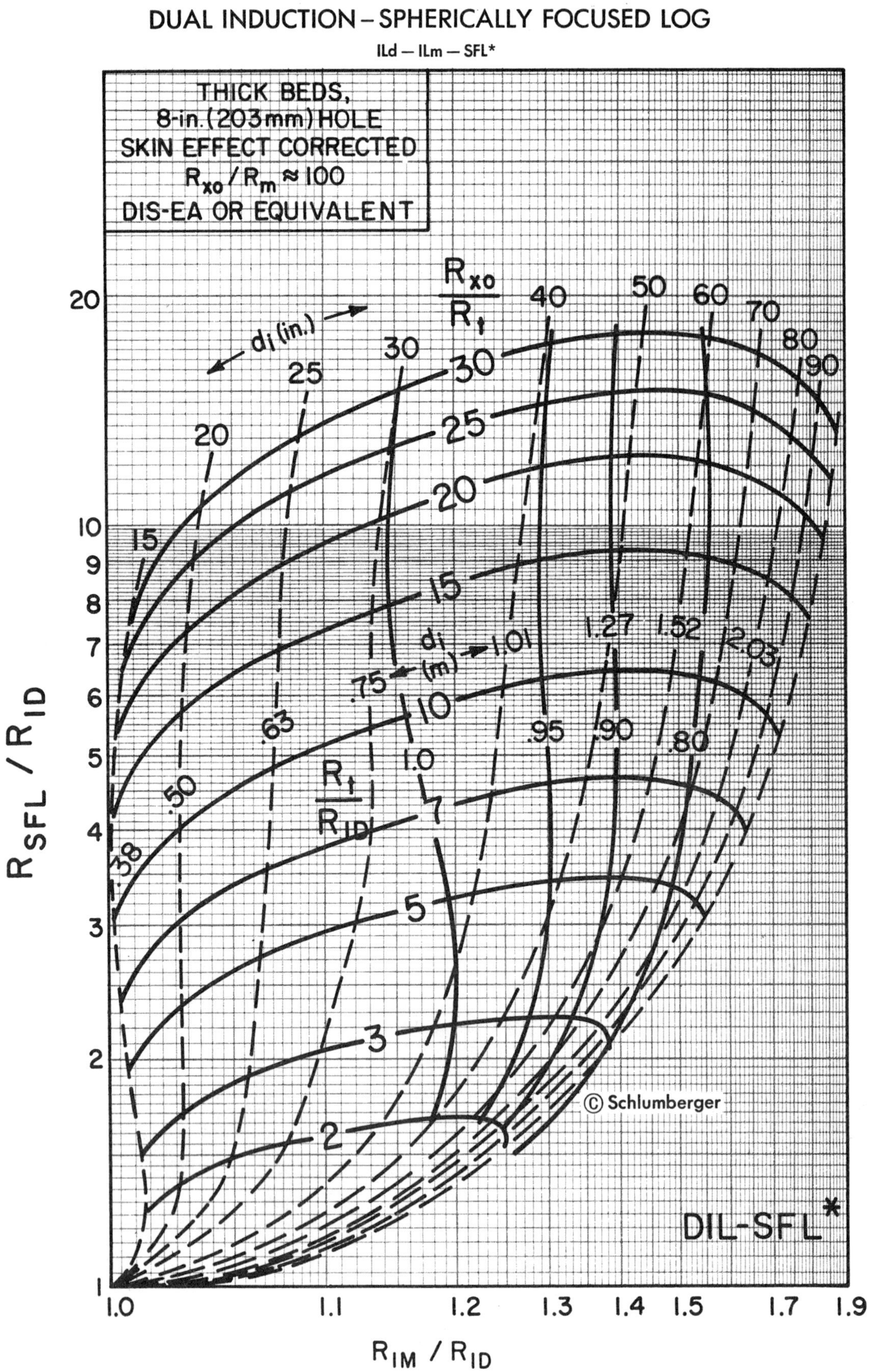
DUAL INDUCTION – SPHERICALLY FOCUSED LOG
ILd — ILm — SFL*
THICK BEDS,
8-in.(203mm) HOLE
SKIN EFFECT CORRECTED
Rxo / Rm ≈ 100
DIS-EA OR EQUIVALENT
Rxo / Rt
di (in.)
di (m)
Rt / RID
© Schlumberger
DIL-SFL*
RSFL / RID
RIM / RID

Figure KK

Solutions to Sample Problems

Problem 1

Solution:

See Figure LL.

First convert R_{mf} @ 93° to 75°.

$$R_2 = R_1 \left(\frac{93 + 6.77}{75 + 6.77}\right) = 0.95 \text{ @ } 75°$$

Since $R_{mf} > 0.1$ @ 75°

$$R_{mfe} = 0.95 \times 0.85 = 0.81$$

$$\frac{R_{we}}{R_{mfe}} = 10^{-(SP/K)}$$

$$K = 60 + 0.133 \times 93$$

$$K = 72.4$$

$$\frac{R_{we}}{R_{mfe}} = 10^{-(-100/72.4)} = 24$$

$$R_{we} = \frac{0.81}{24} = 0.034$$

Since R_{we} @ 75° < 0.12

$$R_w = \frac{(77 \times 0.034) + 5}{146 - (377 \times 0.034)} = 0.057 \text{ @ } 75°$$

Convert to a formation temperature of 93°

$$R_2 = 0.057 \times \left(\frac{75 + 6.77}{93 + 6.77}\right) = 0.047$$

R_w is therefore 0.047 ohm/meters @ 93°.

Problem 2

Solution:

$$SP = -100$$

$$T_f = (BHT - T_o)\ (D_f/TD) + T_o$$

$$T_f = (123 - 70)\ (2{,}506/3{,}300) + 70 = 110$$

$$R_{mf} \text{ @ } 75° = 2.21 \left(\frac{90 + 6.77}{75 + 6.77}\right) = 2.62$$

Since 2.62 > 0.1;

$$R_{mfe} = 2.62 \times 0.85 = 2.23$$

$$\frac{R_{mfe}}{R_{we}} = 10^{-(SP/K)}$$

$$K = 60 + 0.133 \times 110 = 74.6$$

$$\frac{R_{mfe}}{R_{we}} = 10^{-(-100/74.6)} = 21.9$$

$$R_{we} = \frac{2.23}{21.9} = 0.1$$

Since 0.1 @ 75° < 0.12

$$R_w = \frac{(77 \times .1) + 5}{146 - (377 \times .1)} = 0.12 \text{ @ } 75°$$

$$R_w \text{ @ } T_f = 0.12 \times \left(\frac{75 + 6.77}{110 + 6.77}\right) = 0.084$$

GR_{log} (zone of interest) = 53

$$V_{sh} = \frac{GR_{log} - GR_{clean}}{GR_{shale} - GR_{clean}}$$

$$V_{sh} = \frac{53 - 28}{150 - 28} = 0.2$$

Φ_d (from the log) = 0.28

Laminated shale model:

$$\Phi_e = \Phi_d - (\Phi_{sh} \times V_{sh})$$

$$\Phi_e = 0.28 - (0.2 \times .02) = 0.24$$

INDUCTION ELECTRIC LOG

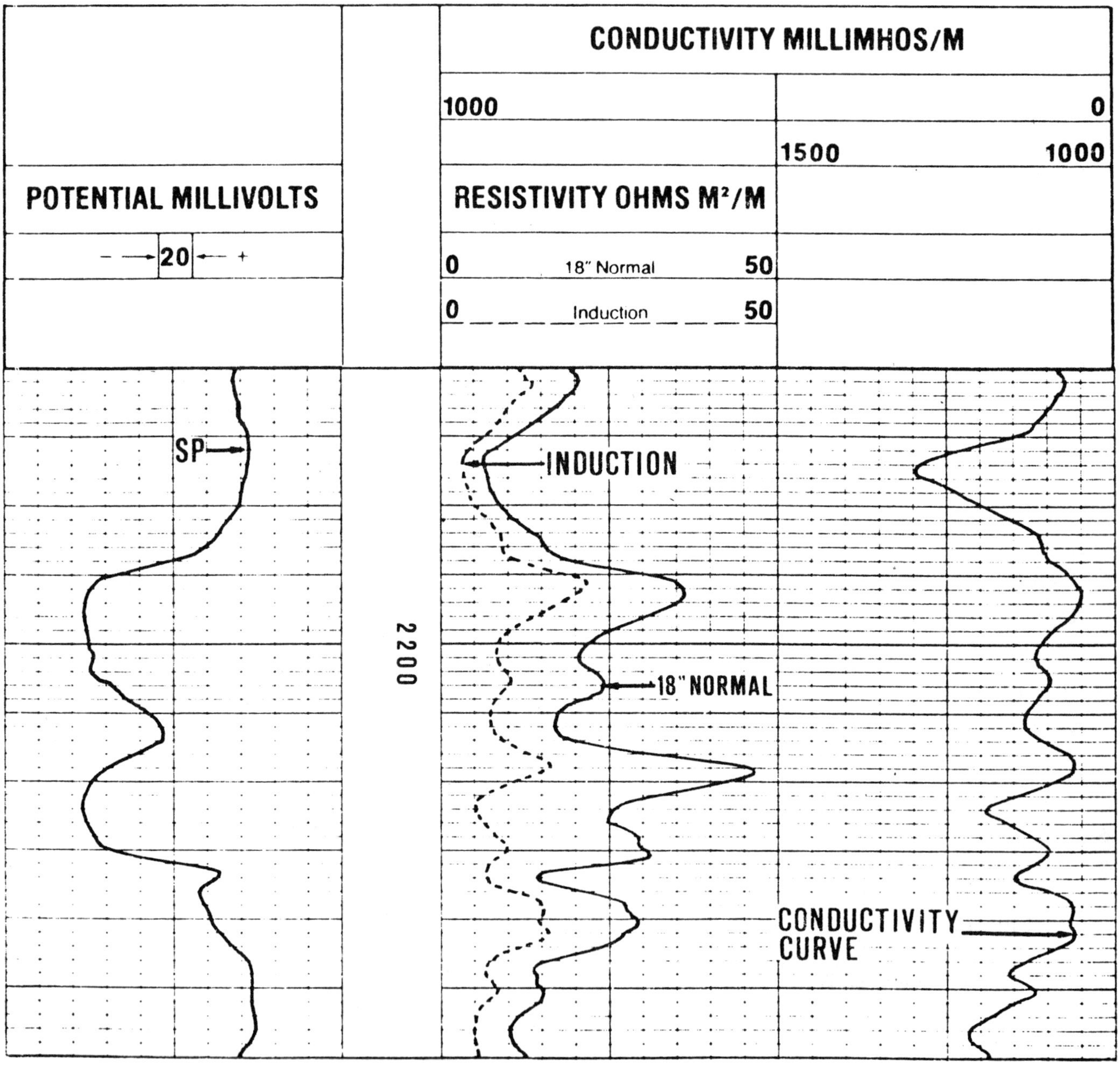

Figure LL

$$F = \frac{1}{.24^2} = 17.4$$

$$R_o = 17.4 \times 0.084 = 1.46$$

$$R_t = 130 \text{ (from the log)}$$

$$S_w = \sqrt{\frac{1.4}{130}} = 0.104 \text{ or } 10.4\%$$

Problem 3

Solution:

$$R_t = R_d \times (R_t/R_d)$$

$$R_{xo} = R_t \times (R_{xo}/R_t)$$

	Zone A	Zone B	Zone C
R_{sg}	800	20	11
R_{med}	61	4.0	3.0
R_d	48	2.2	1.6
R_{sg}/R_d	16.7	9.09	6.87
R_{med}/R_d	1.27	1.82	1.87
Di	30″	56″	67″
R_t/R_d	.98	.86	.81
R_t	47	1.89	1.29
R_{xo}/R_t	30	14	10.5
R_{xo}	1,410	26.5	13.5

Discussion:

R_{xo} for Zone A exhibits a lower resistivity reading than that of the shallow guard. This is due to mud filtrate invasion. The contrasting values signify moveable hydrocarbons.

Problem 4

Solution:

$$SP = -135$$

$$R_{mf} @ 75° = 2.04 \text{ (Arps Equation)}$$

Since R_{mf} @ 75° = 0.1:

$$R_{mfe} = 0.85 \times 2.04 = 1.73$$

$$R_{mfe}/R_{we} = 10^{-(-SP/K)}$$

$$K = 60 + .133 \times 171 = 82.7$$

$$R_{mfe}/R_{we} = 42.9$$

$$R_{we} = 1.73/42.9 = 0.04$$

Since R_{we} = 0.1:

$$R_w = \frac{(77 \times 0.04) + 5}{146 - (377 \times R_{we})} = 0.062 @ 75°$$

$$R_w @ T_f = 0.029 \text{ (Arps Equation)}$$

	Zone A	Zone B	Zone C
R_{sg}	20	19	60
R_{med}	11	8	6.5
R_d	9.5	5.0	3.25
R_{sg}/R_d	2.1	3.8	18.5
R_{med}/R_d	1.15	1.6	2.0
Di	29″	54″	64″
R_t/R_d	0.985	0.88	0.805
R_t	9.4	4.4	2.61
R_{xo}/R_t	3	5.7	29
R_{xo}	28.2	25.1	76
Δt log	78	77	74
Using Wylie Time-Average Equation			
	0.16	0.15	0.13
Note: Since the shale ΔT value is 100, no compaction correction is needed			
Using the Humble Equation			
F	31	35	48
R_o	0.9	1.02	1.39
S_w%	30.9%	48.2%	73%

Problem 5

Solution:

	Zone A			
Depth	4,347	4,350	4,352	4,354
Φ_d	0.28	0.3	0.27	0.23
Φ_n	0.32	0.3	0.3	0.28
Φ_{xp} (crossplot)	0.3	0.3	0.28	0.25
F	11.1	11.1	12.8	16.0
R_o	1.78	1.78	2.05	2.56
R_t	10	16	16	15
S_w%	42.2%	33.4%	35.8%	41.3%

The mixture is lime and dolomite.

Problem 6

Solution:

ρ_{ma} is 2.65 because it is a sandstone matrix.

$$\Phi_d = \frac{\rho_{ma} - \rho_b}{\rho_{ma} - \rho_f}$$

$$\rho_b = 2.37$$

$$\Phi_d = \frac{2.65 - 2.37}{2.65 - 0.2} = 0.17$$

$$\Phi_{dcorr} = \Phi_d \times 0.7 = 0.17 \times .7 = 0.119$$

F = 60 (Humble Equation for sandstone)

$$R_o = 60 \times .15 = 9$$

$$R_t = 55$$

$$S_w\% = 40.5\%$$

Problem 7

Solution:

Depth	Φ_d	F	R_o	R_t	S_w
5,031–36	0.07	31	1.4	14	32
5,036–38	---- shale	------	------	------	------
5,038–42	0.11	18	0.82	18	21
5,042–47	0.15	12	0.55	19	17
5,047–55	0.07	31	1.40	30	22
5,055–59	0.10	20	0.90	37	16
5,059–62	0.08	26	1.20	35	18
5,062–64	0.10	20	0.90	35	16.1
5,064–70	0.082	25	1.10	24	22
5,070–78	0.131	14	0.65	11	25
5,078–83	0.103	19	0.88	12	27
5,083–87	0.14	13	0.6	15	20
5,087–90	0.09	23	1.00	28	19
5,090–95	0.11	18	0.82	28	17
5,095–01	0.08	26	1.20	28	21

Problem 8

Solution:

First, read the resistivity-porosity values and write them down.

Depth	R_t	$\Phi_d\%$
2,210	7	11.8
2,212	7	11.0
2,214	10	10.0
2,216	15	8.0
2,218	16	8.2
2,220	10	8.0
2,222	6	9.0
2,224	5	9.0
2,226	6	8.0
2,228	8	7.0
2,230	10	4.0
2,232	7	2.0
2,234	6	2.0
2,236	10	−2.0
2,238	15	−2.0
2,240	15	0.0
2,242	16	0.0
2,244	11	1.0
2,246	6	1.0
2,248	6	2.0

Second, plot the points on the crossplot, as shown on Figure MM.

Next, determine the 100% water line (R_o line).

Finally, choose any point which falls on the line and calculate R_w as follows: for this example, the point corresponding to a resistivity of 4 ohms and a porosity of 10%.

$$F = \frac{.62}{\Phi^{2.15}} = \frac{.62}{0.1^{2.15}} = 87$$

$$R_o = F \times R_w$$

and transposing the equation:

$$R_w = \frac{R_o}{F} = \frac{4}{87} = 0.046$$

$$R_w = 0.046 \text{ @ } T_f$$

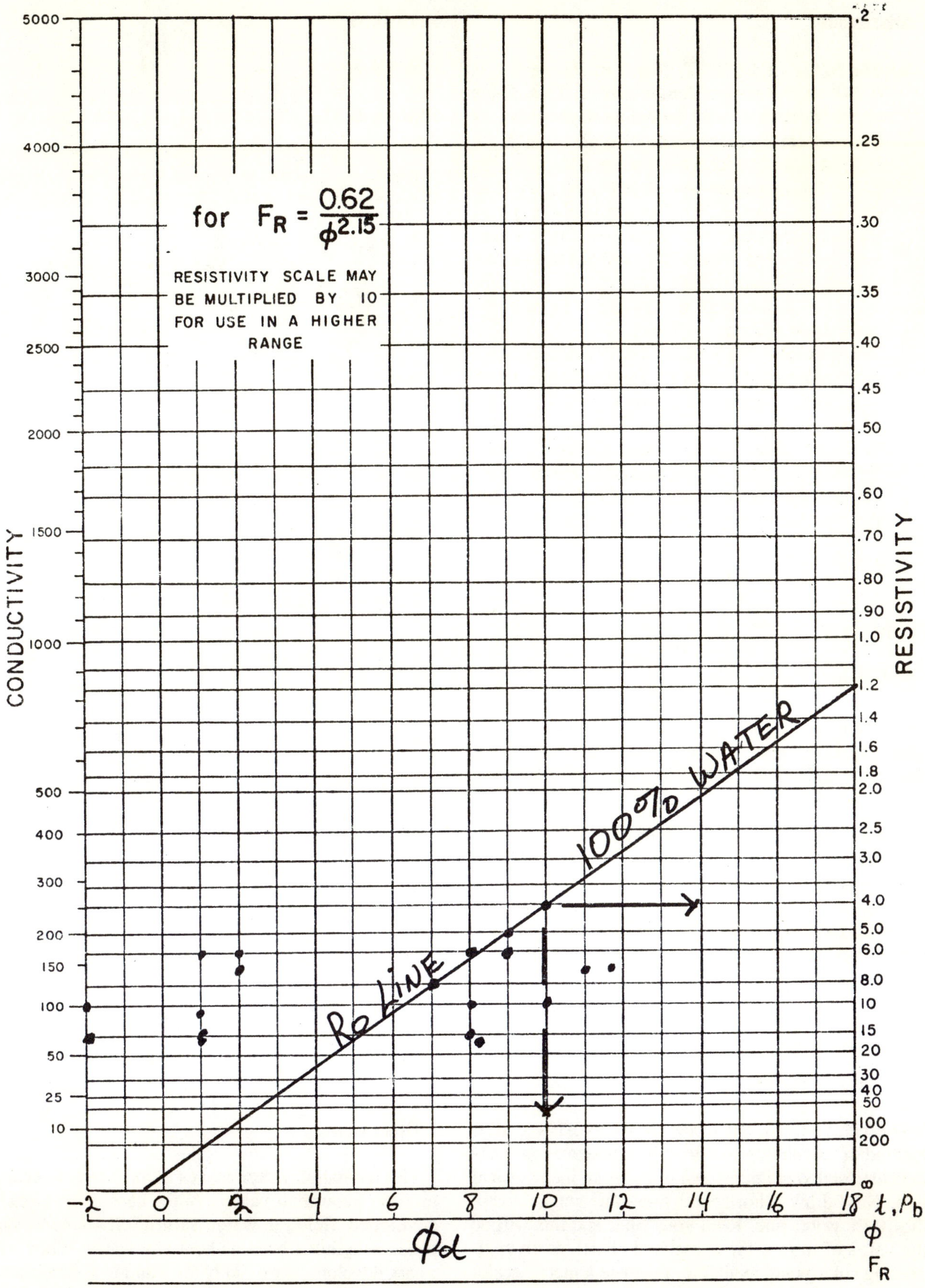

5000
4000
3000
2500
2000
1500
1000
500
400
300
200
150
100
50
25
10
CONDUCTIVITY
for $F_R = \frac{0.62}{\phi^{2.15}}$
RESISTIVITY SCALE MAY BE MULTIPLIED BY 10 FOR USE IN A HIGHER RANGE
.2
.25
.30
.35
.40
.45
.50
.60
.70
.80
.90
1.0
1.2
1.4
1.6
1.8
2.0
2.5
3.0
4.0
5.0
6.0
8.0
10
15
20
30
40
50
100
200
∞
RESISTIVITY
100% WATER
Ro LINE
-2
0
2
4
6
8
10
12
14
16
18
t, Pb
φ
FR
φd

Figure MM

Discussion:

1. The value of R_w @ T_f found in Problem 1 is 0.047 at formation temperature. This value agrees quite well with the value found here.
2. There seem to be two sets of points which are separated on the chart by some distance. The points falling on the bottom left-hand portion of the chart (−2 to +2%) are most likely changing lithology and thus read rather low. These points must be ignored because they are not representative of 100% water. The R_o or 100% water line is chosen as the most northwesterly point of the second set of data.
3. Whether this step is necessary depends on what data are available. If an R_w for this field is available, then a comparison of the known R_w and the SP-derived R_w would suffice for accuracy. If this is not possible, then this step is highly recommended, because the formation contains shale. Shale could diminish the amplitude of the SP; thus we have no guarantee of a static SP.

Problem 9

Solution:

Construct the chart scale that best fits the data (in some cases, the use of 4 or 5 cycles is required). A resistivity scale of 0.01–10.0 is indicated and shown on the plot.

Plot all indicated data and determine which southwesterly points are representative of 100% water. Draw a line through the points, extending it to the 100% porosity mark. In this case, the line intersects 100% porosity at a resistivity of 0.015, therefore (see Figure NN):

$$0.015 = a\, R_w$$

$$a = \frac{0.015}{R_w} = \frac{0.015}{0.026} = 0.58$$

The value of m is the slope of the line. The easiest method of calculating the slope is to take an engineering ruler scaled in tenths, and draw a one-inch vertical line. Now draw a horizontal line until you intercept the 100% water line. Read your ruler and this will be the value of m (see the completed chart). If an engineering ruler is not available, then use a metric scale.

$$a = 0.58$$

$$m = 1.3$$

Finding S_w:

$$F = \frac{0.58}{0.128^{1.3}} = .84$$

$$R_o = 8.4 \times 0.026 = 0.22$$

$$S_w = \sqrt{\frac{0.22}{7.0}} = 0.177 \text{ or } 17.7\%$$

When

$$a = 1$$

$$m = 2$$

$$F = \frac{1}{0.128^2} = 61$$

$$R_o = 61 \times 0.026 = 1.59$$

$$S_w = \sqrt{\frac{1.59}{7.0}} = 0.477 \text{ or } 47.7\%$$

then

$$a = 0.62$$

$$m = 2.15$$

$$F = \frac{0.62}{0.128^{2.15}} = 51.5$$

$$R_o = 51.5 \times 0.026 = 1.34$$

$$S_w = \sqrt{\frac{1.34}{7.0}} = 0.438 \text{ or } 43.8\%$$

Discussion:

The formation factor makes a considerable change in water saturation values. In the case of an already-developed field, a water saturation of 43% may produce hydrocarbons without any water. When such a case develops, most likely the a and m values chosen

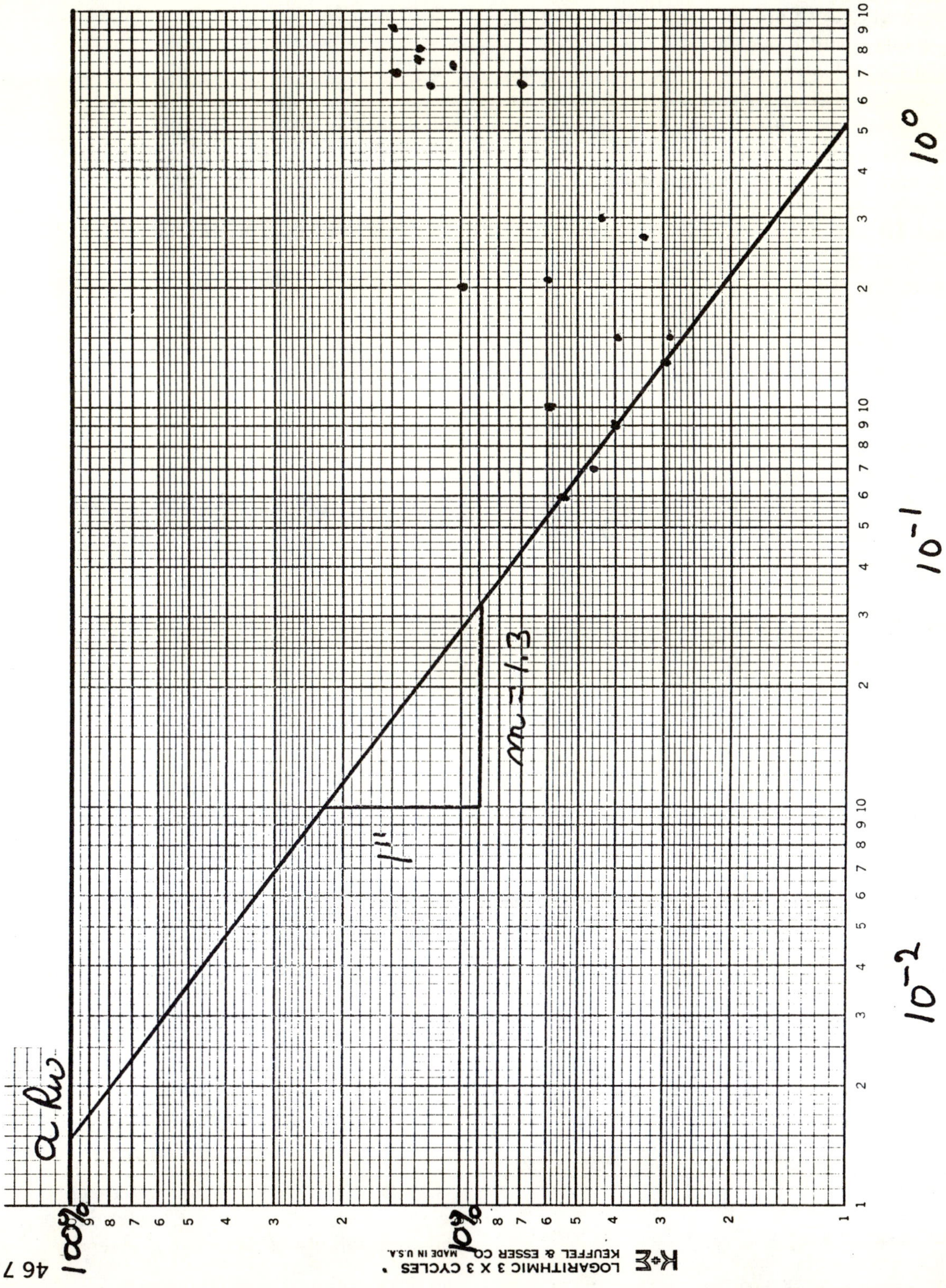
m = 1.3
1''
100%
10%
10^{-2}
10^{-1}
10^{0}
K+E LOGARITHMIC 3 X 3 CYCLES
KEUFFEL & ESSER CO. MADE IN U.S.A.
46 7

Figure NN

were not optimum. The point here is that it is known that 43% will produce no water and thus we have a standard by which to base an analysis. It is advantageous for everyone to use the same values so that, right or wrong, there is a standard.

In the case of a wildcat, accurate values are crucial. In this example, one may have abandoned the well for fear of excessive water production, when in fact, the reservoir is at irreducible water!

Problem 10

Solution:

Zone	A	B	C	D	E
GR =	66	62	53	38	56
I_{GR} =	0.23	0.2	0.133	0.022	0.156
V_{sh} =	0.09	0.077	0.049	0.0075	0.058
Φ_n =	0.17	0.17	0.17	0.19	0.15
$\Phi_{n\ corr}$ =	0.138	0.142	0.152	0.187	0.13
Φ_d =	0.14	0.15	0.15	0.15	0.12
$\Phi_{d\ corr}$ =	0.126	0.139	0.143	0.149	0.111
Φ_{xp} =	0.132	0.14	0.148	0.168	0.121
F =	47	41.7	38	29.6	55.7
R_o =	2.86	2.54	2.31	1.80	3.4
R_t =	8.0	9.5	9.0	6.5	11.0
S_w% =	59.8	51.7	50.7	52.7	55.6

Problem 11

Solution:

See Figures OO and PP.

Problem 12

Solution:

The first step is to compute a residual hydrocarbon saturation. Before this can be done, the value of R_{xo} must be determined.

Using the dual-induction SFL tornado chart (see Figure QQ):

$$R_{SF}/R_d = 156/67.4 = 2.3$$

$$R_m/R_d = 89.7/67.4 = 1.33$$

Plotting 2.3 and 1.33 on the tornado chart, we obtain the following ratios:

$$R_t/R_d = 0.9$$

$$R_{xo}/R_t = 3$$

Solving for R_t and R_{xo}:

$$R_t = R_d \times 0.90 = 67.4 \times 0.90 = 60.7$$

$$R_{xo} = R_t \times 3 = 60.7 \times 3 = 182$$

The saturation of the flushed zone is given by:

$$S_{xo} = \sqrt{\frac{R_{mf}\,F}{R_{xo}}}$$

and

$$F = \frac{1}{\Phi^2}$$

using the density porosity

$$F = \frac{1}{0.12^2} = 69$$

thus:

$$S_{xo} = \sqrt{\frac{0.95 \times 69}{182}} = 0.60 \text{ or } 60\%$$

Residual saturation is then:

$$S_{hr} = 1 - S_{xo} = 1 - 0.6 = 0.40 \text{ or } 40\%$$

Next, we determine the density of the hydrocarbon using the *estimation of hydrocarbon density* chart.

$$\Phi_{CNL}/\Phi_d = 4.8/12.0 = 0.40$$

Now, plot $S_{hr} = 0.40$ and 0.40 on the chart, to obtain $\rho_h = 0.15$, which is quite obviously gas.

The density of the formation water can be found on the salinity-density table of Chapter 11 (page 155). Thus, the salinity of the water is given by:

$$\text{Salinity (PPM)} = 10^b$$

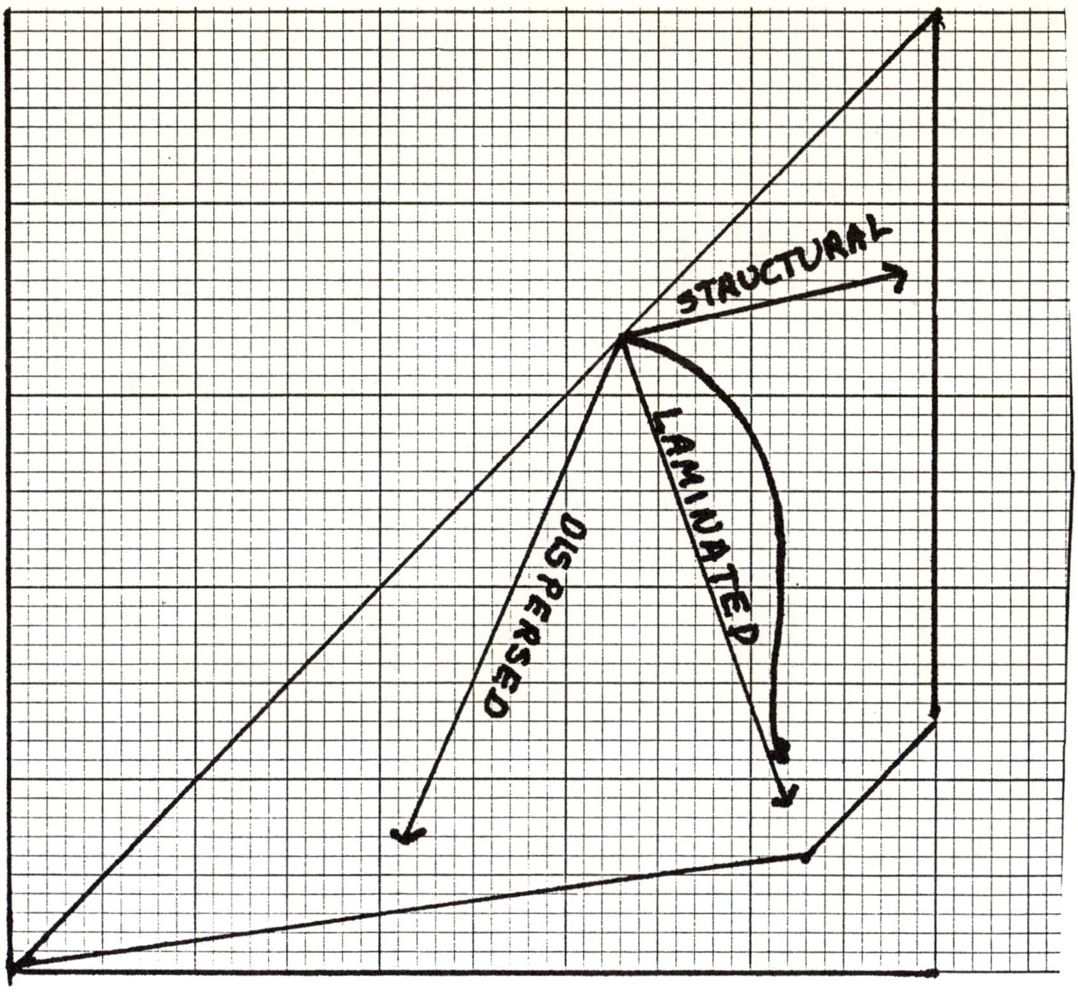

Figure OO

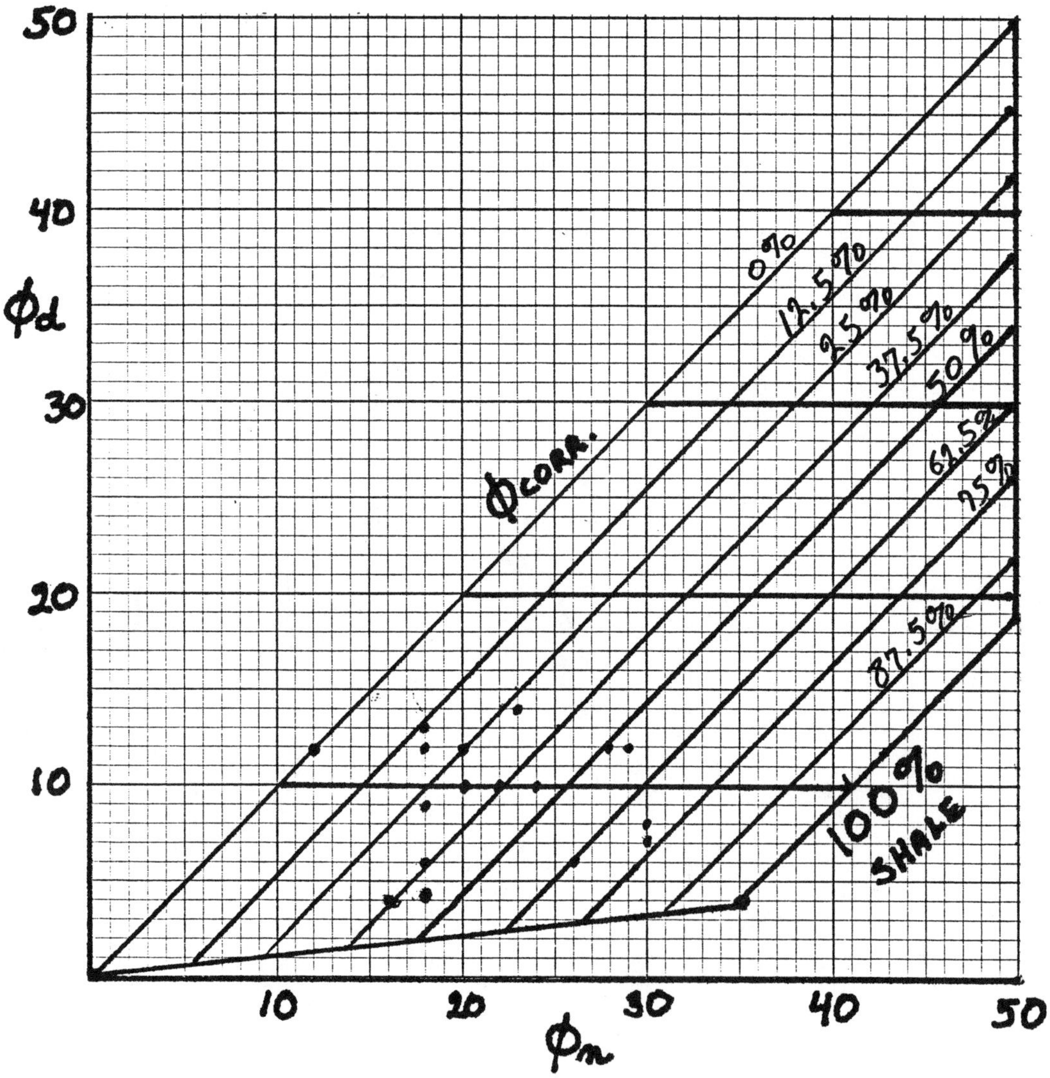
50
40
30
20
10
ϕd
0%
12.5%
25%
37.5%
50%
62.5%
75%
87.5%
100% SHALE
ϕCORR.
10
20
30
40
50
ϕn

Figure PP

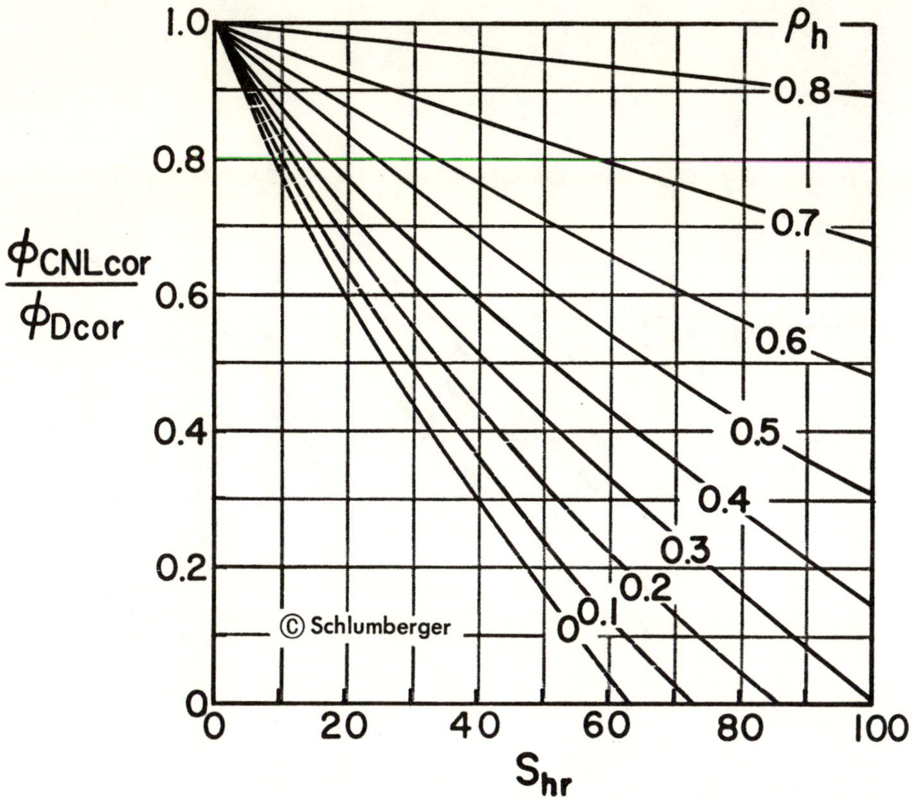

Courtesy of Schlumberger.

Figure QQ

where

$$b = \frac{3.562 - \log (R_w - 0.0123)}{0.955}$$

thus

$$b = \frac{3.562 - \log (0.05 - 0.0123)}{0.955}$$

$$b = 5.22$$

and

$$\text{salinity} = 10^{5.22} = 166{,}182 \text{ PPM}$$

From the salinity-density table of Chapter 11, this salinity corresponds to a water density of 1.109.

Solving for permeability using the resistivity gradient chart:

$$\rho_w - \rho_h = 1.109 - 0.15 = 0.96$$

and

$$a = \frac{\Delta R}{\Delta D} \times \frac{1}{R_o}$$

where

$$\frac{1}{R_o} = \frac{1}{F \times R_w} = \frac{1}{69 \times .05} = 0.29$$

$$\frac{\Delta R}{\Delta D} = \frac{67.4 - 47}{6{,}800 - 6{,}792} = 2.5$$

Thus:

$$a = 2.5 \times 0.29 = 0.725$$

Plotting 0.96 and 0.725 on the gradient chart, we obtain a permeability of 60md (Figure RR).

Discussion:

This type of permeability estimate does not require knowledge of a zone to be at irreducible water. Consequently, this is preferred because it eliminates one more "guesstimate." Also, the porosity curves used were not corrected for gas. The density porosity should be 8.4% instead of 12%. This will affect the formation factor F as well as R_o and the residual saturations. Having determined the hydrocarbon is gas, we may go back and recompute the parameters.

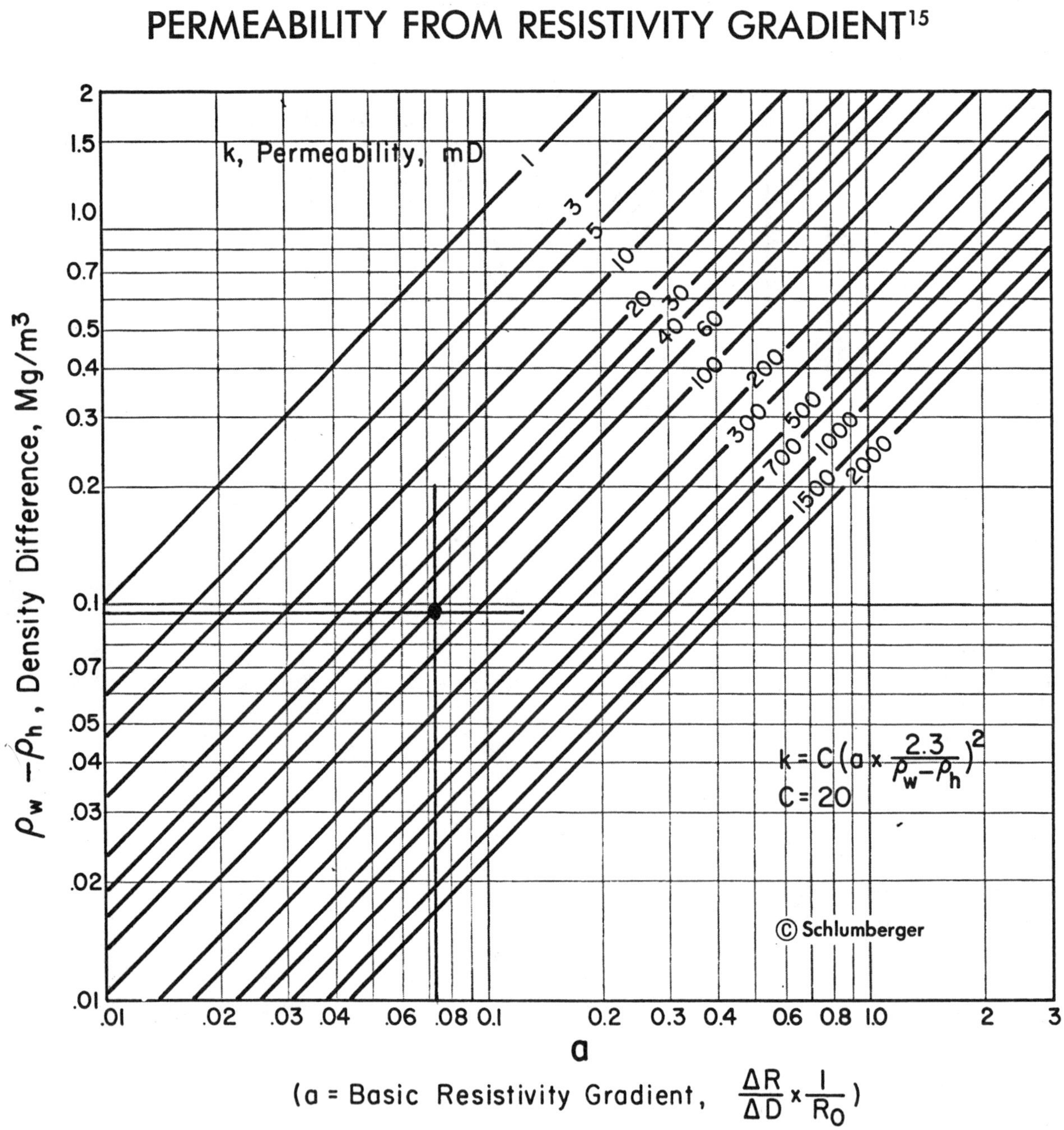

Courtesy of Schlumberger.

Figure RR

Expanded Glossary

(This addendum to *Applied Open-Hole Log Analysis* supersedes the glossary on pages 277–279.)

Angularity—The shape of the rock grain is angular or pointed rather than smooth or round.

Annulus—The area of the formation between the flushed zone and the uninvaded zone. This area contains a mixture of virgin formation fluid and mud filtrate.

Apparent—What a parameter appears to be as it is read from the log. For example, apparent resistivity.

Bimetallism—The interaction of two metals of different composition reacting in an electrolyte to produce an electrical current. In essence, bimetallism is the action which takes place in a common battery.

Borehole compensated sonic or BHC—A method which employs two transmitters and two receivers to compensate the sonic signal for tool tilt or formation washouts.

Bulk density—Primary measurement of the density tool expressed in grams per cubic centimeter.

Cementation—The process of precipitation or growth of a binding material around grains or fragments of rocks.

Cementation exponent—The portion of the formation resistivity factor which mathematically accounts for the rock cement. This value is indicated by the letter m in general terms.

Compaction—The reduction in pore space in a formation resulting from the weight of formations above it.

Compressive wave—Sonic sound wave which forces the rock to compress by its sound pressure.

Compton scattering—Occurs when a gamma ray collides with an atom and dislodges an electron, causing a reduction of energy.

Connate water—Water entrapped in the interstices of the rock (either sedimentary or extrusive igneous) at the time the rock was deposited.

Conductivity—The ability of a substance to conduct electrically charged particles. Conductivity is the inverse of resistivity and its unit of measure is the *mho*.

Darcy—The unit of measure of permeability. The permeability of a porous medium that will allow a flow of 1 ml/s of fluid of 1 cp through 1 cm^2 under a pressure gradient of 1 atm/cm.

Delta t (Δt)—The amount of time it takes a sonic sound to travel through the formation expressed in microseconds per foot.

Delta t (Δt) fluid—The amount of time it takes a sonic sound wave to travel through a fluid. The value for water is 189 microseconds per foot.

Density log—A nuclear log used to determine the electron density of a matrix, inferring a porosity from it.

Delta t (Δt) matrix—The amount of time it takes the sonic sound wave to travel through solid matrix material.

Electromagnetic propagation—The conduction of electrically produced magnetic fields through a substance or medium.

Fissure—A crack or fracture in a subsurface formation.

Fluid waves—The sonic sound signal traversing the mud column.

Flushed zone—The area of formation which is saturated by mud filtrate.

Focusing (guard)—The concentration of current into a thin sheet in order to obtain better resolution in resistivity readings.

Football effect—The pattern or shape of the neutron-density porosity curves in a zone containing gas.

Formation resistivity factor—Commonly called *formation factor,* it is the mathematical model which describes the characteristics of a formation rock.

Fracture—A break, parting, or separation in brittle rock.

Gamma ray—An electromagnetic particle of energy.

Geothermal gradient—The rate of change in temperature per 100 feet of depth.

Hydrocarbon—An organic compound composed of hydrogen and carbon. Typically used to refer to oil or gas.

Hydrocarbon saturation—The percentage of oil or gas in a reservoir.

Ion—An electrically charged chemical particle.

Ionic current—Electrical current caused by the conduction of ions as opposed to electrons.

Impervious bed—Any rock bed capable of impeding the flow of a fluid. Usually a bed which contains no permeability such as a shale.

Induction—Electromagnetic fields produced by an electrical current flowing through a coil. An electromagnet is a good example of induction.

Invasion—The liquid portion of the mud system which penetrates the formation because of the hydrostatic pressure in the mud column.

Irreducible fluid saturation—The lowest possible saturation of a fluid which can be obtained upon its displacement by another fluid with which it is immiscible.

Liquid junction potential—Voltage produced by two chemical solutions of different salinity. This is the smaller portion of the total SP signal.

Matrix—(1) The solid framework of rock that surrounds pore volume. (2) The smaller-size grains in a rock in which certain grains are distinctly larger than others.

Membrane potential—Voltage caused by the cationic membrane potential of shales. The largest voltage of the total SP signal.

Mho—The unit of measure for conductivity. The *mho* is the inverse of the *ohm.*

Microresistivity—The resistivity of a formation very close to the borehole wall.

Mud—Drilling fluid used to maintain borehole pressure and remove cuttings.

Mudcake—The solid portion of the mud which adheres to the borehole wall by the hydrostatic pressure.

Mud filtrate—The liquid portion of the mud.

Neutron—Subatomic particle having no electrical charge and constituting a part of the atomic nucleus.

Packing—The physical configuration of individual rock grains making up the rock.

Permeability (absolute permeability)—A measure of the ability of a rock to conduct a fluid through its interconnected pores when that fluid is at 100% saturation. (Measured in darcies or millidarcies.)

Porosity—The ratio of void space to the bulk volume of rock containing that void space. Porosity can be expressed as a fraction or percentage of pore volume in a volume of rock.

Relative permeability—The ratio of the amount of a specific fluid that will flow at a given saturation, in the presence of other fluids, to the amount that would flow at a saturation of 100%, other factors remaining the same.

Shale—A clay type of material which has porosity but no permeability.

Shear waves—Sonic sound waves generated by the rock as it decompresses from compressive wave energy.

Sonic—Sound wave frequency in the range of 15,000 to 40,000 cycles per second.

Sonic tool—A porosity tool which uses sonic sound to measure the speed of sound traveling through the formation. The speed of the sound is a function of formation density which is then translated to porosity.

Spontaneous potential (SP)—Current which causes a voltage drop in the formation as a result of two chemically unbalanced saline solutions through a cationic membrane.

Static SP (SSP)—A theoretical concept used to describe the action of the SP.

Track—One linear or logarithmic scale containing small divisions horizontally and depth divisions vertically on a log.

True resistivity—The resistivity measured in the virgin zone after it has been corrected.

Uninvaded zone—The area of the formation which remains uncontaminated by drilling fluids such as mud filtrate.

Vugs—Small caverns in the rock grain existing as secondary porosity.

Virgin zone—See Uninvaded zone.

Water cut—The percentage of water that will be produced with a hydrocarbon.

Water saturation—The percentage of water occupying the pore space.

Zone of transition—See Annulus.

Glossary

Angularity—The shape of the rock grain is angular or pointed rather than smooth or round.

Annulus—The area of the formation between the flushed zone and the uninvaded zone. This area contains a mixture of virgin formation fluid and mud filtrate.

Apparent—What a parameter appears to be as it is read from the log. For example, apparent resistivity.

Bimetallism—The interaction of two metals of different composition reacting in an electrolyte to produce an electrical current. In essence, bimetallism is the action which takes place in a common battery.

Borehole compensated sonic or BHC—A method which employs two transmitters and two receivers to compensate the sonic signal for tool tilt or formation washouts.

Bulk density—Primary measurement of the density tool expressed in grams per cubic centimeter.

Cementation—The chemical or physical process which solidifies loose particles such as sand grains into rocks.

Cementation exponent—The portion of the formation resistivity factor which mathematically accounts for the rock cement. This value is indicated by the letter m in general terms.

Compaction—The reduction in pore space in a formation resulting from the weight of formations above it.

Compressive wave—Sonic sound wave which forces the rock to compress by its sound pressure.

Compton scattering—Occurs when a gamma ray collides with an atom and dislodges an electron, causing a reduction of energy.

Connate water—Water which adheres to the matrix material by surface tension.

Conductivity—The ability of a substance to conduct electrically charged particles. Conductivity is the inverse of resistivity and its unit of measure is the *mho.*

Darcy—The unit of measure for permeability.

Delta t (Δt)—The amount of time it takes a sonic sound to travel through the formation expressed in microseconds per foot.

Delta t (Δt) fluid—The amount of time it takes a sonic sound wave to travel through a fluid. The value for water is 189 microseconds per foot.

Delta t (Δt) matrix—The amount of time it takes the sonic sound wave to travel through solid matrix material.

Density log—A nuclear log used to determine the electron density of a matrix, inferring a porosity from it.

Electromagnetic propagation—The conduction of electrically produced magnetic fields through a substance or medium.

Fissure—A long and narrow crack or cleft in a rock.

Fluid waves—The sonic sound signal traversing the mud column.

Flushed zone—The area of formation which is saturated by mud filtrate.

Focusing (guard)—The concentration of current into a thin sheet in order to obtain better resolution in resistivity readings.

Football effect—The pattern or shape of the neutron-density porosity curves in a zone containing gas.

Formation resistivity factor—Commonly called *formation factor,* it is the mathematical model which describes the characteristics of a formation rock.

Fracture—A split or cracked rock.

Gamma ray—An electromagnetic particle of energy.

Geothermal gradient—The rate of change in temperature per 100 feet of depth.

Hydrocarbon—An organic compound composed of hydrogen and carbon. Typically used to refer to oil or gas.

Hydrocarbon saturation—The percentage of oil or gas in a reservoir.

Ion—An electrically charged chemical particle.

Ionic current—Electrical current caused by the conduction of ions as opposed to electrons.

Impervious bed—Any rock bed capable of impeding the flow of a fluid. Usually a bed which contains no permeability such as a shale.

Induction—Electromagnetic fields produced by an electrical current flowing through a coil. An electromagnet is a good example of induction.

Invasion—The liquid portion of the mud system which penetrates the formation because of the hydrostatic pressure in the mud column.

Irreducible water—See **Connate water.**

Liquid junction potential—Voltage produced by two chemical solutions of different salinity. This is the smaller portion of the total SP signal.

Matrix—A formation composed of one or more rock types or minerals.

Membrane potential—Voltage caused by the cationic membrane potential of shales. The largest voltage of the total SP signal.

Mho—The unit of measure for conductivity. The *mho* is the inverse of the *ohm.*

Microresistivity—The resistivity of a formation very close to the borehole wall.

Mud—Drilling fluid used to maintain borehole pressure and remove cuttings.

Mudcake—The solid portion of the mud which adheres to the borehole wall by the hydrostatic pressure.

Mud filtrate—The liquid portion of the mud.

Neutron—Subatomic particle having no electrical charge and constituting a part of the atomic nucleus.

Packing—The physical configuration of individual rock grains making up the rock.

Permeability—The ability of a fluid flowing through a material. Zero permeability means no fluid can flow.

Porosity—The ratio of solid matter to void space expressed as a percentage.

Relative permeability—The permeability of one fluid relative to another.

Residual oil saturation—Oil in the invaded zone which was not pushed back by mud filtrate.

Resistance—The opposition of electrical current flow. Resistance is an absolute quantity measured in *ohms.*

Resistivity—The resistance of electrical current flow through a given volume. Resistivity is measured volumetrically as *ohms* per cubic unit of volume. In log interpretation the cubic meter is used.

Secondary porosity—Porosity created by fractures, leaching, or other chemical/physical means which adds pore space to the original porosity.

Index

N

O–P

Q–R

U–Z